Progress in Physics
Vol. 7

Edited by
A. Jaffe, G. Parisi,
and D. Ruelle

Birkhäuser
Boston · Basel · Stuttgart

Scaling and Self-Similarity in Physics

Renormalization in Statistical Mechanics and Dynamics

Jürg Fröhlich, editor

1983

Birkhäuser
Boston • Basel • Stuttgart

Editor:

Jürg Fröhlich
Department of Theoretical Physics
ETH - Hönggerberg
8093 Zürich, Switzerland

Library of Congress Cataloging in Publication Data

Scaling and self-similarity in physics.

(Progress in physics ; v. 7)
1. Renormalization group. 2. Statistical mechanics.
3. Field theory. 4. Dynamics. 5. Phase transformations
(Statistical physics) 6. Critical Phenomena (Physics)
I. Fröhlich, Jürg. II. Title: Self-similarity.
III. Series: Progress in physics (Boston, Mass.) ; v. 7.
Oc20.7.R43S33 1983 530.1 83-15788
ISBN 0-8176-3168-2

CIP-Kurztitelaufnahme der Deutschen Bibliothek

Scaling and self-similarity in physics:
renormalization in statist. mechanics and
dynamics/ ed. by Jürg Fröhlich. - Boston ;
Basel ; Stuttgart : Birkhäuser, 1983.
(Progress in physics ; Vol. 7)
ISBN 3-7643-3168-2 (Basel, Stuttgart)
ISBN 0-8176-3168-2 (Boston)

NE: Fröhlich, Jürg [Hrsg.]; GT

ISBN 0-8176-3168-2
ISBN 3-7643-3168-2
Printed in USA

9 8 7 6 5 4 3 2 1

TABLE OF CONTENTS

FOREWORD

The articles collected in this book have grown out of a series of seminar talks held at the Institut des Hautes Etudes Scientifiques, Bures-sur-Yvette, between spring 1981 and summer 1982. These talks were presented by people who, during that period, happened to be visitors at the Institut or worked in the Paris area. Most of them were (and still are) thinking about problems in theoretical and mathematical physics related, in a general way, to ideas of scaling and self-similarity; problems which nowadays are often studied with the help of what has become known under the name of "renormalization group methods". Their ideas and their results illustrate in which way mathematical physics has interacted with one of the big themes of present-day theoretical physics. To present a brief status report about where these interactions have led us, so far, is a guiding idea behind this volume.

During the past ten to twenty years, various concepts of scaling, self-similarity and renormalization have assumed a well defined shape and proven to provide a very productive framework to think about a surprisingly large class of physical problems. Powerful computational and analytical tools have emerged from them. Most readers will be familiar with one or another among many excellent text books and review articles devoted to scaling and renormalization in the theory of critical phenomena in statistical mechanics and of differentiable dynamical systems. [References to some of these texts may be found in several articles in this book.] Numerous theoretical physicists have contributed very important insights and results to these fields. Thanks to their original and persistent efforts we now do have some considerable qualitative and quantitative understanding of "continuous transitions and critical phenomena", not only in equilibrium statistical mechanics and quantum field theory, but also in dynamics and in the theory of disordered systems.

Precisely because very many people have contributed a large numer of substantial results to the subject addressed in this book, it unfortunately turned out to be necessary to apply rather arbitrary, but strict criteria to choose authors of articles for the present book. Such criteria are usually unsatisfactory. For I still believe in a basic unity of theoretical and mathematical physics. The criteria which we finally adopted are not intended to express any value judgments. [This is no attempt towards "eine Umwertung aller Werte".] Rather, they were imposed upon us by convenience and by consideration of what was within our modest possibilities.
They are:

1. Every author was supposed to have spent time or given a seminar talk at the Institut des Hautes Etudes Scientifiques during the period between spring 1981 and summer 1982.

2. Only contributions were accepted which describe mathematically rigorous results related to the main theme of this book.

3. Every article was supposed to be of relevance to statistical physics and/or dynamical systems theory.

4. Contributions concerning "scaling and self-similarity in mathematics" were not invited.

These criteria may explain why many ideas, results and calculations of enormous theoretical and practical importance to critical phenomena in equilibrium statistical mechanics and dynamics cannot be found in any of the articles published in this volume. Moreover, they make it clear why most of the original creators of many important ideas and methods are not among the contributors.

Criterion 3 explains why "scaling and renormalization in relativistic quantum field theory", for example, is greatly underemphasized in this book. This is most regrettable, for several reasons:

a) Many of the important ideas on "scaling and renormalization" actually originated in relativistic quantum field theory.

b) The discovery of many relations and deep connections, heuristic and mathematical, between the theory of continuous transitions and critical phenomena on the one hand and quantum field theory on the other hand, and the many uses of these connections in actual calculations belong to the great achievements of theoretical physics in recent times and has had a profound impact on the development of statistical physics and quantum field theory. [Various connections and analogies between quantum field theory and fluid dynamics may eventually turn out to be of comparable significance.]

c) Quantum field theory is one of a few basic theories in modern theoretical physics, and it is the most plausible and most efficient theoretical basis for particle physics. As such it is of eminent importance, and it still attracts many among the most brilliant theoreticians.

The theory of disordered systems is not represented in this book, at all. This, too, is really quite regrettable; especially since, during the past few years, nice mathematically rigorous results about disordered systems, related to the themes of this book, have started to appear. Moreover, the theory of disordered systems bears quite a lot of promise for the near - and not so near - future.

Criterion 4 has excluded all developments related to scaling, self-similarity and renormalization in mathematics from this book. That there are many and important such developments is presumably clear to many or most readers. In any event, it is a fact! I had not sufficiently appreciated this fact, but learnt it (not always without pain) by attending numerous seminars devoted to mathematical aspects of dynamical systems theory, held at the I.H.E.S. and elsewhere in the south of Paris, and trying to follow (unfortunately rather unsuccessfully) an inspiring course on hyperbolic geometry taught by Dennis Sullivan. Striking examples in Fourier analysis were pointed out and explained to us by Giovanni Gallavotti, and James Glimm told us about equally striking examples in the mathematics of fluid dynamics and related topics. In joint work with David Brydges, Alan Sokal and Thomas Spencer, exploiting Symanzik's random walk representation of lattice field theories and spin systems, we were often struck by the beautiful scaling- and self-similarity properties of various random walk processes, in particular of Brownian motion. Tom Spencer patiently explained to me how those properties were related, at least heuristically, to critical properties of lattice field theories. The mathematicians have used them in many different and more rigorous ways.

These few examples - which may stand for many others - might indicate how high a price had to be paid by excluding "scaling and self-similarity in mathematics" from this book: To say the least, some very beautiful and convincing illustrations of those general concepts within mathematics had to be omitted, and an opportunity for physicists to appreciate how useful and inspiring more interaction with mathematicians would be has been missed.

The reader will easily understand that the volume of this book would have increased almost without bounds if the three topics mentioned above had been included, as well. Moreover, criteria 1 and, in the former two cases, 2 would have largely forbidden that anyway. Unfortunately, our application of the first criterion prevented inviting contributions from several colleagues whose thoughts on, and mathematically rigorous contributions to, "scaling and self-similarity in physics (renormalization in statistical mechanics and dynamics)" have been most important, or even seminal. One must mention P. Bleher, R. Dobrushin, Ya.G. Sinai, and other colleagues in the Soviet Union; D. Brydges, J.-P. Eckmann, P. Federbush, M. Feigenbaum, J. Glimm and A. Jaffe, R. Griffiths, O.E. Lanford III, and many others, the large number of whom motivated adopting the very awkward first criterion. (Some of these and other colleagues satisfied all four criteria and were asked to contribute an article, but, for various respectable reasons, were unable or unwilling to do so.)

Hopefully, some of the shortcomings of the present collection of articles, described above, are already or will be remedied in other books[1].

I believe that I shall not offend any of the authors of articles appearing here when I mention that most of these articles are not really complete, polished and particularly readable accounts of a coherent subject matter. Rather, they are intended to indicate, where the interactions between mathematical physics and an important circle of ideas in theoretical physics have led and might lead us; as already remarked upon above. They might show, moreover, that scientific progress can occur in, among many others, the following ways: By taking a concept literally, sharpening it and applying it to new problems; by reinterpreting a concept in a novel way and within new contexts; by generalizing and abstracting a concept and placing it in a wider perspective. Progress, however, rarely occurs by diluting a concept, or making it vague.

Because of these intentions, and because of criterion 2, this book has turned into a rather mathematical one and is somewhat hard to read (I find). Although that circumstance does not really require any defense, it is tempting to recall, at this point, an opinion expressed by Leonhard Euler, undoubtedly one of the great minds and fathers of mathema-

---

1) For example, there are excellent books by Ruelle, by Glimm and Jaffe, by Sinai, and by Collet and Eckmann, where a great many ideas and results missing in this volume may be found.

tical physics:

"Jede Erkenntnis der Wahrheit ist an sich etwas Herausragendes, auch wenn sie weit vom allgemeinen Gebrauch entfernt zu sein scheint; so sind auch alle Aspekte der Wahrheit, die uns zugänglich sind, so untereinander verbunden, dass keiner grundlos zurückgewiesen werden kann, auch wenn er geradewegs nutzlos scheint. Hinzu kommt, auch wenn irgendein bewiesener Satz nichts zu gegenwärtigem Nutzen beizutragen scheint, dass dennoch die Methode, vermittels der entweder Richtigkeit oder Falschheit herausgefunden wurde, meistens den Weg zu anderen Wahrheiten zu öffnen pflegt, die zu erkennen nützlicher sind."[2)]

I hope and believe the present book contradicts Euler's opinion at most mildly. In fact, several of the general methods described in various articles in this book have, meanwhile, been extended and successfully applied to new and possibly more important physical problems, as the reader may easily convince himself. (Progress has been so encouraging that it would actually be tempting to continue this project.)

I should like to add some observations about mathematical physics which might perhaps be helpful for readers who do not usually read mathematical-physics texts: First, there is often a surprisingly large gap between a convincing intuitive argument in favour of some physical fact and a mathematically rigorous proof thereof. Sometimes a

---

2) Rough translation: Every discovery of truth is outstanding, even if it seems to be far from general use. All aspects of truth accessible to us, are interrelated in such a way that none of them should be dismissed, even if it seems useless. It should be added that, quite generally, even a theorem which does not seem to be of any use, at present, may have been found on the basis of a method that will open the way to other and more useful discoveries of truth.

proof - and it need neither be a bad, nor a very difficult proof - does simply not mirror physical intuition. Second, not every mathematical theorem has only one "natural" proof. In fact, most of them do not, and complementary aspects of a fact may lead to complementary proofs of its truth. Third, sometimes hard analysis cannot be avoided. Finally, the technically simplest proof of a theorem need not be the "best", or most intuitive, or most suggestive proof. (In order to avoid unnecessary controversies, I refrain from giving examples for these obvious claims which, unfortunately, are often ignored.)

This book is divided in two parts. The first and more voluminous part is devoted to mathematical theories of phase transitions and critical phenomena, the second part to classical, dissipative and conservative dynamical systems of finitely many degrees of freedom. Some authors really ought to have contributed papers to both parts. The order of articles is largely, but not entirely accidental.

In getting the project for this book under way I received most substantial support and advice from H. Epstein, G. Gallavotti, G. Jona-Lasinio and E. Seiler. Without their help and encouragement this project could not have been realized! I am greatful for their generous assistance.I also thank the publisher for his faith, valuable advice and infinite patience. [The original deadline was missed by almost one year.]

Of course, I am particularly grateful to all those who invested time, energy and thoughtfulness to contribute a paper to this book. They really have done all the important work.

This book would never have been realized, had we not all enjoyed a most pleasant and inspiring hospitality at the Institut des Hautes Etudes Scientifiques, during various periods. In the name of all of us, I wish to thank my former colleagues at I.H.E.S., its director, N. Kuiper, and its staff for their generosity of offering us their hospitality and help. May this book be a token of continued friendship between them and us.

On a more personal level, I wish to acknowledge my great indebtedness to Res Jost for following my professional career with interest and active support and for his friendship, to David Ruelle for teaching me what a benevolent, reliable, inspiring and pleasant colleague is, and - last but not least - to Tom Spencer for generously teaching me what good mathematical physics might be, for the joy of continued and exceptionally harmonious collaboration, and for a marvellous personal friendship.

Jürg Fröhlich

Zürich, July 1983 .

PART I : Equilibrium statistical mechanics and field theory

LARGE FLUCTUATIONS OF RANDOM FIELDS AND RENORMALIZATION GROUP: SOME PERSPECTIVES

by

G.JONA-LASINIO

Abstract

In this paper we approach the theory of the so called effective potential from a probabilistic point of view. We consider lattice systems and we clarify the connection between effective potential, fluctuations of mean fields (e.g. the magnetization in a large volume) and preasymptotic behaviour of the renormalization group (corrections to limit theorems). We point out how this approach may provide non perturbative calculational tools based on recursion and we discuss the case of hierarchical models as an example.

1. Introduction

In this paragraph we discuss some background and motivation for the point of view that will be developed in the following sections. Any approach to constructive statistical mechanics begins by assigning a Gibbs measure to the random variables associated to a finite volume in an infinite cubic lattice Z^d. The first object of study is then the partition function in terms of which one proves the existence of the free energy density or the pressure in the thermodynamical limit. If we deal for example with a ferromagnetic system of spins, we obtain the free energy density as a function of temperature and the external field. This is not however the quantity in terms of which were expressed the first successful theories of phase transitions, i.e. mean field theories and their modern generalisations and improvements.

The most general mean field approach to phase transitions, notably Landau's theory /1/, was based on the notion of a non-equilibrium free energy $F(|\Lambda|,T,\phi)$ where besides the volume $|\Lambda|$ and the temperature T, appears the value of the order parameter ϕ in the given non equilibrium situation. From the thermodynamical point of view $F(|\Lambda|,T,\phi)$ is the minimal work necessary to create in a reversible way the fluctuation ϕ. In our magnetic example the non equilibrium state would correspond to an arbitrary value of the magnetization, i.e. the mean spin in the volume $|\Lambda|$, for zero external field.

So far there has been no attempt to put such a theory on a firm basis by connecting it to the only rigorous approach to the problem of phase transitions known today, that is the theory of infinite volume Gibbs measures. It is one of the aims of the present paper to formulate explicitely such a connection and to show that it seems possible to accomodate in a unique scheme both Landau ideas and their modern counterparts, i.e. renormalization group theories of the critical point. In addition this scheme is sufficiently precise that a completely rigorous version of it should follow in a reasonable time.

Conceptually the main ingredient of our approach is the theory of large fluctuations for dependent random variables supplemented by the wisdom gained during over ten years of activity with the renormalization group (R.G.).

Let us illustrate the starting point of our argument. A typical prediction of the theory of large fluctuations tells us that the probability that the mean spin in a region Λ, $\psi_\Lambda = \sum_{i\in\Lambda} \psi_i/|\Lambda|$ be larger than a given value $\phi > 0$ is of the order

(1.1) $$P(\psi_\Lambda \geq \phi) \simeq e^{-|\Lambda|\Gamma(|\Lambda|,\phi)}$$

for $|\Lambda|$ sufficiently large. $\Gamma(|\Lambda|,\phi)$ is given by

(1.2) $$\Gamma(|\Lambda|,\phi) = \sup_{\theta>0} \{ \theta\phi - \frac{1}{|\Lambda|} \ln E(e^{\theta \sum_{i\in\Lambda} \psi_i}) \}$$

and is the Legendre transform of $\frac{1}{|\Lambda|} \ln E(e^{\theta \sum_{i\in\Lambda} \psi_i})$

$E(\cdot)$ denotes expectation with respect to the given Gibbs measure. The meaning of $\approx$ has to be taken in a logaritmic sense i.e.

(1.3) $$\lim_{|\Lambda|\to\infty} \Gamma(|\Lambda|,\phi) = V_{eff}(\phi) = - \lim_{|\Lambda|\to\infty} \frac{1}{|\Lambda|} \ln P(\psi_\Lambda \gtrsim \phi)$$

$\Gamma(\infty, \phi) = V_{eff}(\phi)$ is what physicists usually call the effective potential. But $\Gamma(|\Lambda|, \phi)$ is also the basic object of Landau theory and in fact can be identified with $\beta F / |\Lambda|$. Landau theory is then developed on the assumption that near the critical point the expansion

$$\lim_{|\Lambda|\to\infty} \frac{\beta F(|\Lambda|,\beta,\phi)}{|\Lambda|} = V_{eff}(\phi) = \frac{1}{2} a \phi^2 + \frac{1}{4} b \phi^4 + \ldots..$$

is valid.

A formula like (1.3) or (1.1) in probability theory would be used to estimate P in terms of Γ. Here we want to adopt the opposite attitude and use (1.3) as a definition of $V_{eff}(\phi)$. One reason for doing so is that in (1.3) a limit theorem is involved and this allows to make direct contact with R.G. methods. Actually there is a deep connection between the theory of large fluctuations and the R.G. that will be illustrated in Section 3. This connection can be usefully exploited for example to study $P(\psi_\Lambda \gtrsim \phi)$ near the critical point by R.G. methods. This seems to be the way followed by Bleher in a recently announced work /3/. However also a literal exploitation of

(1.3) is possible and this opens the way to a direct and in principle non perturbative calculation of the effective potential /4/.

The effectiveness of this point of view will be illustrated within the context of hierarchical models where exact recursion relations are available. It will be shown in particular that a quantity like the critical temperature becomes more accessible from our point of view while being only an indirect byproduct of the usual R.G. approach which is more concerned with model independent properties like critical indices. Actually it is the critical surface that can be studied directly in this way by varying the model within a given class. One can say that our approach while being inspired by the R.G. and especially by the idea of recursion is somehow complementary to it in so far as it tries to keep all the information about the initial model.

We now outline the content of the following sections. Section 2 contains a short course on large fluctuations and parallels the well established treatment in the case of independent random variables. In writing it I was strongly influenced by a lecture of S.R.S. Varadhan on that subject. Section 3 discusses the general connection between the theory of large fluctuations and the R.G. in its probabilistic interpretation /5/. Some earlier results by Bleher are briefly commented in the light of the present approach. In Section 4 hierarchical models are introduced and sample computations are made to illustrate the approach in a concrete case. Finally in Section 5 we make some remarks on the general case and indicate directions for possible generalizations.

2. Large Fluctuations /6/

For definiteness we consider a ferromagnetic lattice system of unbounded spins ψ_i in a cubic region Λ described by a Gibbs measure μ_Λ^β . We assume that $E(\psi_i) = 0$. We now sketch an argument leading to the estimate (1.1). This will be obtained by establishing upper and lower bounds.

<u>Upper Bound</u> - From the exponential Chebysheff inequality we have for $\theta > 0, \phi > 0$.

$$P(\psi_\Lambda \geqslant \phi) \leqslant e^{-|\Lambda|\theta\phi} E(e^{\theta \sum_{i\varepsilon\Lambda} \psi_i})$$

If we optimize with respect to θ we obtain

$$(2.1) \qquad P(\psi_\Lambda \geqslant \phi) \leqslant e^{-|\Lambda| \Gamma(|\Lambda|, \phi)}$$

with Γ given by (1.2).

<u>Lower Bound</u> - This is slightly more difficult. Let θ be defined by

$$(2.2) \qquad \begin{aligned} \Gamma(|\Lambda|,\phi) &= \sup_{\theta>0} \left(\theta\phi - \frac{1}{|\Lambda|} \ln E(e^{\theta \sum_{i\varepsilon\Lambda} \psi_i})\right) \\ &= \theta^* \phi - \frac{1}{|\Lambda|} \ln E(e^{\theta^* \sum_{i\varepsilon\Lambda} \psi_i}) \end{aligned}$$

From this it follows that

$$(2.3) \qquad \phi = \frac{1}{|\Lambda|} \frac{E(\Sigma_{i\varepsilon\Lambda} \psi_i e^{\theta^* \sum_{i\varepsilon\Lambda} \psi_i})}{E(e^{\theta^* \sum_{i\varepsilon\Lambda} \psi_i})} = \frac{1}{|\Lambda|} \frac{E_\theta(\theta^*)}{E(\theta^*)}$$

where $E_\theta(\theta^*)$ is the derivative of E calculated for $\theta = \theta^*$. Define the new probability measure (Cramer transform)

(2.4) $$dG_\Lambda = \frac{1}{E(\theta^*)} e^{\theta^* \sum_{i\in\Lambda} \psi_i} d\mu_\Lambda^\beta$$

Eq. (2.3) says that with respect to G_Λ the mean spin in Λ is equal to ϕ. Consider now the obvious chain of inequalities

$$P(\psi_\Lambda \geq \phi) = E(\theta^*) \int_{\sum_{i\in\Lambda} \psi_i/|\Lambda| \geq \phi} e^{-\theta^* \sum_{i\in\Lambda} \psi_i} dG_\Lambda \geq$$

$$\geq E(\theta^*) \int_{\phi \leq \sum_{i\in\Lambda} \psi_i/|\Lambda| \leq \phi+\varepsilon} e^{-\theta^* \sum_{i\in\Lambda} \psi_i} dG_\Lambda \geq$$

(2.5)

$$\geq E(\theta^*) \quad e^{-|\Lambda| \, \theta^*(\phi + \varepsilon)} \int_{\phi \leq \sum_{i\in\Lambda} \psi_i/|\Lambda| \leq \phi + \varepsilon} dG_\Lambda =$$

$$= e^{-|\Lambda| (\Gamma(|\Lambda|,\phi)+\theta^*\varepsilon)} \int_{\phi \leq \sum_{i\in\Lambda} \psi_i/|\Lambda| \leq \phi + \varepsilon} dG_\Lambda$$

The integral in the last expression is a number $p(|\Lambda|, \varepsilon) < 1$ that we expect to be close to 1/2 as much as we want if, for fixed ε, we take $|\Lambda|$ sufficiently large. In conclusion we have the estimate

(2.6) $$P(\psi_\Lambda \geq \phi) \geq e^{-|\Lambda| (\Gamma(|\Lambda|,\phi) + \alpha(\varepsilon,|\Lambda|)}$$

where α , due to the arbitrariness of ε , for $|\Lambda|$ sufficiently large can be made as small as we wish. Combining (2.1) and (2.6) we obtain the expression (1.3) for the effective potential $V_{eff}(\phi)$.

Some comments are in order. First of all it is easy to see that the above discussion would have remained essentially unchanged if instead of $P(\psi_\Lambda \geq \phi)$ we had considered $P(\phi \leq \psi_\Lambda \leq \phi+\varepsilon)$. This means, if P admits a sufficiently regular density $p(|\Lambda|,\phi)$, that in addition to (1.3) we have the relationship

(2.7) $$V_{eff}(\phi) = -\lim_{|\Lambda|\to\infty} \frac{1}{|\Lambda|} \ln p(|\Lambda|,\phi)$$

This is the form of the large fluctuation theorem that we shall use in Section 4.

From (1.3) it follows that $V_{eff}(\phi)$ is a monotonic function of ϕ, for ϕ of constant sign.

It is clear that various properties, that must be verified for each particular model, were assumed in the above argument. In particular for the lower bound differentiability of $E(\theta)$ and the law of large numbers for the mean spin played an important role.

In our reasoning the thermodynamic limit coincided with the limit $|\Lambda| \to \infty$. We could have started from an infinite volume Gibbs state and nothing essential would have changed for $\beta<\beta_{cr}$. For $\beta>\beta_{cr}$ the analysis is more complicated due to the possibility of phase coexistence.

Obviously large fluctuation theorems are not restricted to the mean spin. The following is another example which makes contact with mathematical objects widely used by physicists.

Consider for a given lattice vector $\underline{k}$, products of random variables $\psi_{\underline{i}}\,\psi_{\underline{i+k}}$ and try to estimate the probability

(2.8) $$P\left(\frac{1}{|\Lambda|}\sum_{i\varepsilon\Lambda}(\psi_{\underline{i}}\,\psi_{\underline{i+k}} - E(\psi_{\underline{i}}\,\psi_{\underline{i+k}})) \geq \Delta_{\underline{k}}\right)$$

An argument of the same type as before now gives

$$(2.9) \qquad P \approx e^{-|\Lambda|\,\Gamma(|\Lambda|,\Delta_{\underline{k}})}$$

where

$$(2.10) \qquad \Gamma(|\Lambda|,\Delta_{\underline{k}}) = \sup_{\mu>0} \{ \mu\Delta_{\underline{k}} - \frac{1}{|\Lambda|} \ln E\,(e^{\mu \sum_{i\varepsilon\Lambda} \psi_{\underline{i}}\psi_{\underline{i}+\underline{k}}}) + \frac{\mu}{|\Lambda|} \sum_{i\varepsilon\Lambda} E(\psi_{\underline{i}}\psi_{\underline{i+k}}) \}$$

This Γ is a special case of what physicists call the second Legendre transform /7/.

As a final remark we mention a more general form of the large fluctuation estimates. Suppose A is a measurable set on the line. Then

$$(2.11) \qquad e^{-|\Lambda|(V(A)+h)} \lesssim P(\psi_\Lambda \in A) \lesssim e^{-|\Lambda|(V(\bar{A})-h)}$$

where

$$(2.12) \qquad V(A) = \inf_{\phi\varepsilon A} V_{eff}(\phi)$$

$\bar{A}$ is the closure of A and h is an error term.

3. Connection with the Renormalization Group

To illustrate the connection with the R.G. it is convenient to consider first the case of independent random variables. It is now well established the connection between the R.G. and limit theorems of probability theory /5/.

A classical problem in limit theorems for independent variables is the estimate of the correction terms to the central limit theorem when the argument of the limit distribution becomes very large. A well known result in this domain is for example the following /8/: suppose we have a sequence of zero mean independent identically distributed random variables ψ_i and we want to estimate $P(\sum_{i=1}^{n} \psi_i/\sqrt{n} \geqslant x)$ when $x = o(\sqrt{n})$. Then under the general condition that the initial distribution admit a Laplace transform i.e. $E(e^{\theta\psi_i}) < \infty$ for $0 \leqslant \theta < a$ (Cramer condition)

(3.1) $$P\left(\sum_{i=1}^{n} \psi_i/\sqrt{n} \geqslant x\right) \approx e^{-n\sum_{k=2}^{s} \Gamma_k \left(\frac{x}{\sqrt{n}}\right)^k}$$

if $x \to \infty$ and $\lim_{n\to\infty} x^{s+1}/n^{\frac{1}{2}(s-1)} = 0$

The function $\Gamma(z) = \sum_{k=2}^{\infty} \Gamma_k z^k$ is the Legendre transform of the logarithm of $E(e^{\theta\psi_i})$. The sign $\approx$ has to be understood in the usual sense of logarithm dominance. If $x = x_0\sqrt{n}$ then the whole function $\Gamma(z)$ contributes and we are back to the type of large deviation estimates considered in the previous Section. (3.1) shows that the central limit theorem is only a first approximation when x is large. Let us go back to the dependent variables case.

By extending, under a natural condition that will be explained in a moment, the large fluctuation discussion of the previous Section we expect a result like (3.1) to hold true for reasonable Gibbs distributions which satisfy the central limit theorem away from the critical point. We then see that higher order vertex functions i.e. the coefficients of an expansion of $\Gamma(|\Lambda|,\phi)$ in powers of ϕ determine the corrections due to large fluctuations to the central limit theorem for the one-block distribution.

More interesting is the situation right at the critical point. Suppose first that the one-block limit distribution is Gaussian but with an anomalous normalization as it is the case for example in hierarchical models for appropriate choices of the parameter. Then instead of (3.1) we expect an estimate of the form

$$(3.2)\qquad P\Big(\sum_{i\in\Lambda} \psi_i/|\Lambda|^{\rho} \geqslant x\Big) \approx e^{-|\Lambda| \sum_{k=2}^{s} \Gamma_k(|\Lambda|)\left(\frac{x}{|\Lambda|^{1-\rho}}\right)^k}$$

with $\rho > \frac{1}{2}$, $x \to \infty$, $\lim_{|\Lambda|\to\infty} x^{s+1}/|\Lambda|^{(1-\rho)s-\rho} = 0$

We see immediately that for the distribution to be a Gaussian the coefficient Γ_2 must vanish for $|\Lambda|\to\infty$ (as expected) with the following law

$$(3.3)\qquad \Gamma_2 \quad \alpha \; |\Lambda|^{1-2\rho}$$

The study of the $|\Lambda|$ dependence of the Γ_k is interesting because it may provide a way to a rigorous foundation of the so called finite size scaling. Similar considerations can be developed for a non-gaussian limit distribution. In this case if we write the limit distribution in the form $e^{-g(x)}$ the large fluctuation theory will give for large enough x

$$(3.4)\qquad e^{-g(x) + \gamma(|\Lambda|,x)} \approx e^{-|\Lambda|\Gamma\left(|\Lambda|,\frac{x}{|\Lambda|^{1-\rho}}\right)}$$

The correction $\gamma(|\Lambda|,x)$ will have the form

(3.5) $$\gamma(|\Lambda|,x) = g(x) - |\Lambda|\Gamma(|\Lambda|,\frac{x}{|\Lambda|^{1-\rho}})$$

and must vanish when $|\Lambda| \to \infty$. Notice that now we have to consider the full $\Gamma(|\Lambda|,\phi)$ as the Γ_k will diverge for $|\Lambda| \to \infty$ and k sufficiently large as we know from the conventional R.G. wisdom.

The main point we have learnt from the above discussion is that the fixed point effective potential i.e. g(x) is related to the asymptotic behaviour of $\Gamma(|\Lambda|,\phi)$ when the argument ϕ is properly rescaled by a power of $|\Lambda|$. In other words $\Gamma(|\Lambda|,\phi)$ is interesting as a function of two variables as it determines in different limits either $V_{eff}(\phi)$ or g(x). We now come to the condition which insures the validity of (3.1), (3.2), (3.4). By arguing along the lines of Section 2 it is easily seen that what is required now are laws of large numbers which guarantee that the distribution of $(\sum_{i\epsilon\Lambda}\psi_i - \bar{\phi}|\Lambda|)/|\Lambda|^{\alpha}$ is concentrated at zero when $|\Lambda| \to \infty$ for any α satisfying $1/2 < \alpha \leq 1$ or $\rho < \alpha \leq 1$ according to which case applies. $\bar{\phi}$ is the expectation of ψ_i.

We finally come to a basic question. To what extent the connections revealed by the large fluctuation theory can be used as a computational scheme for $\Gamma(|\Lambda|,\phi)$? From the discussion of this Section it is clear that to the extent that one has control over pre-asymptotic terms in the R.G. recursions one has informations on $\Gamma(|\Lambda|,\phi)$ for small ϕ. In fact much of the work of Bleher /9/ on the hierarchical model and in particular his results on the relationship between the magnetic field and the magnetization near the critical point constitute a calculation of $V_{eff}(\phi)$ via direct large fluctuation estimates.

In the next Section we will show that for hierarchical models, which have a built in recursive structure, an equation like (2.7) provides an effective and more direct approach to the calculation of $V_{eff}(\phi)$.

4. Direct Calculation of $V_{eff}(\phi)$ for Hierarchical Models

We recall the basic recursion relation of Dyson type hierarchical models /10/. We have a countable set, Z for definiteness, and a decreasing sequence of partitions $P_0 \quad P_1 \ldots$ satisfying 1) P_0 is the partition of Z into separate points 2) any element of the partition P_n consists of 2 elements of the partition P_{n-1}. To each point i of Z we associate a spin variable ψ_i and we consider the mean spin of an element of the partition P_n $\psi_n = \Sigma_1^{2^n} \psi_i / 2^n$. If we define

$$g_n(\phi)\, d\phi = P(\phi \leq \psi_n < \phi + d\phi) \tag{4.1}$$

the models are defined recursively by

$$g_n(\phi) = L_n\, e^{\beta c^n \phi^2} \int d\phi'\, g_{n-1}(\phi')\, g_{n-1}(2\phi-\phi') \tag{4.2}$$

where $1 < c < 2$ and L_n is fixed by the normalization condition.

For each initial distribution $g_0(\phi)$ of the spins ψ_i we obtain a different model. Of special interest is the ϕ^4 hierarchical model defined by the distribution

$$g_0(\phi) = L_0\, e^{-\frac{1}{2}\phi^2 - \frac{\lambda}{4}\phi^4} \tag{4.3}$$

We now define the effective potential at level n

$$V_n(\phi) = -\frac{1}{2^n} \ln g_n(\phi) \tag{4.4}$$

which by (4.2) satisfies the recursion equation

$$V_n(\phi) = -\frac{1}{2^n} \ln L_n - \beta \left(\frac{c}{2}\right)^n \phi^2 - \frac{1}{2^n} \ln \int d\phi' \, e^{-2^{n-1}(V_{n-1}(\phi') + V_{n-1}(2\phi-\phi'))} \tag{4.5}$$

To understand the difference between (4.5) and the usual R.G. recursion for hierarchical models we consider first the trivial gaussian case obtained by setting $\lambda = 0$ in (4.3). This can be easily solved exactly and gives for β sufficiently small

$$V_n(\phi) = \frac{1}{2} \left[1 - \beta c \sum_{k=0}^{n} \left(\frac{c}{2}\right)^k\right] \phi^2 + \nu_n \tag{4.6}$$

Therefore going to the limit $n \to \infty$

$$V_{eff}(\phi) = \frac{1}{2} \left[1 - \frac{2\beta c}{2-c}\right] \phi^2 \tag{4.7}$$

from which we obtain the critical temperature

(4.8) $$\beta_{cr} = \frac{1}{c} - \frac{1}{2}$$

If we had considered the usual recursion for the renormalized density distribution

(4.9) $$f_n(x) = g_n(c^{-n/2}x)$$

the critical point would have been defined as the only β for which f_n converges to the gaussian fixed point $f_\infty(x) = A\exp(-\beta\frac{c}{2-c}x^2)$ and an easy calculation shows that this corresponds to the β_{cr} for which the difference between two divergent expressions

(4.10) $$2\beta\sum_{k=0}^{n}\left(\frac{2}{c}\right)^k - \left(\frac{2}{c}\right)^{n+1}$$

is finite. It is immediately seen that this β_{cr} coincides with (4.8).

The recursion (4.5) converges for all $\beta < \beta_{cr}$ and for each n one can calculate an approximate $\beta_{n\,cr}$ as the β for which the coefficient of ϕ^2 vanishes.

We consider now the more interesting case (4.3) and we assume λ small. The first recursion can be performed analytically and we obtain

$$V_1(\phi) = \frac{1}{2}\phi^2 + \frac{\lambda}{4}\phi^4 - \beta\frac{c}{2}\phi^2 - \frac{1}{2}\frac{(1+3\lambda\phi^2)^2}{4\lambda} -$$

(4.11)

$$-\frac{1}{4}\ln(1+3\lambda\phi^2) - \frac{1}{2}\ln K_{1/4}\left(\frac{(1+3\lambda\phi^2)^2}{4\lambda}\right) + \nu_1$$

where K_ν are the modified Bessel functions. For λ small we can use the asymptotic expansion of $K_{1/4}$ and we obtain uniformly in ϕ

$$V_1(\phi) = \frac{1}{2}\phi^2 + \frac{\lambda}{4}\phi^4 - \beta\frac{c}{2}\phi^2 +$$

(4.12)

$$+\frac{1}{4}\ln(1+3\lambda\phi^2) + \nu_1 + \ldots\ldots$$

We would like to emphasize that this is not a perturbative expansion in λ as can be seen from the fact that the argument of $K_{1/4}$ is $\frac{(1+3\lambda\phi^2)^2}{4\lambda}$

The structure of the logarithmic term reminds of the one-loop resummations of conventional field theory. The next iteration is more difficult. However if we concentrate on the coefficient of ϕ^2 which is the one which determines the critical temperature, a straightforward calculation gives

(4.13) $$V_2(\phi) = \frac{\phi^2}{2}\left(1 - \beta c - \beta\frac{c^2}{2} + 2\lambda f(\lambda,\beta)\right) + O(\phi^4) + \ldots\ldots$$

where $f(\lambda,\phi)$ can be easily obtained either numerically or analytically in terms of an asymptotic series in λ. For example we have

$$(4.14) \qquad f(0,\beta) = \frac{3}{4} \left(1 + \frac{1}{2} \frac{1}{1 - \beta c}\right)$$

and this determines the correction of order λ to the critical temperature of the second iteration. We do not pursue here the study of the higher iterations as it seems to us that our sample calculation indicates clearly the path that one can follow to obtain the effective potential.

We remark that to the extent that the scheme can be implemented for different initial distributions it provides direct information on the structure of the critical surface.

5. Concluding Remarks

So far ve have limited our explicit calculations to hierarchical models. Hierarchical models, besides being one of the most successful areas of application of rigorous R.G. ideas, seem to provide if properly constructed, substantial information about more complex theories as some recent work in constructive field theory indicates /11/. Within our context there are features of an equation like (4.5) that we expect to hold in more general situations. In (4.5) the interaction is represented by the term $\beta \left(\frac{c}{2}\right)^n$ which goes to zero with increasing n (remember that $c < 2$). Consider now a more realistic case with short range forces: if we try to compose the distributions of the mean spins belonging to contiguous cubes (for example 4 squares of side L to give a square of side 2L in two dimensions), the interaction term should be of the order of the ratio between the surface of the blocks and their volume. This at least away from the critical point. One can then envisage that approximate recursions can be set up once the size of the blocks becomes sufficiently large. For blocks up to some critical size the distribution should be computed directly. This is not impossible and examples of direct numerical calculations can be found, although in a different perspective, in a recent article by Shulman /12/.

In this paper we have been concerned only with one-block distributions. Actually estimates similar to those of Section 2 can be obtained for simultaneous fluctuations in different blocks. These generalizations will not be discussed here.

Acknowledgments

During the preparation of this paper I benefitted from discussions with M.Cassandro, J.P.Eckmann, E.Seiler and K.Yoshida. I should mention that originally my interest in large fluctuations for Gibbs fields was considerably stimulated by a series of seminars given by C.Boldrighini and L.Triolo at the University of Rome during the academic year 1978-79. Finally I would like to express my gratitude to Prof. M.Mebkhout for the warm hospitality at the University of Marseille-Luminy where part of this work was written.

References

/1/ L.Landau, E.Lifchitz, "Physique Statistique" MIR, Moscou 1967. A more complete exposition of Landau theory can be found in A.Patashinsky, V.Pokrowsky, "Fluctuation Theory of Phase Transitions" 2nd edition, Nauka, Moscow 1982 (Russian).

/2/ The effective potential first appeared in a perturbative context in J.Goldstone, A.Salam, S.Weinberg, Phys. Rev. 127, 965 (1962). It is a special case of the more general notion of effective action introduced in G.Jona-Lasinio, Nuovo Cimento 34, 1790 (1964) and developed in H.Dahmen, G.Jona-Lasinio, Nuovo Cimento A52, 807 (1967). The physical meaning of the effective action was especially discussed in an appendix to the last paper. See also S.Coleman "Secret Symmetries" in Laws of Hadronic Matter" ed A.Zichichi, Acad. Press, New York 1975.

/3/ P.M.Bleher, "Large Deviations Theorem near the Critical Point of the ψ^4-Hierarchical Model" Abstracts of the 1981 Vilnius Conference on Probability Theory and Mathematical Statistics.

/4/ I became aware after the completion of the present work that aa somewhat related point of view was proposed by R.Fukuda, Progr. Theor. Phys. 56, 258 (1976) within the usual field theoretic heuristic. I am indebted to K.Symanzik for informing me about Fukuda's work.

/5/ Probabilistic concepts in R.G. theory first appeared in P.M.Bleher, Ya. G.Sinai, Comm. Math. Phys. 33, 23 (1973). The general connection between the R.G. and limit theorems for random fields and the basic notion of stable (equivalently: automodel, self-similar) random field was introduced in G.Jona-Lasinio, Nuovo Cimento B26, 99 (1975), G.Gallavotti, G.Jona-Lasinio, Comm. Math. Phys. 41, 301 (1975) and independently in Ya. G.Sinai, Theory of Prob. and its Appl. XXI, 64 (1976). The notion of self-similar random field is the probabilistic counter part of the fixed point in the usual R.G.terminology: the physicist concept of universality corresponds in probability to that of domain of attraction. For an informal exposition of the connection between R.G. and probability see M.Cassandro, G.Jona-Lasinio, Advances in Physics 27, 913 (1978) where one can find additional references. A more recent perspective which includes field theory is in J.Frölich, T.Spencer, "Some Recent Rigorous Results in the Theory of Phase Transitions and Critical Phenomena" Seminaire Bourbaki, 34e anneé, 1981/82, n° 586.

/6/ Results on large deviations for Gibbs random fields are still very scanty. See however S.K.Pogosian, Uspehi, Mat. Nauk. 36, 201 (1981) (Russian) and the interesting preprint by R.Ellis, "Large Deviations and Other Limit Theorems for a class of Dependent Random Variables with Applications to Statistical Mechanics".

/7/ See for example H.D.Dahmen, G.Jona-Lasinio in Ref. /2/.

/8/ I.A.Ibragimov, Yu.V.Linnik, "Independent and Stationary Sequences of Random Variables" Groningen, Wolter Noordhoff Publ. 1971.

/9/ P.M.Bleher, Works of the Moscow Mathematical Society 33, 155 (1975) (Russian).

/10/ A systematic and rigorous discussion of hierarchical models is given in P.Collet, J.P.Eckmann, "A Renormalization Group Analysis of the Hierarchical Model", Lect. Notes in Physics 74, Springer, Berlin-Heidelberg-New York, 1979.

/11/ This trend was initiated by G.Gallavotti and his collaborators and is reviewed by G.Gallavotti in "Quantum Fields - Algebras, Processes" ed. L.Streit, Wien 1980. For more recent developments see K.Gawedsky, A.Kupiainen, contribution to this volume.

/12/ L.S.Shulman, J.Phys. A. 13, 237 (1980).

G.Jona-Lasinio
Istituto di Fisica
Università di Roma
GNSM and INFN - Roma, Italy.

THE BEREZINSKII-KOSTERLITZ-THOULESS TRANSITION
(ENERGY-ENTROPY ARGUMENTS AND RENORMALIZATION IN DEFECT GASES)

Jürg Fröhlich[1] and Thomas Spencer[2*]

Contents.

---

[1] Theoretical Physics, ETH-Hönggerberg, CH-8093 Zürich

[2] Courant Institute of Mathematical Sciences, New York University, 251 Mercer Street, New York, N.Y. 10012, U.S.A.

* Work supported in part by the NSF under grant DMR 81-00417

1. Introduction: The General Context, a Description of Some Models and Summary of the Main Results

The study of phase transitions and of the approach to critical points in physical systems in thermal equilibrium is an important part of equilibrium statistical mechanics, but its significance goes beyond statistical physics proper: It is crucial for condensed matter physics and for relativistic quantum field theory.

Phase transitions have been analyzed since the early days of statistical mechanics, as formulated by Maxwell, Boltzmann, Gibbs, Einstein and others. The first example of a phase transition which was understood in much detail and quite rigorously was the Bose-Einstein condensation exhibited by ideal, monatomic quantum (Bose) gases. However, the notion of phase transition remained somewhat vague and a mathematical theory of phase transitions rudimentary untill the period between 1936 and 1944. In 1936, Peierls invented his famous argument to establish the existence of a phase transition and spontaneous magnetization at low temperature in the two-dimensional Ising model [1]. In 1944, Onsager solved the two-dimensional Ising model [2], in the sense that he explicitly calculated the free energy of this model. Onsager's work had an important impact on the course of events in understanding phase transitions. The full power of Peierls' ideas was recognized only relatively late [3]. [In fact, as recently as in 1980/1981, we found it still possible to reinterpret the Peierls argument and put it into a more general context in such a way that new results which will be described in some detail in this article emerged.]

Quantitative theories of the type (or: the order) of phase transitions and of the approach to the critical point have remained somewhat incomplete and heuristic, untill now. Although the simplicity of the picture provided by Landau theory is appealing, its quantitative predictions, concerning critical exponents, for example, turn out to be incorrect in dimension $d \leq 4$; (superstructures used to "derive" it, like catastrophe theory, cannot change this fact). A proof that many of the predictions of Landau theory (mean field theory) are correct for certain lattice spin systems in dimension $d \geq 5$ forms the contents of other review articles in this volume. If and where applicable the renormalization group, in particular the ε-expansion, makes predictions which are in excellent agreement with numerical data and experiments. But the only mathematically rigorous information on the precise value of critical exponents in four or fewer dimensions comes from exact solutions of two-dimensional models and from exact results for Dyson's hierarchical model. [Qualitative information about the approach to critical points and rigorous bounds on exponents are reviewed in other articles in this volume.]

In spite of over forty years of intensive work on phase transitions and critical phenomena, our understanding of these matters is still rather unsatisfactory. If we exclude relativistic quantum field theory we know of just one very simple model of a continous system, the Widom-Rowlinson model, for which a phase transition with order parameter and symmetry breaking has been proven to exist [4], with the help of a Peierls-type argument. For a regularized two-dimensional Coulomb gas in the continuum the existence of a Kosterlitz-Thouless transition and of an interval of critical points can be established by using the methods of [5] which we review in detail in this article. However, we are still far from a mathematical understanding of realistic models describing crystalline solids and their melting, liquid crystals, ferromagnets, metal-insulator transitions, etc.

In spite of this somewhat discouraging state of affairs there has been some progress in our mathematical understanding of phase transitions and critical phenomena:

There are basically three, or - with some optimism - four, essentially different ways of rigorously establishing the existence of phase transitions in (lattice) systems in thermal equilibrium.

(a) Exact solution of a model. This can be achieved only in a very limited class of models such as the two-dimensional Ising model, the six - and the eight vertex models. If a model can be solved exactly one usually gains very detailed quantitative information, but sometimes basic physical mechanisms remain obscure. See [6] for reviews.

(b) Energy-entropy-, or Peierls arguments. They apply to a wide variety of models which can be interpreted as gases of discrete excitations such as the Ising model, the classical XY model ($\tilde{=}$ gas of vortices of integer vorticity), etc. They give detailed, quantitative information at very low temperatures and have a lot of intuitive appeal. See [1,3,7,8,5] and this article.

(c) A rigorous version of spin wave theory, based on infrared bounds, [9]. This method can be used to analyze a large class of lattice models with continuous symmetry groups, classical and quantum, including the classical Heisenberg model [9] and the quantum anti-ferromagnet [10]. It is the only presently known method permitting to analyze such models, but is restricted to systems with "reflection-positive" interactions [9]. When applicable it tends to give fairly accurate bounds on things like the critical temperature. For reviews of (b) and (c), see also [11,12,13].

(d) Rigorous analysis of the flow of coupling constants (or thermodynamic parameters) under renormalization group transformations. This method (used without any additional techniques) has so far been limited to hierarchical models [14], but appears to make rapid progress nowadays. It tends to be rather involved analytically, but has a lot of potential. See the contributions of Gawedzki and Kupiainen and others to this volume, and refs. given there.

In this article we describe in some detail methods to establish the existence of (Berezinskii-Kosterlitz-Thouless) transitions in two-dimensional models [15,5], as well as related results for higher dimensional models [16,17,18]. These methods have emerged from a combination of approaches (b) and (d).

Next, we describe the models which we have in mind. These models are classical lattice spin systems and systems dual or related to them, e.g. the Coulomb gas, Higgs models and the like. A classical lattice system is defined as follows: As our lattice we choose, for simplicity, Z^d, $d = 2,3,4,\dots$. At each site $x \varepsilon Z^d$ there is a classical spin $\vec{S}_x = (S_x^1,\dots,S_x^N)$, with a priori distribution $d\lambda(\vec{S}_x)$, where $d\lambda$ is some finite measure on R^N . Let $\Lambda \subset Z^d$ be some bounded region. The Hamilton function for a ferromagnetic spin system confined to the region Λ is given, for example, by

$$H_\Lambda = - \sum_{\substack{x,y\varepsilon\Lambda \\ |x-y|=1}} \vec{S}_x \cdot \vec{S}_y - h \sum_{x\varepsilon\Lambda} S_x^1 , \tag{1.1}$$

where h is an external magnetic field. The equilibrium state of this system at inverse temperature β is given by

$$d\mu_{\beta,h,\Lambda}(\vec{S}) = Z_{\beta,h,\Lambda}^{-1} e^{-\beta H_\Lambda(\vec{S})} \prod_{x\varepsilon\Lambda} d\lambda(\vec{S}_x) \ , \tag{1.2}$$

where $Z_{\beta,h,\Lambda}$ is the partition function.

When $N=2$ it is convenient to introduce polar coordinates:

$$\vec{S}_x = r_x(\cos\theta_x, \sin\theta_x) \tag{1.3}$$

We shall focus our attention on the classical rotor (XY) model for which

$$d\lambda(\vec{S}_x) = \delta(|\vec{S}_x|^2 - 1)d^2S_x \ , \tag{1.4}$$

which corresponds to $r_x = 1$ and Lebesgue measure $d\theta_x$. We shall also consider the Z_n clock models characterized by

$$\vec{S}_x \varepsilon \ Z_n \ , \tag{1.5}$$

with n large. The model with $n=2$ is the Ising model.

From our results for the rotor- (and the Z_n -)models one can derive results for models with more general distributions $d\lambda$ with the help of correlation inequalities [19,20].

We define the free energy

$$f_\Lambda(\beta,h) = -\frac{1}{\beta|\Lambda|}\log Z_{\beta,h,\Lambda}, \tag{1.6}$$

$|\Lambda| \equiv \#$ sites in Λ, and the correlation functions

$$<\prod_{i=1}^{n} S_{x_i}^{\alpha_i}>(\beta,h) \equiv \int d\mu_{\beta,h,\Lambda}(\vec{S}) \prod_{i=1}^{n} S_{x_i}^{\alpha_i} \ . \tag{1.7}$$

For $N=1,2$ the thermodynamic limit, $\Lambda\uparrow Z^d$, can be shown to exist, for a class of $d\lambda$ containing the ones defined in (1.4) and (1.5), by means of correlation inequalities [19]. To denote the limiting quantities we shall drop the subscript Λ . In zero magnetic field, $h=0$, we write $f(\beta)$ for $f(\beta,0)$ and $<(\cdot)>(\beta)$ for $<(\cdot)>(\beta,0)$. Finally, unless specified otherwise, $d\lambda(\vec{S})$ will be as in (1.4).

We are interested in the behaviour of $f(\beta,h)$ as a function of β and h , and in the behaviour of correlation functions at large distances.

In the classical rotor model, $f(\beta,h)$ is analytic in h as long as $\mathrm{Re}h\neq 0$; it is concave in β and smooth for real $h\neq 0$.

It is a standard consequence of high temperature expansions that, for sufficiently small β and all h ,

$$<(\cdot)>(\beta,h) \quad \text{is } \underline{\text{unique}} \ , \tag{1.8}$$

and correlations exhibit <u>exponential clustering</u>. If $h\neq 0$, there is a unique translation-invariant equilibrium state in the thermodynamic limit [21], exhibiting exponential cluster properties. When $h=0$, and at values of β at which $f(\beta)$ is continuously differentiable, all translation-invariant states of the classical rotor model in the thermodynamic limit can be labelled by an angle $\phi\varepsilon[0,2\pi)$: Let $<(\cdot)>^0(\beta)$ be the state obtained by letting $h\downarrow 0$. Then any other translation-invariant state is of the form

$$<(\cdot)>^{\phi}(\beta) \equiv <R_{\phi}(\cdot)>^0(\beta) \ , \tag{1.9}$$

where R_{ϕ} rotates every spin through an angle ϕ . See [22] and refs. given there. In two dimensions,

$$<(\cdot)>^{\phi}(\beta) = <(\cdot)>^0(\beta) \ , \quad \text{for all } \beta \ .$$

In fact, every state of the two-dimensional rotator in 0 magnetic field is invariant under rotations, R_{ϕ} , of all the spins. This is a special case of a general theorem, due to Dobrushin and Shlosman [23]. It can be understood in terms of spin wave theory [24]:

Let

$$R^N_{\phi,x} = \begin{cases} R_\phi \ , & |x|\leq N \\ R_{(2-|x|/N)\phi} \ , & N\leq|x|\leq 2N \\ 1 \ , & |x|\geq 2N \ . \end{cases} \tag{1.10}$$

Given a configuration, $\vec{S}$, of spins, let $\vec{S}_N$ be the configuration obtained from $\vec{S}$ by rotating $\vec{S}_x$ by $R^N_{\phi,x}$, for all x. Let $<(\cdot)>(\beta)$ be an arbitrary equilibrium state in the thermodynamic limit. Then in two, but not in three or more, dimensions

$$<|H(\vec{S}) - H(\vec{S}_N)|>(\beta) \leq \text{const.} \ , \tag{1.11}$$

uniformly in N. It follows - by an elementary application of Jensen's inequality - that $<(\cdot)>(\beta)$ and $<R^N_{\phi,\cdot}(\cdot)>(\beta)$, with $R^N_{\phi,x}$ given by (1.10), are absolutely continuous with respect to each other, i.e. not "orthogonal", even when $N\to\infty$. Thus (choosing $<(\cdot)>(\beta)$ to be an extremal state, w.l.o.g.) we conclude that

$$\lim_{N\to\infty} <R^N_{\phi,\cdot}(\cdot)>(\beta) = <R_\phi(\cdot)>(\beta) = <(\cdot)>(\beta) \ . \tag{1.12}$$

For details of this argument, see the first two refs. in [24].

In dimension $d\geq 3$, the argument sketched above does not work, and the conclusion is wrong. The original proof [9] of the fact that at low temperatures there is non-zero spontaneous magnetization made use of a rigorous version of spin wave theory: Let Ω be a large, finite (hyper) cube in $Z^d, d\geq 3$, and

$$\vec{S}_\Omega = \frac{1}{|\Omega|}\sum_{x\in\Omega} \vec{S}_x \ , \quad \vec{S}_x = (S^1_x,\dots,S^N_x) \ , \ N \geq 1 \ .$$

One expects that for large β and large $|\Omega|$

$$\vec{S}_\Omega = M\vec{e}_1 + \delta\vec{S}_\Omega, \quad M \equiv |<\vec{S}_0>(\beta,h)| \tag{1.13}$$

when $h>0$; ($\vec{e}_1 \equiv$ unit vector in 1-direction). The fluctuation, $\delta\vec{S}_\Omega$, of $\vec{S}_\Omega$ around $M\vec{e}_1$ is expected to be approximately Gaussian, with variance $\lesssim\beta^{-1}|\Omega|^{(2-d)/d}$. These ideas can be formalized as follows: Let $G^c(k)$ be the Fourier transform of

$<\delta\vec{S}_0 \cdot \delta\vec{S}_x>(\beta,h)$ in x ; $d\lambda$ as in (1.4). By using the transfer matrix method one can prove that

$$0 \leqq G^c(k) \leqq N\beta^{-1}[2d-2\sum_{\alpha=1}^{d} \cos k_\alpha]^{-1} . \tag{1.14}$$

The upper bound is called infrared bound and is what spin wave theory predicts for large β and N . Let

$$I_d \equiv (2\pi)^{-d}\int d^dk[2d-2\sum_{\alpha=1}^{d} \cos k_\alpha]^{-1} . \tag{1.15}$$

Note that $I_d < \infty$, for $d \geqq 3$. By (1.14) and (1.15)

$$0 \leqq < |\delta\vec{S}_0|^2 > (\beta,h) \leqq N\beta^{-1} I_d ,$$

hence, using (1.13) and $<|\vec{S}_0|^2>(\beta,h) = 1$,

$$M^2 \geqq 1-N\beta^{-1}I_d$$

which is positive for $\beta > NI_d$, for arbitrary $h>0$. Thus

$$\lim_{h\downarrow 0} <\vec{S}_0>(\beta,h) \neq 0 , \tag{1.16}$$

for $\beta > NI_d$. Since for small $\beta, <\vec{S}_0>(\beta,0) = 0$, see (1.8), we conclude that there is a phase transition, accompanied by spontaneous symmetry breaking and spontaneous magnetization. By (1.15)

$$\delta\vec{S}_\Omega \lesssim \sqrt{N\beta^{-1}|\Omega|^{(2-d)/d}} .$$

This becomes small for large Ω when $d \geqq 3$, but is divergent when $d \leqq 2$, in accordance with the argument sketched in (1.10) - (1.12).

Next, we define the exponential decay rate of correlations (inverse correlation length), m , and the magnetic susceptibility χ :

$$m(\beta) \equiv \lim_{|x|\to\infty} \frac{1}{|x|} \log<\vec{S}_0 \cdot \vec{S}_x>(\beta) , \tag{1.17}$$

(and $x \to \infty$ in the direction of a lattice axis). Since for $\beta > NI_d, d \geqq 3$,

$$<\vec{S}_0 \cdot \vec{S}_x>(\beta) \to M^2 > 0 ,$$

there exists some finite $\beta_c \leqq NI_d$ with the property that

$$m(\beta) = 0 , \quad \text{for all} \quad \beta \geqq \beta_c .$$

It is rigorously known that for $N=1,2$

$$\left.\begin{aligned} &\lim_{\beta \uparrow \beta_c} m(\beta) = 0 , \\ &\text{with} \quad m(\beta) \leqq \text{const.} \, (\beta_c - \beta)^{1/2} . \end{aligned}\right\} \qquad (1.18)$$

See [25] and refs. given there. The magnetic susceptibility χ is given by

$$\chi(\beta) = \sum_x <\vec{S}_0 \cdot \vec{S}_x>^c(\beta) \equiv \sum_x \{<\vec{S}_0 \cdot \vec{S}_x>(\beta) - M^2\} . \qquad (1.19)$$

(Note, for $h \neq 0$, $\chi(\beta,h) \equiv \sum_x <\vec{S}_0 \cdot \vec{S}_x>^c(\beta) = - \dfrac{\partial^2(\beta f(\beta,h))}{\partial h^2}$.)

It has been proven in [26] that

$$\lim_{\beta \uparrow \beta_c} \chi(\beta)^{-1} = 0 ,$$

with (1.20)

$$\chi(\beta) \geqq \text{const.} \, (\beta_c - \beta)^{-1} .$$

See also [27] for some details.
Results at very large values of β (asymptoticity of the low temperature expansion) for the classical rotor model may be found in [28].

When $N=1$, i.e. for the Ising model, the results summarized above also hold in two dimensions, since β_c(2-dim Ising) is finite. However, models with a continuous symmetry $(N \geqq 2)$ do not exhibit transitions accompanied by spontaneous magnetization in two dimensions; see (1.12). One of the main results reviewed in this article is the following result for the two-dimensional rotor model

Theorem A.

In the classical rotor model (N=2) there exists a finite constant β_0 such that for $\beta > \beta_0$

$$\text{const.}\,(1+|x-y|)^{-(1/2\pi\beta')} \leq \langle \vec{S}_x \cdot \vec{S}_y \rangle(\beta)$$

$$\leq \text{const.}_\varepsilon\,(1+|x-y|)^{-(1/(2\pi+\varepsilon)\beta)}, \quad (1.21)$$

for $\varepsilon>0$ arbitrarily small, $\frac{1}{2\pi} \leqq \beta'(\beta) \leq \beta$, with

$$\frac{\beta'(\beta)}{\beta} \to 1, \text{ as } \beta \to \infty . \quad (1.22)$$

The upper bound in (1.21) (Mc Bryan-Spencer upper bound) has been proven in [29]; see also (2.43). (The result is predicted by spin wave theory, the original proof employs "complex rotations of spins", a technique which is also used in Sect. 5 and extends to some quantum mechanical models.) The proof of the lower bound in (1.21) and of (1.22) is considerably more intricate; see [5]. By [30],

$$\beta_c(\text{rotor}) \geqq 2\beta_c(\text{Ising}) . \quad (1.23)$$

Theorem A asserts that $\beta_c(\text{rotor})<\infty$, in $d\geqq 2$. By [31]

$$|x|\langle \vec{S}_0 \cdot \vec{S}_x \rangle(\beta) \underset{|x|\to\infty}{\longrightarrow} +\infty, \text{ for } \beta \geqq \beta_c . \quad (1.24)$$

It is believed that

$$\langle \vec{S}_0 \cdot \vec{S}_x \rangle(\beta_c) \sim |x|^{-(1/4)}, \quad (1.25)$$

as $|x|\to\infty$, up to possible logarithmic corrections. (It is not particularly challenging to give a heuristic explanation of (1.25); see e.g. [32] and (3.10), but a proof seems hard.) The results (1.18) and (1.20) remain true in the two-dimensional, classical rotor model, i.e.

$$\lim_{\beta\uparrow\beta_c} m(\beta) = \lim_{\beta\uparrow\beta_c} \chi(\beta)^{-1} = 0 \quad (1.26)$$

Heuristically, one expects that $m(\beta) \sim \exp[-\text{const.}/(\beta_c-\beta)]$; [15,32]. It is a challenge to enthusiasts of renormalization group methods (method (d)!) to prove (1.25) and to prove the following

Conjecture.

For $N \geq 3$, $d = 2$,

$$m(\beta) > 0 \text{ , for all } \beta < \infty .$$

See [33]. It has been shown in [12] that $m(\beta) \leq \exp[-\text{const. } \beta/N]$.

Next, we consider some further lattice models mathematically closely related to the rotor model: Let $\phi(x)$ be an integer-valued "classical spin", $x \in Z^d$, $d = 2,3,\ldots$. The discrete Gaussian (DG) model is the model with an equilibrium state given by

$$\mu_\beta^{DG}(\phi) = \lim_{\Lambda \uparrow Z^d} Z_{\beta,\Lambda}^{-1} \prod_{\langle x,y\rangle \subset \Lambda} e^{-\frac{\beta}{2}(\phi(x)-\phi(y))^2} , \tag{1.27}$$

where $\langle x,y\rangle$ indicates that x and y are nearest neighbors. The solid-on-solid (s-o-s) model has an equilibrium state given by

$$\mu_\beta^{s\text{-}o\text{-}s}(\phi) = \lim_{\Lambda \uparrow Z^d} Z_{\beta,\Lambda}^{-1} \prod_{\langle x,y\rangle \subset \Lambda} e^{-\beta|\phi(x)-\phi(y)|} . \tag{1.28}$$

The DG- and the s-o-s model serve to analyze the so called roughening transition. Let $\langle(\cdot)\rangle(\beta)$ denote the expectation in $d\mu_\beta^{DG}$, or $d\mu_\beta^{s\text{-}o\text{-}s}$. Let us interpret ϕ as a height function describing the surface of a crystal or an interface. We can prove the following result [5]:

Theorem B.

In the two-dimensional DG- and s-o-s models, for β sufficiently small,

$$\langle(\phi(x)-\phi(y))^2\rangle(\beta) \geq \text{const.}_\beta \log|x-y| . \tag{1.29}$$

The proof of this result is very closely related to the proof of Theorem A: By a duality transformation, i.e. a Fourier transformation in the angular variables $\theta_x, x \in Z^2$, the rotor model is mapped onto a model very similar to the DG- and s-o-s models.

By means of a Peierls argument one can show that, for β large enough,

$$<(\phi(x)-\phi(y))^2>(\beta) \leqq \text{const.} , \qquad (1.30)$$

uniformly in x and y, for both the DG- and the s-o-s model. The infrared bound, see (1.14), also holds for the DG- and the s-o-s models and yields

$$<(\phi(x)-\phi(y))^2>(\beta) \leqq \begin{cases} \text{const.}_\beta \log |x-y| , & d = 2 \\ \text{const}'_\beta , & d \geqq 3 , \end{cases}$$

with $\text{const.}_\beta \sim \text{const.}'_\beta \sim \beta^{-1}$ in the DG model. See [34] and refs. given there.

Next we summarize our main results for the two-dimensional Z_n - models, see (1.2), (1.5), with n large.

<u>Theorem C.</u>

<u>When</u> n <u>is sufficiently large, there exist positive, finite numbers</u> $\underline{\beta} \leqq \beta' < \beta'' \leqq \overline{\beta}$ <u>such that</u>

1) $m(\beta) > 0$, <u>and the</u> Z_n - <u>symmetry is unbroken, for</u> $\beta < \underline{\beta}$

2) $m(\beta) = 0$, <u>and at large distances the</u> Z_n - <u>symmetry is enhanced to a full</u> $U(1)$ - <u>symmetry, for</u> $\beta' < \beta < \beta''$.

3) $m(\beta) > 0$ <u>and the</u> Z_n - <u>symmetry is spontaneously broken, for</u> $\beta > \overline{\beta}$.

It is expected that $\underline{\beta} = \beta'$, $\overline{\beta} = \beta''$ and that $\beta' < \beta''$, for $n \geqq 5$.

The methods on which the proofs of Theorems A-C are based and which we outline in Sects. 2-6 have proven to be rather general and powerful. Variants of these methods have enabled us to establish the existence of a transition and spontaneous magnetization at low temperature in the one-dimensional Ising model with $^1/r^2$ interaction energy [35], in the d-dimensional, classical XY model, $d \geqq 3$, and in abelian lattice gauge theories such as the four-dimensional $U(1)$ theory [16,17]. The unifying feature of these different problems which permits one to analyze them with very similar techniques is explained in the next section. Rather surprisingly, some of the basic ideas described in Sects. 4 and 6 have played a fairly important role in our recent proof [36] of absence of diffusion in Anderson's tight binding model for large disorder, but this topic cannot be included here.

2. Energy-Entropy Arguments In Defect Gases

In this section we sketch some features of a general theory the purpose of which it is to reformulate some class of spin systems, gauge theories and other physical systems as gases of interacting "topological" defects which one can analyze with the help of "generalized Peierls arguments". As a heuristic guide as to what results are predicted by that theory, the topological classification of defects in ordered media (see [37,38] and refs. given there) and simple energy-entropy considerations applied to individual defects, or "defect molecules" [5,35], play very useful roles.

The archetype of a mathematically rigorous energy-entropy argument applied to a "gas of defects" is the Peierls argument which we now sketch very briefly.

We consider an ordered medium (a simple charicature of a ferromagnet) with order parameter $\sigma_x \varepsilon \{-1,1\}$, $x \varepsilon Z^d$. The system is ordered completely if $\sigma_x = +1$, or $\sigma_x = -1$, for all x. Defects in this system represent obstructions against complete ordering. A bond $b \equiv \langle x,y \rangle \subset Z^d$ is "frustrated" (i.e. produces disorder) if

$$\sigma_b \equiv \sigma_x \sigma_y = -1 . \tag{2.1}$$

Let p be an arbitrary unit square (called "plaquette") of the lattice and let ∂p be its boundary. By (2.1)

$$\prod_{b \varepsilon \partial p} \sigma_b = 1 , \quad \text{for all } p . \tag{2.2}$$

Let b be a frustrated bond, in the sense of (2.1). Let $c \equiv c(b)$ be the $(d-1)$ - cell in the dual lattice, $(Z^d)^*$, passing through b. By (2.2) the subset, Γ, of $(Z^d)^*$ defined by

$$\Gamma \equiv \{c(b) \subset (Z^d)^* : \sigma_b = -1\} \tag{2.3}$$

consists of closed (d-1) - dimensional surfaces, γ , which are called Peierls contours. Separating regions where $\sigma_x = 1$ from regions where $\sigma_x = -1$, the contours γ are really wall defects (Bloch walls) and measure the amount of disorder. If the disorder is weak a typical contour γ has small volume (or area, length), $|\gamma| \equiv \#\{c\varepsilon\gamma\}$. If the disorder is large then there exist infinite contours with positive probability. The quality of ordering can be measured by means of the expectation value of the "order parameter", $\langle\sigma_x\rangle$, where $\langle(\cdot)\rangle$ describes the state of the system. Similarly one can define a "disorder parameter", (an operator creating a Bloch wall) which measures the strength of disorder; see e.g. [39,40].

We now briefly consider the example of the ferromagnetic Ising model, with Hamilton function given by

$$H(\sigma) = - \sum_{\langle x,y\rangle} \sigma_x \sigma_y ; \tag{2.4}$$

the magnetic field h is set=0 . The state of the system is assumed to be an equilibrium state at inverse temperature β , as defined in (1.2). (The measure $d\lambda(\sigma_x)$ assigns probability 1/2 to 1 and -1.) We impose + b.c. , i.e. $\sigma_x = +1$, for all $x \notin \Lambda$. We propose to estimate the probability, $p_-(\beta)$, that the spin at the origin is down, i.e. $\sigma_0 = -1$: If σ is a configuration for which $\sigma_0 = -1$ then there must exist at least one contour, γ , surrounding the origin, because $\sigma_x = +1$, for all $x \notin \Lambda$. We choose the outermost contour, denoted $\gamma(\sigma)$. Given a configuration σ with $\gamma(\sigma) = \gamma_0$, γ_0 an arbitrary, but fixed contour surrounding 0 , there exists a configuration σ' , with $\gamma(\sigma') \neq \gamma_0$, obtained by flipping all spins "inside γ_0" .

By (2.4), the energy of a contour γ , relative to the totally ordered state, is

$$E(\gamma) = 2 |\gamma| . \tag{2.5}$$

It is a well known fact that the total number of contours γ surrounding the origin and of volume $|\gamma|=2n$ is bounded by

$$c_d^{2n} \quad , \quad \text{for } d \geqq 2 \ , \tag{2.6}$$

where c_d is some finite, geometrical constant. Thus, the a priori probability, w_n, of choosing a particular contour of length $2n$ surrounding 0, is at least c_d^{-2n}. Defining an entropy S_n, as usual, by

$$S_n = - \ln w_n \ , \tag{2.7}$$

we conclude, using (2.6), that

$$S_n \leqq \text{const. } n \ .$$

Thus

$$\begin{aligned}
p_{-}(\beta) &= \sum_{\gamma_0 \text{ surrounding } 0} Z_\beta^{-1} \sum_{\sigma:\gamma(\sigma)=\gamma_0} e^{-\beta H(\sigma)} \\
&\leqq \sum_{\gamma_0} \left\{ \sum_{\sigma:\gamma(\sigma)=\gamma_0} e^{-\beta H(\sigma')} \right\}^{-1} \sum_{\sigma:\gamma(\sigma)=\gamma_0} e^{-\beta H(\sigma)} \\
&\leqq \sum_{\gamma_0} e^{-\beta E(\gamma_0)} \\
&= \sum_{n=4,6,8,\ldots} e^{S_n} e^{-2\beta n} \ , \qquad \text{by (2.5)} \\
&\leqq \sum_{n=4,6,8,\ldots} \left(c_d e^{-2\beta} \right)^n \ , \qquad \text{by (2.7), (2.6)} \ .
\end{aligned}$$

Thus $p_{-}(\beta) \to 0$, as $\beta \to \infty$, uniformly in Λ. It follows that

$$\langle \sigma_0 \rangle(\beta) = p_{+}(\beta) - p_{-}(\beta) = 1 - 2p_{-}(\beta) > 0 \ ,$$

for β large enough, in dimension $d \geq 2$. It is well known that

$$<\sigma_0>(\beta) = 0 \ , \quad \text{for small } \beta \ .$$

This can be understood, heuristically, as a consequence of defect condensation: For β small enough

$$e^{S_n} e^{-\beta E(\gamma : |\gamma| = n)} \uparrow + \infty \ , \tag{2.8}$$

as $n \to \infty$. One expects, therefore, that at high temperature, there exist infinitely long contours with positive probability, and the choice of b.c. becomes irrelevant.

In one dimension, a contour γ consists of a pair of frustrated bonds, i.e. of point defects. Let l be the distance between these frustrated bonds. There are l contours, γ , of size (diameter) l enclosing the origin 0 of the lattice Z . We therefore define the entropy S_l to be

$$S_l = l n l \ . \tag{2.9}$$

We now consider a one-dimensional Ising model with Hamilton function

$$H(\sigma) = \sum_{x,y} \frac{1}{|x-y|^{2+\delta}} (1 - \sigma_x \sigma_y) \tag{2.10}$$

Let σ_γ be the configuration of the one-dimensional Ising system which has precisely one contour γ . We set $E(\gamma) = H(\sigma_\gamma)$. If γ has diameter l then

$$E(\gamma) \sim \begin{cases} \text{const.} \ , & \delta > 0 \\ \text{const. } l n l \ , & \delta = 0 \\ \text{const. } l^{-\delta} \ , & \delta < 0 \ . \end{cases} \tag{2.11}$$

Interpreting $E(\gamma)$ as a two-body potential we find that for $\delta \leq 0$ there is a confining potential between frustrated bonds. Notice that if $\delta > 0$

$$e^{S_l} e^{-\beta E(\gamma:\mathrm{diam}(\gamma)=l)} \nearrow +\infty ,$$

as $l \to \infty$, for arbitrary β . We expect that the system is always disordered and $\langle(\cdot)\rangle(\beta)$ is unique, for all β . This can be proven by using the arguments between (1.11) and (1.12), for example. See [24]. When $\delta<0$ we expect ordering for β sufficiently large. This can be proven for a general class of one-dimensional systems, by using infrared bounds similar to the one in (1.14). See [9]. The most interesting case corresponds to $\delta=0$. The naive energy-entropy argument based on (2.9) and (2.11) predicts a transition, and this is indeed what happens. But the rigorous proof [35] is fairly involved, because of the long range interactions between contours. It employs techniques closely related to the ones in Sects. 4-6. Quite generally, energy-entropy analyses of gases of point defects tend to be much more subtle than energy-entropy analyses of gases of defects of dimension ≥ 1 .

We now turn to the discussion of a more general setting. We temporarily replace the lattice Z^d by a continuum, R^d (i.e. we pass to the scaling limit). This will enable us to use classical topological notions, not directly applicable to lattice systems. We consider a "physical" system whose configurations can be described by a random field $\vec{S}$, (e.g. a classical-spin field), with values in a compact manifold M , e.g. $M=S^N$, $N=0,1,2,\ldots$.

We propose to exhibit the topological defects of a configuration $\vec{S}$, [37,38]. For this purpose one assumes that $\vec{S}$ is continuous, except on surfaces of co-dimension ≥ 1 . Consider, as an example, a configuration $\vec{S}$ which is continuous, except on a hyperplane H^k of dimension $k \leq d-1$. The space of all continuous maps

$$\vec{S} : R^d \sim H^k \to M \tag{2.12}$$

can be decomposed into homotopy classes which are labelled by the elements of the homotopy groups, $\pi_{d-k-1}(M)$. [To see this, restrict $\vec{S}$ to the sphere at infinity of the space $R^d \sim H^k$, where $\vec{S}$ is continuous by hypothesis. That sphere has dimension $d-k-1$.] A configuration $\vec{S}$, continous except on a k-dimensional surface Σ^k

and labelled by a non-trivial element, g, of $\pi_{d-k-1}(M)$, is called a topological defect of dimension k and type $[g]$, [where $[g]$ is the conjugacy class of g in $\pi_{d-k-1}(M)$. When this group is abelian, in particular when $k<d-2$, $[g] = g$.]

The idea is now to interpret configurations of the field $\vec{S}$, distributed according to an equilibrium state $d\mu_\beta(\vec{S})$, as equilibrium configurations of a gas of interacting topological defects. Let us assume that all homotopy groups of M are discrete (and that there are only finitely many non-trivial such groups). When all defects have dimension ≥ 1 (i.e. $\pi_{d-1}(M) = \{e\}$), the main features of the statistical mechanics of such a defect gas can be inferred from an energy-entropy argument of the sort explained above for the Ising model: Consider a defect of type $[g]$, $g \in \pi_{d-k-1}(M)$, $g \neq e$, located on a k-dimensional surface Σ^k. Using $d\mu_\beta(\vec{S})$, try to calculate a specific self-energy, $\varepsilon([g])$, such that the self-energy of a defect of type $[g]$ with locus Σ^k is bounded below by

$$E([g],\Sigma^k) \geqq \varepsilon([g])|\Sigma^k| , \tag{2.13}$$

where $|\Sigma^k|$ is the k-dimensional volume of Σ^k. After introducing some course graining, one can argue (see e.g. [36] for details) that the entropy S_n of all defects of a given type whose loci contain a given point, 0, and have k-dimensional area $|\Sigma_k|=n, n=2,4,6,\ldots$, behaves like

$$S_n \approx c(k,d)n , \tag{2.14}$$

for some geometrical constant $c(k,d)$. The "activity" of defects of dimension k, area n and type $[g]$ at inverse temperature β is then expected to behave like

$$z([g],n) \approx e^{-\beta\varepsilon([g])n+c(k,d)n} , \tag{2.15}$$

where we have used (2.13) and (2.14), and the assumption is made that interactions between different defects are, in some sense, weak. Formula (2.15) suggests that when β decreases below some value

$\beta([g]) \approx c(k,d)/\varepsilon([g])$, defects of dimension k and type $[g]$ condense, and there are, with positive probability, infinitely extended defects of type $[g]$. One can thus expect that there are transitions at values of the inverse temperature $\approx \max_{g\in\pi_{d-k-1}(M)} \beta([g])$, $0<k\leq d-1$. [We do not know of any example of a system to which the above considerations can be applied, but for which the conclusions are qualitatively wrong.] The arguments outlined above do not apply to gases of point defects which are more difficult to analyze, as already mentioned in the discussion of the one-dimensional Ising model: Transitions in such gases depend very much on the interactions between different defects. If these interactions are confining order-disorder transitions can generally be expected (see Sects. 3-7), otherwise disorder remains prevalent even at low temperature, (Sect. 8).

We now must ask what remains of the topological considerations sketched above when one studies lattice models, i.e. what means "topology" in lattice models? If one thinks about the question briefly, but not enough, the answer will be "nothing". This answer is, however, unsatisfactory, as the discussion of the Ising model clearly demonstrates. In fact, lattice models of defect gases in arbitrary dimension d , with defects of dimension k characterized by a homotopy group

$$\pi_{d-k-1}(M) = Z_2$$

have first been proposed and analyzed by Wegner [39]. These models are generalizations of the Ising model with, for $k<d-1$, a local (gauge) Z_2 invariance. They exhibit interesting phase diagrams, with continuous, as well as, presumably, first order transitions.

Models closely related to Wegner's can be constructed also when

$$\pi_{d-k-1}(M) = \begin{cases} Z_n , & n=3,4,5,\ldots \\ Z , & \end{cases} \tag{2.16}$$

e.g. $M=S^1$, $k=d-2$ (which describes vortices in a system of classical rotors), or $M=U(1)$, $k=d-3$ (which describes "magnetic monopoles" in a $U(1)$ gauge theory, namely in compact QED), etc. The analysis of transitions and their order becomes, however, considerably more difficult. See [17].

We now wish to explain how a modified defect gas analysis, explained in general terms above, can be used to study the classical rotor model.*) For simplicity we consider the two-dimensional model, but our arguments can be generalized to arbitrary dimension and defects of co-dimension > 2 ; ($k < d-2$ and $M = U(1)$ in (2.16).) Given a configuration $\vec{S}$ of rotors on the lattice Z^2, we imagine interpolating $\vec{S}$ by a configuration $\vec{S}_{cont.}$ defined on R^2 and continuous, except at points located in plaquettes $p \subset Z^2$ which can be interpreted as positions of vortices. [Clearly, there is no preferred choice of an $\vec{S}_{cont.}$, given $\vec{S}$.] Let ψ_{xy} denote the angle between the spin $\vec{S}_x$ at site x and the spin $\vec{S}_y$ at y. Let γ_{xy} be a curve starting at x and ending at y, parametrized by a vector-valued function $u(t) \varepsilon R^2$, $0 \leqq t \leqq 1$, with $u(0) = x$, $u(1) = y$. Then

$$\psi_{xy} = \int_0^1 \frac{d}{dt}\, \theta(u(t))\, dt \;, \tag{2.17}$$

where $\theta(x)$ is the angle corresponding to $\vec{S}_{cont.}(x)$.

Equ. (2.17) defines ψ_{xy} as a real-valued, rather than as an S^1-valued function. It is thus natural to let ψ_{xy} range over R. But there is clearly a constraint: If we add up the angles ψ_{xy} along a loop γ we must obtain an integer multiple of 2π ; (the angle between $\vec{S}_x$ and $\vec{S}_x$ vanishes!) Hence

$$\sum_{xy \varepsilon \gamma} \psi_{xy} = 2\pi\, n_\gamma ,\; n_\gamma \varepsilon Z \;, \tag{2.18}$$

where xy denotes the bond from x to y, assumed to be nearest neighbors, and γ is an arbitrary loop in Z^2. By Stokes' theorem,

$$n_\gamma = \sum_{\substack{p \varepsilon S, \\ \partial S = \gamma}} n_p \;,$$

where $$2\pi n_p = \sum_{b \varepsilon \partial p} \psi_b \;. \tag{2.19}$$

*) We thank K. Gawedzki and E. Seiler for discussions on this topic.

Here S is the set of plaquettes bounded by γ. If the angle differences ψ_b are defined as in (2.17), with $\gamma_{xy}=b$, we see that n_p really measures the total vorticity of $\vec{S}_{cont.}$ inside the unit square p.

We now need some more mathematical formalism to develop these ideas: Let Λ be an arbitrary, finite sublattice of Z^d with trivial homology. A site in Λ is called a 0-cell, an oriented bond b a 1-cell, an oriented plaquette a 2-cell, etc. A function ω defined on oriented k-cells with values in Z, or R (or C) is called a k-chain (or k-form), and

$$\omega(c_k^{-1}) = -\omega(c_k) , \tag{2.20}$$

where c_k^{-1} denotes the k-cell obtained from c_k by reversing the orientation. We define

$$(\delta\omega)(c_{k-1}) = \sum_{c_k : \partial c_k \ni c_{k-1}} \omega(c_k) , \tag{2.21}$$

and

$$(d\omega)(c_{k+1}) = \sum_{c_k \in \partial c_{k+1}} \omega(c_k) . \tag{2.22}$$

A scalar product of two k-forms is defined by

$$(\omega,\phi) \equiv \frac{1}{2} \sum_{c_k \subset Z^d} \overline{\omega(c_k)}\, \phi(c_k) \tag{2.23}$$

With respect to this inner product, d and δ are adjoints. We have,

Lemma 2.1

1) $\delta^2 = d^2 = 0$.

2) If α is a k-form such that $\delta\alpha=0$ then there exists a (k+1)-form β such that

$$\alpha = \delta\beta .$$

If α is integer-valued β can be chosen to be integer-valued and such that suppβ is contained in the smallest hypercube containing supp α, and

$$|\beta(c_{k+1})| \leqq \Sigma|\alpha(c_k)| .$$

(It is assumed that $\Sigma\alpha(x)=0$ when $k=0$.) Similar statements hold with δ replaced by d . See [41,16,17,42] and refs. given there. By (2.19) and Lemma 2.1, 2) there exists an integer-valued 1-chain m and a real-valued function θ such that

$$\psi_b = (d\theta)_b + 2\pi m_b ,$$

with (2.24)

$$(dm)_p = n_p .$$

(We shall usually suppose that the total vorticity vanishes, i.e. $\sum_{p\in\Lambda} n_p=0$.) One checks that the multiplicity of solutions of the equation $dm=n$ is independent of n . One can, in fact, select a unique solution to that equation by enforcing a gauge condition. One obtains all configurations $\psi = \{\psi_b\}_{b\in Z^2}$ satisfying the constraint (2.19), with constant multiplicity, by choosing all possible, integer-valued 1-chains m and requiring that

$$\theta_x \in [-\pi,\pi] . \tag{2.25}$$

Let

$$-\Delta \equiv \delta d + d\delta \tag{2.26}$$

be the Laplace operator (with suitable b.c. at $\partial\Lambda$). Equation (2.19) can also be solved in the real numbers as follows:

$$\psi_b = (d\phi)_b + \overset{0}{\psi}_b , \tag{2.27}$$

where

$$\overset{0}{\psi}_b = - 2\pi(\delta\Delta^{-1}n)_b \tag{2.28}$$

[Note, by Lemma 2.1, 1), $d(d\phi)=0$, and

$$d\psi^{0} = -2\pi d\delta\Delta^{-1}n \ .$$

Now, $d\Delta^{-1}n=\Delta^{-1}dn=0$, since $n=\frac{1}{2\pi}d\psi$. Hence

$$d\psi^{0} = -2\pi(d\delta+\delta d)\Delta^{-1}n = 2\pi n \ .]$$

Next, we define a Hamilton function the choice of which is inspired by standard spin wave theory:

$$H_{\Lambda}(\psi) \equiv \frac{1}{2} \sum_{b\subset\Lambda} \psi_b^{2} \tag{2.29}$$

The equilibrium state of a system of classical rotors in Λ with Hamilton function given by (2.29) is defined by

$$d\mu_{\beta,\Lambda}(\psi) = Z_{\beta,\Lambda}^{-1} e^{-\beta H_{\Lambda}(\psi)} \prod_{p}\delta((d\psi)_p-2\pi n_p)\prod_{b}d\psi_b \ , \tag{2.30}$$

where n is an arbitrary integer-valued 2-chain, and

$$\prod_{p\subset\Lambda} \delta((d\psi)_p-2\pi n_p)$$

imposes the constraint (2.19).

Let A be a periodic function of ψ .

Then

$$\langle A\rangle_{\Lambda}(\beta) \equiv \int A(\psi)d\mu_{\beta,\Lambda}(\psi)$$

$$= \frac{1}{Z_{\beta,\Lambda}} \sum_{n}\int A(\psi)\delta(d\psi-2\pi n)\prod e^{-\frac{\beta}{2}\psi_b^{2}}d\psi_b \ ;$$

by (2.24), (2.25)

$$\langle A\rangle_{\Lambda}(\beta) = \frac{1}{Z_{\beta,\Lambda}} \sum_{m} \int_{-\pi}^{\pi}\cdots\int_{-\pi}^{\pi} A(d\theta+2\pi m) \ .$$

$$\cdot \prod_{b} e^{-\frac{\beta}{2}((d\theta)_b+2\pi m_b)^2} \prod_{x} d\theta_x \ .$$

Since A is assumed to be periodic,

$$A(d\theta+2\pi m) = A(d\theta) .$$

We define a function

$$g_\beta(\theta) = \sum_{m=-\infty}^{\infty} e^{-\frac{\beta}{2}(\theta+2\pi m)^2} . \tag{2.31}$$

It follows that

$$\begin{aligned} \langle A\rangle_\Lambda(\beta) &= \frac{1}{Z_{\beta,\Lambda}} \int_{-\pi}^{\pi}\cdots\int_{-\pi}^{\pi} A(\theta)\prod_b g_\beta((d\theta)_b)\prod_x d\theta_x \\ &\equiv \langle A\rangle'_\Lambda(\beta) . \end{aligned} \tag{2.32}$$

The r.s. of (2.32) defines the equilibrium expectation of the <u>Villain model</u> closely related to the classical rotor model defined in Sect. 1: $\exp \beta\vec{S}_x\cdot\vec{S}_y = \exp \beta\cos(\theta_x-\theta_y)$ is replaced by $g_\beta(\theta_x-\theta_y)$.

Next, using (2.27) and (2.28),

$$\begin{aligned} \langle A\rangle_\Lambda(\beta) = \frac{1}{Z_{\beta,\Lambda}} \sum_n \int_{-\infty}^{+\infty}\cdots\int_{-\infty}^{+\infty} & A(d\phi+\overset{0}{\psi})\cdot \\ & \cdot e^{-\frac{\beta}{2}(d\phi+\overset{0}{\psi},d\phi+\overset{0}{\psi})} \prod_{x\in\Lambda} d\phi_x , \end{aligned}$$

with $\phi_x=0$, $x\notin\Lambda$. Since $\overset{0}{\psi}=-2\pi\delta\Delta^{-1}n$,

$$\delta\overset{0}{\psi} = 0 ,$$

by Lemma 2.1, 1). Thus

$$(\overset{0}{\psi},d\phi) = (d\phi,\overset{0}{\psi}) = (\phi,\delta\overset{0}{\psi}) = 0 . \tag{2.34}$$

Let

$$\tilde{A}(n) \equiv \frac{1}{Z^{s.w.}_{\beta,\Lambda}} \int \prod_{x\in\Lambda} d\phi_x e^{-\frac{\beta}{2}(d\phi,d\phi)} A(d\phi+\overset{0}{\psi}) , \tag{2.35}$$

where

$$Z^{s.w.}_{\beta,\Lambda} = \int \prod_{x\varepsilon\Lambda} d\phi_x \, e^{-\frac{\beta}{2}(d\phi,d\phi)} \tag{2.36}$$

is the partition function of the pure spin wave theory. The partition function of the vortex gas is then given by

$$Z^{v}_{\beta,\Lambda} \equiv Z_{\beta,\Lambda}/Z^{s.w.}_{\beta,\Lambda} \ . \tag{2.37}$$

By (2.33) - (2.35),

$$\begin{aligned} \langle A\rangle_\Lambda(\beta) &= (Z^{v}_{\beta,\Lambda})^{-1} \sum_n \tilde{A}(n) e^{2\pi^2\beta(n,\Delta^{-1}n)} \\ &\equiv \langle\tilde{A}\rangle''_\Lambda(\beta) \ , \end{aligned} \tag{2.38}$$

where we have used that

$$(\overset{0}{\psi},\overset{0}{\psi}) = -4\pi^2(n,\Delta^{-1}n) \ .$$

<u>We have thus reformulated the Villain model as a gas of point defects, the vortices</u> $\{n_p\}_{p\subset\Lambda}$, <u>interacting via a two-body Coulomb potential,</u> $(-\Delta)^{-1}_{xy}$ $[\approx \frac{1}{2\pi} \ln \frac{1}{|x-y|}$, for large $|x-y|$ when $d=2$].

If

$$A = \vec{S}_0 \cdot \vec{S}_x = \mathrm{Re} \exp\,(i \sum_{b\varepsilon\gamma_{0x}} \psi_b) \tag{2.39}$$

we obtain

$$\begin{aligned} \langle\vec{S}_0\cdot\vec{S}_x\rangle_\Lambda(\beta) &= \langle e^{i(\theta_0-\theta_x)}\rangle'_\Lambda(\beta) \\ &= e^{(2\beta)^{-1}((\delta_0-\delta_x),\ \Delta^{-1}(\delta_0-\delta_x))} \ . \\ &\quad \cdot\langle\exp(i \sum_{b\varepsilon\gamma_{0x}} \overset{0}{\psi}_b)\rangle''_\Lambda(\beta) \ , \end{aligned} \tag{2.40}$$

with $\overset{0}{\psi}$ given by (2.28).

Let $f_{\gamma_{0x}}(b) = 1$, $b \varepsilon \gamma_{0x}$, and $= 0$, otherwise. Then

$$\sum_{b \varepsilon \gamma_{0x}} \psi_b^0 = -2\pi (f_{\gamma_{0x}} , \delta \Delta^{-1} n)$$
$$\equiv (\sigma, n) ,$$

where

$$\sigma = -2\pi d \Delta^{-1} f_{\gamma_{0x}} . \tag{2.41}$$

Thus

$$<e^{i(\theta_0 - \theta_x)}>'_\Lambda(\beta) = e^{(2\beta)^{-1}((\delta_0 - \delta_x), \Delta^{-1}(\delta_0 - \delta_x))} \cdot$$
$$\cdot <\exp i(\sigma, n)>''_\Lambda(\beta) \tag{2.42}$$

The first factor on the r.s. of (2.42) is the spin wave contribution to the spin-spin correlation, the second factor is the vortex contribution. It clearly follows from (2.42) that

$$<\vec{S}_0 \cdot \vec{S}_x>_\Lambda(\beta) = <e^{i(\theta_0 - \theta_x)}>'_\Lambda(\beta)$$
$$\leqq \exp(2\beta)^{-1}((\delta_0 - \delta_x) , \Delta^{-1}(\delta_0 - \delta_x))]$$
$$\leqq \text{const. } |x|^{-(1/2\pi\beta)} , \tag{2.43}$$

in two dimensions. This is the Mc Bryan-Spencer bound for the Villain model. It is the purpose of subsequent sections to discuss the proof of a lower bound on $<\vec{S}_0 \cdot \vec{S}_x>_\Lambda(\beta)$ by proving a lower bound on

$$<\exp i(\sigma, n)>''_\Lambda(\beta) .$$

We emphasize that nothing in the calculations between (2.17) and (2.42) depended on the assumption that $d=2$, (except that in (2.19) the surface S is not unique when $d>2$.) Thus, all conclusions, in particular (2.32), (2.38) and (2.42) hold in general. We just must remember that, by (2.19), n is constrained to satisfy

$$dn = 0 . \tag{2.44}$$

Thus vortices form (d-2)-dimensional, closed surfaces. [It is sometimes convenient to set $n = {*\lambda}$, $*$ is the Hodge $*$ operation, associating with k-chains on Z^d the dual (d-k)-chains on $(Z^d)^*$. Then the constraint (2.44) becomes

$$\delta\lambda = 0 .$$

This corresponds to the homogeneous Maxwell equations, in the magnetic interpretation of these models.]

Formula (2.38) shows that the energy of a vortex n is

$$\begin{aligned} E(n) &= -2\pi^2(n,\Delta^{-1}n) \\ &\geq \frac{\pi^2}{2d}(n,n) \geq \frac{\pi^2}{2d}|\Sigma_{d-2}(n)| , \end{aligned} \tag{2.45}$$

where $\Sigma_{d-2}(n)$ is the (d-2)-dimensional, closed surface dual to the support of n. For $d>2$, the energy-entropy arguments explained in (2.13) - (2.15) can thus be applied and predict a phase transition. This prediction is, of course, correct, and it is possible to convert the heuristic energy-entropy argument into a rigorous proof [17].

In two dimensions, the arguments are more complicated [5]; see Sects. 3-7.

The ideas developed in this section extend to a general class of models with U(1)-symmetry [17], including the U(1) lattice gauge theory [41,16,42].

3. Kosterlitz-Thouless Transition In the Two-Dimensional Villain Model

The purpose of Sects. 3-7 is to explain the main ideas of the proof [5] of the following result.

Theorem 3.1.

In the two-dimensional Villain model there is a finite constant β_0 and a function $\beta'=\beta'(\beta)$, with the properties

$$\beta'(\beta) \geqq \frac{1}{2\pi} \quad \text{for} \quad \beta > \beta_0 , \quad \lim_{\beta\to\infty} \beta'(\beta)/\beta = 1 , \tag{3.1}$$

such that

$$\begin{aligned} \text{const.}' \, (1+|x-y|)^{-1/2\pi\beta'} &\leqq \langle \vec{S}_x \cdot \vec{S}_y \rangle(\beta) \\ &\leqq \text{const.} \, (1+|x-y|)^{-1/2\pi\beta} . \end{aligned} \tag{3.2}$$

Remarks.

1) This result is Theorem A of Sect. 1 for the Villain model. It also holds for the classical rotor model, but the proof is technically more complicated, (because the Fourier transform of $\exp(\beta\cos\theta)$ is more difficult to handle than $\exp(-(2\beta)^{-1}k^2)$, $k\varepsilon Z$ - the Fourier transform of $g_\beta(\theta)$. See however [5]. All the basic ideas of the proof remain unchanged).

2) It is well known that $\langle \vec{S}_x \cdot \vec{S}_y \rangle(\beta)$ decays exponentially in $|x-y|$, for β sufficiently small ($\beta<\beta_c$). Thus there is a phase transition, and - as already remarked in Sect. 1 - all $\beta \geqq \beta_c$ are critical points.

3) Let p and p' be two distinct plaquettes. Let $\{p_n\}_{n=0}^N$ be a path of adjacent plaquettes joining p to p', with $p_0=p$, $p_N=p'$. Let $b_j=p_{j-1}\cap p_j$. A disorder operator $D^\xi_{pp'}$ may be defined by

$$D^\xi_{pp'} = \prod_{j=1}^{N} \frac{g_\beta((d\theta)_{b_j}+2\pi\xi)}{g_\beta((d\theta)_{b_j})} , \quad 0 < \xi < 1 . \tag{3.3}$$

The expectation value, $\langle D^\xi_{pp'} \rangle(\beta)$, is independent of the path $\{p_n\}_{n=0}^N$. One can prove (see [5]) that

$$\langle D^\xi_{pp'} \rangle(\beta) \leqq \text{const.} \, \text{dist}(p,p')^{-\beta''} , \tag{3.4}$$

where $\beta''=\beta''(\xi,\beta)$ is positive for $\beta>\beta_0(\xi)$, and $\beta'' \sim \text{const.}\ \xi^2\beta$, for ξ small and $\beta\to\infty$. By a high temperature expansion,

$$\langle D^{\xi}_{pp'}\rangle(\beta) \geq \text{const.} ,$$

uniformly in p and p', for small β.

The proof of (3.4) is closely related to the proof of the lower bound in (3.2). See Sect. 5 of [5].

<u>An outline of the proof of Theorem 3.1</u>

We recall that the upper bound in (3.2) has been proven in Sect. 2; see (2.43). The physical intuition in the proof of the lower bound is that, at low temperatures, all vortices are bound in small molecules of zero total vorticity (neutrality!) which form a dilute gas. Thus spin wave theory should be close to exact, i.e. the spin wave contribution,

$$\exp(1/2\beta)((\delta_0-\delta_x), \Delta^{-1}(\delta_0-\delta_x)) \approx |x|^{-1/2\pi\beta} , \tag{3.5}$$

to $\langle\vec{S}_0\cdot\vec{S}_x\rangle(\beta)$ and $\langle\vec{S}_0\cdot\vec{S}_x\rangle(\beta)$ are expected to have the same long distance behaviour, except for a renormalization of the exponent $\frac{1}{2\pi\beta}$ on the r.s. of (3.5): $\beta\to\beta_{ren} \equiv \varepsilon^{-1}\beta$.

To make this more precise, we may adapt the energy-entropy argument, sketched for the one-dimensional Ising model with $1/r^2$ interaction energy, to the vortex gas:

Consider a system of vortices confined to a square Λ of size l. We impose 0 Dirichlet data at $\partial\Lambda$ on the interaction potential, $(-\Delta)^{-1}_{xy}$, for a pair of vortices located at x and y. Physically this is the correct choice of b.c. (Incidentally, from now on we shall usually identify plaquettes p in Z^2 with sites x of $(Z^2)^* \cong Z^2$.) The self-energy, E_l, of a free vortex in Λ (in a system of small neutral molecules of vortices) behaves like

$$E_l \approx \varepsilon^{-1}\pi\, l n l , \tag{3.6}$$

where $\varepsilon \equiv \beta/\beta_{ren.} > 1$ is a dielectric constant of the vortex gas. The number of positions of the free vortex is l^2, thus the entropy is given by

$$S_l = 2\, l n l . \tag{3.7}$$

The activity of a free vortex in Λ is therefore

$$z_l \approx e^{S_l - \beta E_l} \approx l^2 e^{-\beta\varepsilon^{-1}\pi\, ln l} \tag{3.8}$$

and thus tends to 0, as $l \to \infty$, when

$$\beta\varepsilon^{-1} > 2/\pi . \tag{3.9}$$

Hence we expect a Kosterlitz-Thouless transition to occur when $\beta = \beta_c \approx 2\varepsilon/\pi$. (Since $\beta_c \geqq 2\beta_c(\text{Ising})$, $\varepsilon \geqq 1.38$) Since we expect that

$$\begin{aligned} <\vec{S}_0 \cdot \vec{S}_x>(\beta) &\approx e^{-(1/2\pi\beta_{ren}) ln|x|} \\ &= |x|^{-(\varepsilon/2\pi\beta)} \end{aligned}$$

we predict that

$$<\vec{S}_0 \cdot \vec{S}_x>(\beta=\beta_c) \underset{|x|\to\infty}{\approx} |x|^{-1/4} , \tag{3.10}$$

as mentioned in Sect. 1.
(Note that (3.10) does not depend on the values of β_c and ε !)

In order to convert the energy-entropy arguments sketched above into a rigorous proof, we shall use the sine-Gordon representation of the Coulomb gas [43,5]. We recall that, by (2.38) and (2.42), the lower bound in (3.2) follows from a lower bound on

$$\begin{aligned} &<\exp i(\sigma,n)>''_{\Lambda}(\beta) \\ &= (Z^V_{\beta,\Lambda})^{-1} \sum_n \exp i(\sigma,n) \exp[2\pi^2\beta(n,\Delta^{-1}n)] \end{aligned} \tag{3.11}$$

uniformly in Λ, where Δ is the finite difference Laplacian with 0 Dirichlet data at $\partial\Lambda$, and

$$\sigma = -2\pi\, d\, \Delta^{-1} f_{\gamma_0 x} ; \tag{3.12}$$

see (2.41) .

Let $d\mu_\beta(\phi)$ be the Gaussian measure with mean 0 and covariance $-\beta\Delta^{-1}$. Its characteristic functional is given by

$$\int e^{i\phi(f)} d\mu_\beta(\phi) = \exp[(\beta/2)(f,\Delta^{-1}f)] , \tag{3.13}$$

where $\phi(f) \equiv \sum_{x\varepsilon\Lambda} \phi_x f_x$, $\Lambda \subset (Z^2)^*$.

Hence

$$\exp[2\pi^2\beta(n,\Delta^{-1}n)] = \int e^{i\phi(2\pi n)} d\mu_\beta(\phi) \tag{3.14}$$

We insert (3.14) on the r.s. of (3.11) and interchange summation over n and integration over $d\mu_\beta(\phi)$; (this can be justified easily!) This yields

$$\langle \exp i(\sigma,n)\rangle''_\Lambda (\beta) = (Z^V_{\beta,\Lambda})^{-1} Z_{\beta,\Lambda}(\sigma) , \tag{3.15}$$

where

$$Z_{\beta,\Lambda}(\sigma) \equiv \int \prod_{x\varepsilon\Lambda} [1+2\sum_{n=1}^{\infty} \cos(2\pi n\phi_x+n\sigma_x)] d\mu_\beta(\phi) ; \tag{3.16}$$

and $Z_{\beta,\Lambda}(0) \equiv Z^V_{\beta,\Lambda}$. $\qquad$ (3.17)

Identities (3.15) - (3.17) are known under the name of <u>sine-Gordon representation</u> (of the Coulomb gas.) In order to establish a lower bound, we would like to apply Jensen's inequality to $Z_{\beta,\Lambda}(\sigma)$, with the purpose of finding a lower bound in terms of $Z^V_{\beta,\Lambda}$. However, the series $\sum_{n=1}^{\infty}\cos(2\pi n\phi_x+n\sigma_x)$ isn't absolutely convergent, and σ isn't small. To overcome these difficulties we shall prove a <u>basic identity</u> (our main technical result!):

Let Λ be some rectangular array of sites, $j \in Z^2$. We confine the positions of all the vortices to be inside Λ . A vortex density ρ is an integer-valued function on Λ whose value, $\rho(j)$, at some site $j \in \Lambda$ indicates the total vorticity concentrated at j . We say that ρ is <u>neutral</u> (or of total vorticity 0) iff

$$Q(\rho) \equiv \sum_x \rho_x = 0 \ .$$

An ensemble, E, is a collection of vortex densities, ρ, whose supports are mutually disjoint (and are contained in Λ).

We may now state our basic identity.

Theorem 3.2.

For arbitrary, real σ and all $\beta > \beta_0$, where β_0 is some finite constant, there exists a family $F = F_\Lambda$ of ensembles N of vortex densities such that

$$Z^V_{\beta,\Lambda} \langle \exp i(\sigma,n) \rangle''_\Lambda(\beta) \equiv Z_{\beta,\Lambda}(\sigma)$$

$$= \int \prod_{x\in\Lambda} [1+2 \sum_{n=1}^{\infty} \cos(2\pi n\phi_x + n\sigma_x)] \, d\mu_\beta(\phi) \tag{3.18}$$

$$= \sum_{N\in F_\Lambda} c_N \int \prod_{\rho\in N} [1+z_N(\beta,\rho)\cos(2\pi\phi(\bar{\rho})+\sigma(\rho))] \, d\mu_\beta(\phi) \ ,$$

where $\phi(\rho) = \sum_x \phi_x \rho_x$, etc.;

$c_N > 0$, for each ensemble $N \in F_\Lambda$;

the coefficients $z_N(\beta,\rho)$ are "renormalized (or: effective) activities" of vortex molecules with vortex density ρ, which, for $\beta > \beta_0$, are bounded by

$$0 < z_N(\beta,\rho) \leqq \exp[-\beta(c\|\rho\|_2^2 + d \ ln \ \mathrm{diam}(\mathrm{supp}\ \rho))] \ , \tag{3.19}$$

where c and d are positive constants, and

$$\|\rho\|_2^2 = \sum_j |\rho_j|^2 \ ;$$

finally $\bar{\rho} = \bar{\rho}_N$ is a renormalized vortex density which depends linearly on ρ, and

$$Q(\bar{\rho}) = Q(\rho) \ . \tag{3.20}$$

[$\bar{\rho}$ need not be integer-valued and depends on N.]

Remark. We emphasize that, in (3.18), F_Λ , N , c_N , $z_N(\beta,\rho)$ and $\bar{\rho}$ are all independent of σ .

For technical reasons we shall henceforth replace the Gaussian measure $d\mu_\beta = d\mu_{\beta,\Lambda}$ with Dirichlet boundary conditions at $\partial\Lambda$ by the infinite volume Gaussian measure which we shall also denote by $d\mu_\beta$. Theorem 3.2 remains true after this slight change, but, in addition, all vortex densities ρ of all ensembles $N \varepsilon F_\Lambda$ are neutral, (i.e. $Q(\rho) = Q(\bar{\rho}) = 0$.) This is a (non-trivial!) consequence of the fact, special to dimension 2, that

$$\int e^{i\phi(f)} d\mu_\beta(\phi) = \begin{cases} 0 \quad \text{if} \quad \sum_x f_x \equiv \hat{f}(0) \neq 0 \; ; \\ \\ \exp[\frac{\beta}{2}(f,\Delta^{-1}f)] \quad \text{if} \quad \hat{f}(0) = 0 \; . \end{cases} \tag{3.21}$$

(Here and below, $d\mu_\beta$ is the infinite volume Gaussian measure.)

The main advantage of free b.c. will be seen in the course of the renormalization transformations carried out in Sect. 5, but they are also more convenient in the proof of a lower bound on $Z_{\beta,\Lambda}(\sigma)$, sketched below. The changes required to handle 0 Dirichlet b.c. are explained in Appendix D of [5]. The difference of b.c. is irrelevant in the thermodynamic limit.

Using Theorem 3.2 we can now apply Jensen's inequality and obtain a lower bound on $Z(\sigma)$ which yields Theorem 3.1. By Taylor's theorem with remainder

$$\begin{aligned} &\prod_{\rho\varepsilon N} [1+z_N(\beta,\rho)\cos(2\pi\phi(\bar{\rho})+\sigma(\rho))] \\ &= \exp\left\{\sum_{\rho\varepsilon N} \log[1+z_N(\beta,\rho)\cos(2\pi\phi(\bar{\rho})+\sigma(\rho))]\right\} \\ &= \prod_{\rho\varepsilon N} [1+z_N(\beta,\rho)\cos(2\pi\phi(\bar{\rho}))] \exp O_N(\phi,\sigma)\cdot \\ &\qquad \cdot\exp[-R_N(\phi,\sigma)] \; , \end{aligned} \tag{3.22}$$

where $O_N(\phi,\sigma)$ is odd in ϕ , and

$$R_N(\phi,\sigma) \leqq \sum_{\rho\in N} C\ z_N(\beta,\rho)[\sigma(\rho)|\mathrm{mod}.2\pi]^2 , \tag{3.23}$$

for some constant C which is finite (and $\propto[1-z_N(\beta,\rho)]^{-2}$) , provided β is so large that $z_N(\beta,\rho) < 1$. (To prove these facts one applies Taylor's theorem with remainder to $\log[1+z_N(\beta,\rho)\cos(2\pi\phi(\bar{\rho})+\sigma(\rho))]$, to second order in σ , and notes that, by periodicity,

$$\cos(2\pi\phi(\bar{\rho})+\sigma(\rho)) = \cos(2\pi\phi(\bar{\rho})+\sigma(\rho)|\mathrm{mod}.2\pi).)$$

Since each vortex density $\rho \in N$ is <u>neutral</u>, we may write

$$\sigma(\rho) = \sum_{\mu} (\sigma_{p_\mu} - \sigma_{q_\mu}) , \tag{3.24}$$

where p_μ and q_μ are points in $\mathrm{supp}\ \rho$, and there are no more than $\frac{1}{2}||\rho||_1$ terms on the r.s. of (3.24). Now, $\sigma_{p_\mu} - \sigma_{q_\mu}$ can be rewritten as

$$\sigma_{p_\mu} - \sigma_{q_\mu} = \sum_b (\delta\sigma)_b\ g_\mu(b) ,$$

where $g_\mu(b) = 1$, for b in a path joining q_μ to p_μ , and $g_\mu(b) = 0$, otherwise.

Now, we are really interested in choosing σ as in (2.41), (3.12), i.e.

$$\sigma = -2\pi\ d\ \Delta^{-1}\ f_{\gamma_{0x}} .$$

Then

$$\begin{aligned}(\delta\sigma)_b &= -2\pi(\delta d\Delta^{-1} f_{\gamma_{0x}})_b \\ &= 2\pi f_{\gamma_{0x}}(b) + 2\pi(d\delta\Delta^{-1} f_{\gamma_{0x}})_b \\ &= 2\pi f_{\gamma_{0x}}(b) + 2\pi(d\Delta^{-1}[\delta_x - \delta_0])_b\end{aligned}$$

Thus

$$(\delta\sigma)_b|\text{mod.}2\pi = 2\pi[(\nabla C)(b,0)-(\nabla C)(b,x)] ,$$

where $(\nabla h)(b)$ is the gradient of a function h in the direction of b , and $C(j,j') = -\Delta^{-1}_{jj'}$ is the Green's function of Δ . Thus

$$|(\sigma_{p_\mu} - \sigma_{q_\mu})|\text{mod.}2\pi| \leqq 4\pi\ ||\rho||_1 \cdot \max_{b\subset \text{supp}\,\rho} |(\nabla C)(b,0)-(\nabla C)(b,x)| , \qquad (3.25)$$

where $||\rho||_1 \equiv \sum_j |\rho_j|$.

[This inequality would clearly be wrong if σ were not taken modulo 2π .]

Thus, for the above choice of σ ,

$$[\sigma(\rho)|\text{mod.}2\pi]^2 \leqq (2\pi)^2||\rho||_1^4 \cdot \max_{b\subset \text{supp}\,\rho} |(\nabla C)(b,0)-(\nabla C)(b,x)|^2 .$$

Hence

$$R_N(\phi,\sigma) \leqq \sum_b |(\nabla C)(b,0)-(\nabla C)(b,x)|^2 \cdot \sum_{\substack{\rho\varepsilon N \\ \text{supp}\,\rho \supset b}} (2\pi)^2||\rho||_1^4\, Cz_N(\beta,\rho) \qquad (3.26)$$

$$\leqq - c(\beta)\ (\delta_0-\delta_x,\Delta^{-1}(\delta_0-\delta_x))$$

$$\underset{|x|\to\infty}{\approx} c(\beta)\,\frac{1}{\pi}\,ln\,|x| ,$$

where $c(\beta)$ is a finite constant, for β sufficiently large, thanks to the bound (3.19) on the activities $z_N(\beta,\rho)$. [Actually, $c(\beta)$ tends to 0 exponentially fast, as $\beta \to \infty$.]

From this and (3.19) we conclude that there is some $\beta_1 < \infty$ such that, for all $\beta > \beta_1$,

$c_N > 0$, for all $N \varepsilon F_\Lambda$,

$z_N(\beta,\rho) < 1$, for all $\rho \varepsilon N$ and all $N \varepsilon F_\Lambda$, and

$c(\beta) < \infty$.

Therefore, for $\beta > \beta_1$,

$$\sum_{N\varepsilon F_\Lambda} c_N \prod_{\rho\varepsilon N} [1+z_N(\beta,\rho)\cos 2\pi\phi(\bar{\rho})] \, d\mu_\beta(\phi)$$

is a positive measure. This crucial fact permits us to apply Jensen's inequality to the r.s. of (3.22) which yields

$$Z_{\beta,\Lambda}(\sigma) = \sum_{N\varepsilon F_\Lambda} c_N \int \prod_{\rho\varepsilon N} [1+z_N(\beta,\rho)\cos 2\pi\phi(\bar{\rho})]\cdot \exp O_N(\phi,\sigma)\cdot$$

$$\cdot \exp[-R_N(\phi,\sigma)] \, d\mu_\beta(\phi)$$

$$\geqq \sum_{N\varepsilon F_\Lambda} c_N \exp[-\max_\phi R_N(\phi,\sigma)]\cdot$$

$$\cdot \int \prod_{\rho\varepsilon N} [1+z_N(\beta,\rho)\cos 2\pi\phi(\bar{\rho})] \, d\mu_\beta(\phi) ,$$

by Jensen's inequality,

$$\geqq |x|^{-c(\beta)/\pi} \sum_{N\varepsilon F_\Lambda} c_N \int \prod_{\rho\varepsilon N} [1+z_N(\beta,\rho)\cos 2\pi\phi(\bar{\rho})] \, d\mu_\beta(\phi),$$

by (3.26),

$$= |x|^{-c(\beta)/\pi} Z^V_{\beta,\Lambda} , \tag{3.27}$$

by (3.18), with $\sigma = 0$.

In our application of Jensen's inequality we have used the fact that $O_N(\phi,\sigma)$ is odd in ϕ .

Theorem 3.1 now follows from (2.42), (3.18) and (3.27).

We conclude this section with a digression on the relation between the Villain model in the sine-Gordon representation and the discrete Gaussian (DG) model, introduced in Sect. 1, (1.27). Recall that

$$Z_{\beta,\Lambda}(\sigma) = \int \prod_{x\in\Lambda} [1+2\sum_{n=1}^{\infty} \cos(2\pi n\phi_x+n\sigma_x)]d\mu_\beta(\phi) ,$$

see (3.16). But

$$1+2\sum_{n=1}^{\infty} \cos(2\pi n\phi_x+n\sigma_x) = \sum_{m=-\infty}^{\infty} \delta(\phi_x+\bar{\sigma}_x-m) , \tag{3.28}$$

with $\bar{\sigma}_x \equiv \sigma_x/2\pi$. Inserting this identity into our expression for $Z_{\beta,\Lambda}(\sigma)$ we see that $Z_{\beta,\Lambda}(\sigma)$ is really a shifted $(\phi \to \phi + \bar{\sigma})$ partition function of the DG model, (in particular $Z_{\beta,\Lambda}(0)$ is its partition function). By a simple change of variables, correlations in the DG model, like $\langle e^{\phi(f)}\rangle_\Lambda^{DG}(\beta)$, $\Sigma f_x = 0$, can be expressed in terms of the quotient $Z_{\beta,\Lambda}(\sigma)/Z_{\beta,\Lambda}(0)$: Choosing

$$\sigma_x = - \beta(\delta_x,\Delta^{-1}f)$$

we have

$$\langle e^{\phi(f)}\rangle_\Lambda^{DG}(\beta) = e^{-(\beta/2)(f,\Delta^{-1}f)} Z_{\beta,\Lambda}(\sigma)/Z_{\beta,\Lambda}(0) , \tag{3.29}$$

as one sees by setting $\phi_x = \phi_x'+\sigma_x$.
This establishes an isomorphism between the Villain model (in the sine-Gordon representation) and the discrete Gaussian. This will be used in Sect. 8.

4. An Expansion in Terms of Neutral Vortex Molecules, Entropy Estimates

In this section we prove our main combinatorial theorem which expresses expectations in the two-dimensional Villain model, written in the sine-Gordon representation, as convex combinations of expectations in dilute gases of neutral vortex molecules of variable size. Our principal estimate is on the entropy (or: the bare activity) of the neutral vortex molecules which constitute the resulting gases.

We derive a combinatorial expansion for the generalized partition function $Z_{\beta,\Lambda}(\sigma)$. A corresponding expansion for $\langle \vec{S}_0\cdot\vec{S}_x\rangle_\Lambda(\beta)$ then follows from (2.42) and (3.18).

The measure $d\mu_\beta$ will henceforth be the infinite volume Gaussian measure. The changes necessary to accomodate Dirichlet boundary conditions are explained in Appendix D of [5].

4.1 Notation and a simple identity.

The diameter of the support of a vortex density ρ is denoted by $d(\rho)$, and $\mathrm{dist}(\rho,\rho')$ is the minimum (Euclidean) distance between $\mathrm{supp}\,\rho$ and $\mathrm{supp}\,\rho'$. A vortex density ρ is said to be compatible with an ensemble E if

$$\rho = \sum_{\rho'\varepsilon E} \varepsilon(\rho,\rho')\rho' , \qquad (4.1)$$

where $\varepsilon(\rho,\rho') = 0, \pm 1$. We say that an ensemble E' is a parent of E , $E' \to E$, if every vortex density $\rho \varepsilon E$ is compatible with E' . We say that a vortex density ρ_1 is a constituent of a vortex density ρ , $\rho_1 \subset \rho$, if

$$\mathrm{supp}\,\rho_1 \subset \mathrm{supp}\,\rho , \quad (\rho_1)_j = \rho_j , \text{ for all } j \varepsilon \mathrm{supp}\,\rho_1 .$$

Next, we define an area (or: 2-volume) of a vortex density ρ on a sequence of distance scales: Let $V_n(\rho)$ be the minimal number of $2^n \times 2^n$ squares, i.e. lattice squares with sides parallel to the lattice axes containing 2^n sites, needed to cover $\mathrm{supp}\,\rho$. Clearly $V_0(\rho) = \mathrm{card}(\mathrm{supp}\,\rho)$.

An ensemble $E = E_n$ is called an n -ensemble if for all ρ and ρ' in E_n , $\rho \neq \rho'$,

$$\mathrm{dist}(\rho,\rho') > 2^n .$$

Next, we recall expression (3.16) for $Z_{\beta,\Lambda}(\sigma)$:

$$Z_{\beta,\Lambda}(\sigma) = \int \prod_{j\varepsilon\Lambda} [1+2\sum_{n=1}^{\infty} \cos(2\pi n\phi_j+n\sigma_j)]d\mu_\beta(\phi) .$$

We choose a sequence $\left\{z_q\right\}_{q=1}^{\infty}$ of positive numbers with the property that

$$\sum_{q=1}^{\infty} z_q = 1 \ , \text{ and } \ z_q^{-1} \leqq \text{const. } e^{\delta q^2} \ , \tag{4.2}$$

(e.g. $z_q = \text{const.}' \ e^{-\delta q^2}$), for some finite $\delta > 0$.

We set $\zeta_q = 2 z_q^{-1}$. Then

$$\prod_{j\varepsilon\Lambda} [1+2 \sum_{n=1}^{\infty} \cos(2\pi n\phi_x + n\sigma_x)] \tag{4.3}$$

$$= \sum_{q_\Lambda} z_{q_\Lambda} I(q_\Lambda, \phi^\sigma) \ ,$$

where

$$I(q_\Lambda, \phi^\sigma) \equiv \prod_{j\varepsilon\Lambda} [1+\zeta_{q_j} \cos(q_j \phi_j^\sigma)] \ , \tag{4.4}$$

and

$$q_\Lambda \equiv \left\{q_j\right\}_{j\varepsilon\Lambda} \ , \ z_{q_\Lambda} \equiv \prod_{j\varepsilon\Lambda} z_{q_j} \ , \tag{4.5}$$

$$\phi_j^\sigma \equiv 2\pi\phi_j + \sigma_j \ .$$

4.2 Statement of the main result

Theorem 4.1

For each q_Λ as in (4.5) there exists a family $F = F_{q_\Lambda}$ of ensembles N such that, for arbitrary σ ,

$$I(q_\Lambda, \phi^\sigma) = \sum_{N\varepsilon F} c_N \prod_{\rho\varepsilon N} [1+K(\rho)\cos\phi^\sigma(\rho)] \cdot$$

$$\cdot [1+K(\rho^c)\cos\phi^\sigma(\rho^c)] \ , \tag{4.6}$$

where $c_N > 0$, for all $N \ \varepsilon \ F$, and each $N \ \varepsilon \ F$ satisfies the following

<u>Condition</u> D :

a) <u>All densities</u> ρ <u>in</u> N <u>are neutral</u>.

<u>There exist constants</u> $M > 1$ <u>and</u> $\alpha \in (3/2,2)$ <u>independent of</u> N <u>such that</u>

b) $\text{dist}(\rho_1,\rho_2) \geqq M\,[\min(d(\rho_1),d(\rho_2))]^{\alpha}$,

<u>for all</u> ρ_1 , ρ_2 <u>in</u> N <u>with</u> $\rho_1 \neq \rho_2$.

c) <u>If</u> $\rho' \subset \rho \in N$ <u>is such that</u>

$$\text{dist}(\rho' \,,\, \rho - \rho') \geqq 2M\, d(\rho')^{\alpha}$$

<u>then</u> ρ' <u>is charged, i.e.</u> $|Q(\rho')| \geqq 1$. □

<u>Finally, the constants</u> $K(\rho)$ <u>on the r.s. of</u> (4.6) <u>satisfy</u>

$$0 < K(\rho) = \zeta(\rho)e^{S(\rho)} \text{ , where} \tag{4.7}$$

$$\zeta(\rho) = \prod_j \zeta_{|\rho_j|} \quad \text{and} \quad S(\rho) \leq C\, V(\rho) \text{ ,}$$

<u>with</u>

$$V(\rho) = \sum_{k=0}^{n(\rho)} V_n(\rho) \text{ , and} \tag{4.8}$$

$$n(\rho) = [\text{smallest integer} \geqq \ln_2 (Md(\rho)^{\alpha})] \text{ ,}$$

<u>for all</u> $\rho \neq \rho^c$.

<u>The density</u> ρ^c <u>is charged, i.e.</u> $|Q(\rho^c)| \geqq 1$, <u>and</u> $K(\rho^c) \geqq 0$; $\text{dist}(\rho,\rho^c) \geqq Md(\rho)^{\alpha}$, <u>for all</u> $\rho \in N$.

<u>Remarks.</u>

(1) By (3.21)

$$\int I(q_\Lambda,\phi^{\sigma})d\mu_\beta(\phi)$$

$$= \sum_{N\in F} c_N \int \prod_{\rho\in N} [1+K(\rho)\cos\phi^{\sigma}(\rho)]d\mu_\beta(\phi)$$

if $d\mu_\beta$ is the infinite volume Gaussian measure, because for arbitrary $\varepsilon(\rho) = 0, \pm 1$

$\sum_{\rho\varepsilon N} \varepsilon(\rho)\rho \pm \rho^c$ is charged, so that

$$\int \exp[i\phi^\sigma(\sum_{\rho\varepsilon N} \varepsilon(\rho)\rho)]\cos\phi^\sigma(\rho)d\mu_\beta(\phi) = 0 .$$

We are thus entitled to neglect the factor $[1+K(\rho^c)\cos\phi^\sigma(\rho^c)]$ henceforth. [Other boundary conditions - Dirichlet b.c. at $\partial\Lambda$ - not permitting to neglect that factor, are briefly studied in Appendix D of [5].]

(2) The constants M and α will be chosen in Sects. 5, 6. We shall see that $\frac{3}{2} < \alpha < 2$ is the admissible range of α. Property b) of the ensembles N ensures that the vortex densities $\rho \in N$ are sufficiently sparse to enable us to extract a selfenergy factor $\exp[-\text{const.}\beta\, E_{loc.}(\rho)]$ with the property that

$$z_N(\beta,\rho) \equiv K(\rho)\exp[-\text{const.}\beta\, E_{loc.}(\rho)]$$

is very small, for large β, as claimed in Theorem 3.2. [The extraction of the selfenergy factor is a problem in electrostatics which we solve in Sect. 5.] The quantity $S(\rho)$ occurring in the definition has the interpretation of an entropy of the vortex molecule with density ρ, $z_N(\beta,\rho)$ is its renormalized activity.

(3) The key assertion of Theorem 4.1 is contained in (4.7) and (4.8). The coefficient $K(\rho) = \zeta(\rho)e^{S(\rho)}$ estimated in (4.7) is the bare activity of the molecule with density ρ.

(4) The magnitude of the coefficients c_N is unimportant, only their positivity is essential.

Defining

$$\lambda_N \equiv c_N\, (Z^V_{\beta,\Lambda})^{-1} \langle \prod_{\rho\varepsilon N} [1+K(\rho)\cos 2\pi\phi(\rho)]\rangle_\beta \quad , \tag{4.9}$$

where $\langle(\cdot)\rangle_\beta$ denotes the expectation in $d\mu_\beta$, we conclude from Theorem 4.1 that the expectation in the sine-Gordon representation of the Villain model is given by

$$\sum_{q_\Lambda} \sum_{N\varepsilon F_{q_\Lambda}} \lambda_N \langle(\cdot)\rangle_N(\beta) \;, \quad (\sum_{q_\Lambda} \sum_{N\varepsilon F_{q_\Lambda}} \lambda_N = 1) \;, \tag{4.10}$$

where $\langle(\cdot)\rangle_N(\beta)$ is the state given by the "measure"

$$Z_N^{-1} \prod_{\rho\varepsilon N} [1+K(\rho)\cos 2\pi\phi(\rho)]d\mu_\beta(\phi) .$$

(5) The proof of Theorem 4.1 consists of an inductive construction of the ensembles N which extends over infinitely many distance scales and is designed in such a way that Condition D will be fulfilled.

4.3 A basic lemma

In this section we prove the basic combinatorial identity needed to prove Theorem 4.1.

Lemma 4.2

Let $E = E_n$ *be an* n *-ensemble. Then there is a family* $\{E_\gamma\}_{\gamma=1,2,3,\ldots}$ *of* $(n+1)$ *-ensembles, labelled by an index* γ , *with the property that* E *is a parent of each* E_γ $(E \to E_\gamma)$, *and*

$$\prod_{\rho\varepsilon E} [1+K(\rho)\cos\phi^\sigma(\rho)] = \sum_\gamma c_\gamma \prod_{\rho'\varepsilon E_\gamma} [1+K'(\rho')\cos\phi^\sigma(\rho')] , \qquad (4.11)$$

where, for each γ , $c_\gamma > 0$ *and for each* $\rho' \varepsilon E_\gamma$

a) $\rho' = \sum_{\rho\varepsilon E} \varepsilon(\rho',\rho)\rho$, *i.e.* ρ' *is compatible with* E ;

b) *if* $\rho_1' \subset \rho'$ *is compatible with* E *then*

$\text{dist}(\rho_1',\rho'-\rho_1') \leq 2^{n+1}$;

c) $$0 < K'(\rho') \leq e^{C_1 V_n(\rho')} \prod_{\rho\varepsilon E} K(\rho)^{|\varepsilon(\rho',\rho)|} , \qquad (4.12)$$

for a constant C_1 *independent of* ρ' *and* n .

Remark. In order to simplify our notation we temporarily write $\phi(\rho)$ instead of $\phi^\sigma(\rho)$.

Proof. The lemma follows from iterated application of the identity

$$(1+K_\alpha\cos\phi(\rho_\alpha))(1+K_\beta\cos\phi(\rho_\beta))$$

$$= 1/3(1+3K_\alpha\cos\phi(\rho_\alpha))+1/3(1+3K_\beta\cos\phi(\rho_\beta))$$

$$+ 1/6(1+3K_\alpha K_\beta\cos\phi(\rho_\alpha-\rho_\beta))$$

$$+ 1/6(1+3K_\alpha K_\beta\cos\phi(\rho_\alpha+\rho_\beta)) \; . \qquad (4.13)$$

Note, all densities on the right side of (4.13) are compatible with $\{\rho_\alpha,\rho_\beta\}$ and all coefficients are positive. Identity (4.13) is only applied to a pair $\{\rho_\alpha,\rho_\beta\} \subset E$ if

$$\text{dist}(\rho_\alpha,\rho_\beta) \leqq 2^{n+1} \; . \qquad (4.14)$$

We start by applying (4.13) to any two factors on the left side of (4.11) corresponding to two charge densities ρ_α,ρ_β in E for which (4.14) holds. (If there are no such factors the lemma holds trivially.) The right side of identity (4.13) is then inserted on the left side of (4.11), replacing the two factors labelled by ρ_α,ρ_β by a sum of four terms, and by expanding we obtain a sum of four products. If one of the resulting products contains two factors labelled by densities ρ'_α,ρ'_β satisfying (4.14), we apply identity (4.13) again and expand the resulting expression into a sum of products. We repeat this operation until we obtain a sum over products indexed by ensembles E_γ with the property that, for arbitrary ρ_1,ρ_2 in E_γ, $\text{dist}(\rho_1,\rho_2) > 2^{n+1}$.

As noted already, each application of (4.13) replaces charge densities which are compatible with E by other charge densities compatible with E . Thus $\rho' = \sum_{\rho\varepsilon E} \varepsilon(\rho',\rho)\rho$, for all $\rho' \varepsilon E_\gamma$ and all γ . Moreover, all coefficients c_γ are clearly positive, [since by (4.13) each $c_\gamma = (1/3)^{n_\gamma}(1/6)^{m_\gamma}$, for some positive integers n_γ and m_γ]. This completes the proof of (4.11) and a). Part b) of the lemma then follows directly from (4.14). Thus we are left with proving c).

At intermediate stages of our operations we have identities

$$\prod_{\rho\varepsilon E} (1+K(\rho)\cos\phi(\rho)) = \sum_I c_I \prod_{\rho_\alpha \varepsilon I} (1+K_\alpha \cos\phi(\rho_\alpha)) ,$$

with $E \to I$, (i.e. E is a parent of each I), and $c_I > 0$, for all intermediate ensembles I . In order to prove (4.12) we now consider some density $\rho' \varepsilon E_\gamma$ and an intermediate ensemble I such that ρ' is compatible with I . (If ρ' is not compatible with I , further operations on the factors indexed by densities in I can never produce ρ' , as is immediate to see.) We must keep track of all applications of identity (4.13) necessary to produce the given ρ' , starting from I , i.e. we must consider all possible application of (4.13) to pairs $\{\rho_\alpha, \rho_\beta\} \subset I$ for which either

i) $\pm \rho_\alpha \subset \rho'$, $\rho_\beta \cap \rho' = \emptyset$, or

ii) $\pm \rho_\beta \subset \rho'$, $\rho_\alpha \cap \rho = \emptyset$, or

iii) $\pm \rho_\alpha \subset \rho'$, $\pm \rho_\beta \subset \rho'$.

In case i) the term on the right side of identity (4.13) is chosen in which ρ_β is eliminated. (Suppose not; then ρ_α and ρ_β would have been combined to $\rho_\alpha \pm \rho_\beta$. At later stages, either $\rho_\alpha \pm \rho_\beta$ would have been eliminated, or it would have been kept, so that either $\rho_\alpha \not\subset \rho'$, or $\pm \rho_\beta \subset \rho'$.) The term in which ρ_β is eliminated is $\propto(1+3K_\alpha \cos\phi(\rho_\alpha))$, so the coefficient, $3K_\alpha$, of $\cos\phi(\rho_\alpha)$ is independent of K_β and ρ_β .

Since (4.14) has been imposed, ρ_β contains at least one density $\rho \varepsilon E$ with the property that $\text{dist}(\rho,\rho') \leq 2^{n+1}$. Thus, the total number of applications of identity (4.13) of type i) necessary to produce ρ' is at most

$$\text{card} \left\{\rho \varepsilon E : \rho \cap \rho' = \emptyset , \text{dist}(\rho,\rho') \leq 2^{n+1}\right\} .$$

Case ii) is the same as case i), with α and β interchanged.

In case iii) a term on the right side of identity (4.13) is chosen in which ρ_α and ρ_β are combined to $\rho_\alpha + \rho_\beta$ or $\rho_\alpha - \rho_\beta$.

The coefficient of $\cos\phi(\rho_\alpha \pm \rho_\beta)$ is $3K_\alpha K_\beta$. Given ρ' , there are precisely card $\{\rho \in E : \rho \subset \rho'\}$ - 1 applications of (4.13) in which a term of type iii) is kept which are needed to form ρ' , independently of the order in which the constituents, $\rho \in E$, of ρ' got paired. (The proof of this is an easy combinatorial exercise.)

From the discussion of cases i) - iii) above, we now conclude that

$$K'(\rho') \leq 3^{n_E(\rho')} \prod_{\rho \in E} K(\rho)^{|\varepsilon(\rho',\rho)|} , \qquad (4.15)$$

where $n_E(\rho') = \text{card}\left\{\rho \in E : \text{dist}(\rho,\rho') \leq 2^{n+1}\right\}$. Next, we make use of the fact that E was assumed to be an n -ensemble, i.e. for any two densities ρ_1 , ρ_2 in E , $\rho_1 \neq \rho_2$, $\text{dist}(\rho_1,\rho_2) > 2^n$. From this property it follows immediately that a $2^n \times 2^n$ square cannot intersect more than three different charge densities in E . Let $\overline{V}_n(\rho)$ be the minimal number of $2^n \times 2^n$ squares needed to cover $\left\{j \in \Lambda : \text{dist}(j,\text{supp }\rho) \leq 2^{n+1}\right\}$. It is easy to check that $\overline{V}_n(\rho) \leq 25\, V_n(\rho)$. Thus

$$n_E(\rho') \leq C_2 V_n(\rho') ,$$

for some constant $C_2 \leq 75$.

From this and inequality (4.15) we obtain part c) of the lemma, with $C_1 = C_2 \ln 3$.

4.4 Proof of Theorem 4.1

Theorem 4.1 is proven by an induction in distance scales 2^n , $n = 0,1,2,\ldots$. Each induction step is carried out with the help of Lemma 4.2. The initial ensemble to which Lemma 4.2 is applied is given by

$$E(q_\Lambda) = \left\{\rho^j\right\}_{j \in \Lambda} ,$$

where

$$(\rho^j)_l \equiv q_j \, \delta_{jl} \, .$$

By Lemma 4.2, (4.11) we have

$$I(q_\Lambda, \phi) = \prod_{j \varepsilon \Lambda} [1+\zeta_{q_j} \cos(2\pi q_j \, \phi_j)] = \sum_\gamma c_\gamma \prod_{\rho \varepsilon E_{0,\gamma}} [1+K(\rho)\cos\phi(\rho)] \, , \tag{4.16}$$

where each $E_{0,\gamma}$ is a 0 -ensemble. Next, we apply (4.11) to each term in the sum on the r.s. of (4.16), with $E = E_{0,\gamma}$, for some γ. We end up with an expansion of $I(q_\Lambda, \phi)$ as a sum extending over 1 -ensembles of products of factors $1+K(\rho)\cos\phi(\rho)$.

For $n \geqq n_0$, where n_0 is an integer bounded above by $ln_2 M$, we must start to choose the ensemble E on the l.s. of (4.11) to be a <u>sub-ensemble</u> of some $E_{n,\gamma}$ obtained from previous applications of (4.11), in order to avoid forming unnecessarily large vortex molecules and large bare activities $K(\rho)$. Roughly speaking, we shall apply Lemma 4.2 to sub-ensembles of ensembles $E_{n,\gamma}$ which do <u>not</u> satisfy Condition D of Theorem 4.1. More precisely, those sub-ensembles are constructed, inductively, as follows:

Let Q_n be one of the ensembles obtained after the $(n+1)^{st}$ application of Lemma 4.1. The induction hypotheses for Q_n are as follows:

$$Q_n = E_n \cup N_n \, ,$$

with (4.17)

$$E_n \cap N_n = \emptyset \, ,$$

and N_n is defined, constructively, as follows:

α) $N_n \supseteq N_{n-1}$, for some sub-ensemble N_{n-1} constructed at the previous induction step.

β) For each vortex density $\rho \varepsilon Q_N \sim N_{n-1}$, let $B(\rho)$ be the set of sites within distance $< Md(\rho)^\alpha$ of supp ρ. Let B_n^1 be the family of all those densities $\rho \varepsilon Q_n \sim N_{n-1}$ for which $Q(\rho) = 0$ (i.e. ρ

is neutral) and $B(\rho) \cap \operatorname{supp} \rho' = \emptyset$, for all $\rho' \in Q_n \sim N_{n-1}$, $\rho' \neq \rho$. We define inductively

$$E_n^{(k)} \equiv Q_n \sim (N_{n-1} \cup B_n^1 \cup \ldots \cup B_n^k) , \tag{4.18}$$

and define B^{k+1} to be the family of all those $\rho \in E_n^{(k)}$ which are neutral and for which $B(\rho) \cap \operatorname{supp} \rho' = \emptyset$, for all $\rho' \in E_n^{(k)}$, $\rho' \neq \rho$. This defines $E_n^{(k)}$ and B_n^k for all $k = 1,2,3,\ldots$. We now set

$$\begin{aligned} N_n &= N_{n-1} \cup \left(\bigcup_{k=1}^{\infty} B_n^k \right) , \\ E_n &= Q_n \sim N_n = \bigcap_k E_n^{(k)} . \end{aligned} \tag{4.19}$$

The set N_n is a maximal subset of Q_n for which Condition D holds.

The induction in the distance scale n is started by setting

$$Q_0 = E_{0,\gamma} , \quad N_0 = \emptyset ,$$

where $E_{0,\gamma}$ is one of the ensembles on the r.s. of identity (4.16) and applying Lemma 4.2 to $\prod_{\rho \in E_{0,\gamma}} [1+K(\rho)\cos\phi(\rho)]$. The induction step, $n \to n+1$, is done by applying Lemma 4.2 to a term

$$\prod_{\rho \in E_n} [1+K(\rho)\cos\phi(\rho)] ,$$

where E_n is one of the ensembles defined in (4.19). This yields

$$\prod_{\rho \in E_n} [1+K(\rho)\cos\phi(\rho)] = \sum_{\kappa} c_{\kappa} \prod_{\rho \in I_{n+1,\kappa}} [1+K'(\rho')\cos\phi(\rho')] .$$

By Lemma 4.2, a), E_n is a parent of each $I_{n+1,\kappa}$, and each $I_{n+1,\kappa}$ is an $(n+1)$-ensemble. We choose $Q_{n+1,\kappa} \equiv Q_{n+1}$ to be $I_{n+1,\kappa} \cup N_n$, for some κ . Since $N_n \cap E_n = \emptyset$, and $E_n \to I_{n+1,\kappa}$, it follows that $N_n \cap I_{n+1,\kappa} = \emptyset$, for all κ . We then choose $N_{n+1,\kappa} \equiv N_{n+1}$ and $E_{n+1,\kappa} \equiv E_{n+1}$, for a given κ , as in (4.19), with n replaced by $n + 1$.

Now, for a bounded region Λ , E_n converges either to the empty set or to an ensemble consisting of a single, charged element, as $n \to \infty$.

Thus, we have established identity (4.6) of Theorem 4.1, (by induction in n): Each ensemble N occuring in (4.6) is a limit of a sequence of ensemble N_n , constructed as in (4,19), as $n \to \infty$. By properties α) and β) of the ensembles N_n , each N occuring on the r.s. of (4.6) satisfies Condition D, a) and b).

In order to prove part c) of Condition D, we pick a vortex density ρ in one of the ensembles N and suppose that ρ_1 is a constituent of ρ satisfying

$$\mathrm{dist}(\rho_1,\rho-\rho_1) \geqq R_1 \equiv 2\,Md(\rho_1)^{\alpha} \tag{4.20}$$

In the course of building up ρ out of some collection of densities, $\{\rho_\gamma\}$, with which ρ is compatible, it must have happened that some $\rho_\mu \subseteq \rho_1$ got paired with some $\pm\rho_\nu$, $\rho_\nu \subseteq \rho-\rho_1$ to form $\rho_\mu + \rho_\nu$. Since $\mathrm{dist}(\rho_\mu,\rho_\nu) \geqq R_1$, this happened on scale 2^k of the inductive construction, with $2^{k+1} \geqq R_1$. Since $R_1 \equiv 2Md(\rho_1)^{\alpha} \geqq 2Md(\rho_\mu)^{\alpha}$, ρ_μ is charged. For, if ρ_μ were neutral, $\rho_\mu \in N_k$, as $2^k \geqq (1/2)\,R_1 \geqq Md(\rho_\mu)^{\alpha}$.

Suppose now that $\rho_\mu \subsetneqq \rho_1$. Then at a later induction step some $\rho' \supseteq \rho_\mu + \rho_\nu$ got combined with some non-empty $\rho_\lambda \subseteq \rho_1 - \rho_\mu$. But

$$\mathrm{dist}(\rho',\rho_\lambda) \leqq \mathrm{dist}(\rho_\mu,\rho_\lambda) \leqq d(\rho_1) \ ,$$

and

$$d(\rho_1) = (R_1/2M)^{\alpha^{-1}} < (M^{-1}2^k)^{\alpha^{-1}} < 2^k \ ,$$

for $\alpha > 1$, $M > 1$. However, the expansion on scales $< 2^k$ was already terminated at this point. Thus $\rho_\mu = \rho_1$, and therefore ρ_1 is charged. This completes the proof of part c) of Theorem 4.1.

Now we turn to the proof of the bounds (4.7) and (4.8). We choose some N . Let ρ be some neutral density in N , and let $n(\rho)$ satisfy $2^{n(\rho)} \geqq Md(\rho)^{\alpha}$, as required in (4.8). Let $m > n(\rho)$, and suppose ρ was produced during the induction step on scale 2^m. Since, for arbitrary constituents ρ_μ, ρ_ν of ρ, $\mathrm{dist}(\rho_\mu, \rho_\nu) \leq d(\rho) < 2^{\alpha-1}m < 2^m$, for $M > 1$, $\alpha > 1$, ρ must have been paired with some $\rho'' \neq \rho$ at distance at least 2^m from ρ, and ρ'' got eliminated, i.e. the first term on the right side of identity (4.13) was chosen, (with $\rho_\alpha = \rho$, $\rho_\beta = \rho''$; see cases i), ii) in the proof of Lemma 4.2). However, since ρ is neutral, and $2^m > 2^{n(\rho)} \geq Md(\rho)^\alpha$ this would violate the rules for choosing N_n, namely $N_{n(\rho)}$ would not have been chosen to be a maximal subset of $Q_{n(\rho)}$. Thus, for some $m \leq n(\rho)$, $\rho \in N_m$. By Lemma 4.2, (4.12)

$$K(\rho) \leq e^{C_1 V_m(\rho)} \prod_{\rho_\gamma} K(\rho_\gamma)^{|\varepsilon(\rho,\rho_\gamma)|},$$

where $\rho = \Sigma\, \varepsilon(\rho,\rho_\gamma)\rho_\gamma$, and all densities ρ_γ belong to some $(m-1)$-ensemble. Applying (4.12) again, we obtain

$$K(\rho) \leq \exp\left\{C_1\, [V_m(\rho) + \sum_{\pm\rho_\gamma \subset \rho} V_{m-1}(\rho_\gamma)]\right\} \prod_{\pm\rho_\sigma \subset \rho} K(\rho_\sigma), \quad (4.21)$$

for densities ρ_σ in some $(m-2)$-ensemble.

Next, we make use of the inequality: If $\mathrm{dist}(\rho_\gamma, \rho_{\gamma'}) > 2^k$, for all $\gamma \neq \gamma'$ then

$$\sum_{\rho_\gamma \subset \rho} V_k(\rho_\gamma) \leq C_3 V_k(\rho), \quad (4.22)$$

for some constant $C_3 \leq 3$. (The proof is straightforward; see also end of proof of Lemma 4.2.) By combining (4.21) and (4.22) and a recursion, (4.7) follows, with $C \leq C_1\, C_3 \leq 225\, \ln 3$; (the upper bound on C is vastly larger than what one could presumably obtain by more detailed, combinatorial arguments). This completes the proof of Theorem 4.1.

5. A Sequence of Renormalization Transformations

We have shown in Sect. 4 - see (4.3) - (4.6) and Remark (1) following Theorem 4.1 that

$$Z_{\beta,\Lambda}(\sigma) = \int \prod_{j\varepsilon\Lambda} [1+2 \sum_{n=1}^{\infty} \cos(2\pi n\phi_j+n\sigma_j)]\, d\mu_\beta(\phi)$$

$$= \sum_{q_\Lambda} z_{q_\Lambda} \sum_{N\, F_{q_\Lambda}} c_N \int \prod_{\rho\varepsilon N} [1+K(\rho)\cos(2\pi\phi(\rho)+\sigma(\rho))]\, d\mu_\beta(\phi)\ . \quad (5.1)$$

The aim of this section is to renormalize the vortex densities ρ and the bare activities $K(\rho)$ in such a way that an identity

$$\int \prod_{\rho\varepsilon N} [1+K(\rho)\cos(2\pi\phi(\rho)+\sigma(\rho))]\, d\mu_\beta(\phi)$$
$$= \int \prod_{\rho\varepsilon N} [1+z_N(\beta,\rho)\cos(2\pi\phi(\bar{\rho})+\sigma(\rho))]\, d\mu_\beta(\phi) \quad (5.2)$$

results, where the renormalized vortex densities $\bar{\rho} = \bar{\rho}_N$ and the renormalized activities $z_N(\beta,\rho)$ have all the properties claimed in Theorem 3.2. The first step in the proof of (5.2) is to notice the trivial identity

$$\prod_{\rho\varepsilon N} [1+K(\rho)\cos(2\pi\phi(\rho)+\sigma(\rho))]$$
$$= \sum_{\varepsilon} \prod_{\rho\varepsilon N} (\tfrac{1}{2} K(\rho))^{|\varepsilon(\rho)|}\, e^{i(2\pi\phi(\varepsilon(\rho)\rho)+\sigma(\varepsilon(\rho)\rho))}\ , \quad (5.3)$$

where $\varepsilon : \rho \varepsilon N \to \varepsilon(\rho)$ is a function on N taking the values ± 1 and 0 , and the sum on the r.s. of (5.3) extends over all distinct such functions. We define ensembles

$$E \equiv E_\varepsilon = \left\{\varepsilon(\rho)\rho : \rho \varepsilon N\ ,\ \varepsilon(\rho) \neq 0\right\}\ . \quad (5.4)$$

Now, we recall that

$$\int \prod_{\rho\in E} e^{i2\pi\phi(\rho)} d\mu_\beta(\phi)$$

$$= \exp[-2\pi^2\beta \sum_{\rho,\rho'\in E} (\rho,C\rho')] , \tag{5.5}$$

where C is the Green's function of Δ , i.e. the two-dimensional Coulomb potential, and $(\rho,C\rho') \equiv \sum_{j,l} \rho_j\ C(j-l)\ \rho'_l$. Thus the self-energies of and the interactions between vortex molecules are given by Coulomb two-body forces (between their constituents).

The idea behind our proof of (5.2), (i.e. the renormalization transformation $\rho \to \bar{\rho}$, $K(\rho) \to z_N(\beta,\rho))$, is to replace the vortex densities $\rho \in E$ by densities $\bar{\rho}$ which are not necessarily integer-valued, but <u>neutral</u>, i.e. $\sum_j \bar{\rho}_j = \sum_j \rho_j = 0$, in such a way that the interaction energy between two different renormalized densities, $\bar{\rho}$ and $\bar{\rho}'$, is unchanged, i.e.

$$(\bar{\rho},C\bar{\rho}') = (\rho,C\rho') ,$$

but that the selfenergy of $\bar{\rho}$ is much smaller than the selfenergy of ρ , i.e.

$$(\bar{\rho},C\bar{\rho}) << (\rho,C\rho) .$$

This is a problem in electrostatics which, thanks to Condition D, b) of Theorem 4.1 turns out to be well posed. We solve it by a judicious use of the method of complex translations [29], following Sect. 4 of [5]. The upshot will be that if an ensemble E is given by (5.4) with N an ensemble satisfying Condition D then the Gaussian integrals of $\prod_{\rho\in E} e^{i(2\pi\phi(\rho)+\sigma(\rho))}$ and of

$$\prod_{\rho\in E} e^{i(2\pi\phi(\bar{\rho})+\sigma(\rho))} \exp[-2\pi^2\beta \left\{(\rho,C\rho)-(\bar{\rho},C\bar{\rho})\right\}]$$

are <u>identical</u>. The factor

$$\zeta_N(\beta,\rho) \equiv \exp[-2\pi^2\beta \left\{(\rho,C\rho)-(\bar{\rho},C\bar{\rho})\right\}] \equiv e^{-\beta E_{\text{loc.}}(\rho)} ,$$

where $E_{loc.}(\rho)$ is a "localized" portion of the electrostatic self-energy of ρ, will turn out to be so small that

$$Z_N(\beta,\rho) = \zeta_N(\beta,\rho)\ K(\rho)$$

obeys the bounds claimed in Theorem 3.2.

We now formalize these ideas.

A lattice square with sides containing 2^n sites parallel to the lattice axes will henceforth be called an n -square.

By S_n we denote an arbitrary collection of n -squares, and we define $|S_n|$ to be the total number of different n -squares in S_n. Given a subset D of Z^2, we let $S_n(D)$ be a minimal family of n -squares (i.e. $|S_n(D)|$ is minimal) needed to cover D. We define the volume of D on scale 2^n to be

$$V_n(D) \equiv |S_n(D)| \ , \tag{5.6}$$

and the total "volume" of D to be given by

$$V(D) = \sum_{n=0}^{n(D)} V_n(D) \ , \tag{5.7}$$

where $n(D)$ is the samllest integer such that

$2^{n(D)} \geq M\, \text{diam}(D)^{\alpha}$, where M and α are as in Condition D, Sect. 4.

If D is the support of a vortex density, ρ, we write $S_n(\rho)$, $V_n(\rho)$, $V(\rho)$, $n(\rho)$, instead of $S_n(\text{supp}\ \rho), \ldots$. [The notions introduced here and in the following first appear in [5] and in a slightly different form in [36], except for some trivial changes in notations.]

Given a family S_n of n -squares, we define a sub-family $S_n' \subset S_n$ of "far isolated" n -squares as follows: Let M and α be the constants introduced in Condition D. We define

$$S_n' \equiv S_n'(\alpha,M) \equiv \Big\{ s : s\varepsilon S_n \ , \ \text{dist}(s,s') \geqq 2M\, 2^{\alpha n} \ , \ \text{for all} \ \ s'\varepsilon S_n \ , \ s' \neq s \Big\} \ , \tag{5.8}$$

$$|S_n'| \equiv \# \{ s\varepsilon S_n' \} \ , \ V_n'(D) \equiv |S_n'(D)| \ , \tag{5.9}$$

$$V'(D) \equiv \sum_{n=0}^{n(D)} V_n'(D) \ , \text{ and} \tag{5.10}$$

$V_n'(\rho) \equiv V_n'(\text{supp}\ \rho)$ etc.

The main result of this section is

Theorem 5.2.

Let E *be an ensemble of vortex densities given by* (5.4), *with* N *satisfying* Condition D *of Theorem* 4.1. *Then*

$$\int \prod_{\rho\varepsilon E} e^{i2\pi\phi(\rho)} d\mu_\beta(\phi) = \int \prod_{\rho\varepsilon E} \zeta_N(\beta,\rho)\, e^{i2\pi\phi(\bar\rho)} d\mu_\beta(\phi) \ , \tag{5.11}$$

where

$$\zeta_N(\beta,\rho) = e^{-\beta E_{loc.}(\rho,N)} \ , \ \textit{and}$$

$$E_{loc.}(\rho,N) \geqq c\, V'(\rho) + d||\rho||_2^2 \ , \tag{5.12}$$

with $\quad ||\rho||_2^2 \equiv \sum_x |\rho_x|^2$, *and*

$$\zeta_N(\beta,\rho) = \zeta_N(\beta,-\rho) \ . \tag{5.13}$$

Furthermore, $\bar\rho = \bar\rho_N$ *is a function which depends linearly on* ρ *and satisfies*

i) $\sum_x \bar\rho_x = 0$

ii) $d(\bar\rho) < 4d(\rho)$, and

supp $\bar\rho \subset D(\rho)$, *where* $D(\rho)$ *is a disc of radius* $2d(\rho)$ containing supp ρ ; ($\bar\rho$ depends on the ensemble N).

An *outline of the proof* of Theorem 5.2 occupies the remainder of this section. It relies on a double induction: We order the vortex densities $\rho\varepsilon E$ according to their size:

$$E = \left\{\rho_\alpha\right\}_{\alpha=1}^{N} \text{ , with } d(\rho_\alpha) \leq d(\rho_{\alpha'}) \text{ if } \alpha < \alpha' . \qquad (5.14)$$

Our first induction extends over all possible α . The induction hypothesis is as follows:

Suppose the induction has reached ρ_α , for some $\alpha < N$. Let

$$E_< \equiv \left\{\rho_1,\ldots,\rho_\alpha\right\} \text{ , } E_> \equiv \left\{\rho_{\alpha+2},\ldots,\rho_N\right\} \text{ , and set}$$

$$\rho^* \equiv \rho_{\alpha+1} .$$

Assume that previous renormalizations have led to the identity

$$\begin{aligned} \underline{H_\alpha} : \int \prod_{\rho\varepsilon E} e^{i2\pi\phi(\rho)} d\mu_\beta(\phi) &= \int \prod_{\rho\varepsilon E_<} \zeta_N(\beta,\rho)\, e^{i2\pi\phi(\bar{\rho})}\, e^{i2\pi\phi(\rho^*)} \cdot \\ &\cdot \prod_{\rho\varepsilon E_>} e^{i2\pi\phi(\rho)} d\mu_\beta(\phi) , \end{aligned} \qquad (5.15)$$

where $\bar{\rho}$ and $\zeta_N(\beta,\rho)$ have all the properties required in Theorem 5.2, for all $\rho\varepsilon E_<$.

We have to renormalize $e^{i2\pi\phi(\rho^*)}$ in such a way that $H_{\alpha+1}$ follows. [The induction is started by setting $E_< = \emptyset, \rho^* = \rho_1$, in which case (5.15) is trivial.] We shall now construct $\overline{\rho^*}$ and $\zeta_N(\beta,\rho^*)$ by a second inductive procedure, extending over all possible scales, 2^n, n = 0,1,2,... , of constituents of ρ^* . More precisely, with half of the sites in supp ρ^* and with each square $s\varepsilon S_n'(\rho^*)$ we shall associate a renormalization transformation.

We define

$$\left.\begin{aligned} R(\phi) &\equiv \prod_{\rho\varepsilon E_<} \zeta_N(\beta,\rho)\, e^{i2\pi\phi(\bar{\rho})} , \\ U(\phi) &\equiv \prod_{\rho\varepsilon E_>} e^{i2\pi\phi(\rho)} . \end{aligned}\right\} \qquad (5.16)$$

Let $r_n\rho^*$ denote the vortex density that is obtained by renormalizing ρ^* on all scales 2^m, $m \leq n$. We shall construct $r_{n+1}\rho^*$ by successively performing transformations associated with all squares in $S'_{n+1}(\rho^*)$. We start by constructing $r_0\rho^*$, using the following easy lemma.

<u>Lemma 5.3.</u>

<u>Let</u> $G(\phi)$ <u>be a functional independent of</u> ϕ_{j_0} . <u>Then</u>

$$\int e^{iq\phi_{j_0}} G(\phi)d\mu_\beta(\phi) = e^{-\beta q^2/8} \int e^{iq\bar{\phi}_{j_0}} G(\phi)d\mu_\beta(\phi) , \tag{5.17}$$

<u>where</u> $\bar{\phi}_{j_0} = \frac{1}{4} \sum_{|k-j_0|=1} \phi_k$.

<u>Proof.</u>

The proof relies on the following well known identity for the Gaussian measure $d\mu_\beta(\phi)$ which plays a prominent role in this section:

$$d\mu_\beta(\phi+ia) = e^{i\beta^{-1}(\phi,\Delta a)} e^{-\frac{1}{2}\beta^{-1}(a,\Delta a)} d\mu_\beta(\phi) . \tag{5.18}$$

On the l.s. of (5.17) we change variables:

$$\phi_{j_0} = \phi'_{j_0}+i\beta q/4 , \quad \phi_k = \phi'_k , \quad \text{for } k \neq j_0 .$$

Using now (5.18), we obtain the r.s. of (5.17). ■

To construct $r_0\rho^*$, we decompose $\operatorname{supp}\rho^*$ into two disjoint subsets, Ω_1 and Ω_2 , such that no two sites in Ω_1 or in Ω_2 are nearest neighbors. Let $\rho^1 \equiv \rho^*\chi_{\Omega_1}$, $\rho^2 \equiv \rho^*\chi_{\Omega_2}$, (χ_Ω = characteristic function of Ω). Then

$$e^{i2\pi\phi(\rho^*)} = (\prod_x e^{i2\pi\rho^1_x\phi_x}) \cdot e^{i2\pi\phi(\rho^2)} .$$

We want to apply Lemma 5.3 successively to each factor in $\prod_x e^{i2\pi\rho^1_x\phi_x}$. Thus, let us check that the hypotheses are verified, no matter how Ω_1 , Ω_2 were chosen: By the induction hypothesis H_α ,

$$\operatorname{supp}\bar{\rho} \subset D(\rho) , \quad \operatorname{diam} D(\rho) \leq 4d(\rho) ,$$

for all $\rho \varepsilon E_<$; (use the definition of $D(\rho)$ and (5.15)). Thus

$$\begin{aligned} \mathrm{dist}(\rho^*,\bar{\rho}) &\geqq \mathrm{dist}(\rho^*,D(\rho)) \\ &\geqq \mathrm{dist}(\rho^*,\rho) - 4d(\rho) \\ &\geqq (M-4)d(\rho)^{\alpha} > 2 \ , \end{aligned} \tag{5.19}$$

for $M \geqq 5$, $\alpha > 1$. Furthermore,

$$\begin{aligned} \mathrm{dist}(\rho^*,\rho) &\geqq M\,[\min(d(\rho^*)\ ,\ d(\rho))]^{\alpha} \\ &\geqq M\,d(\rho^*)^{\alpha} \geqq 10 \ , \end{aligned} \tag{5.20}$$

for all $\rho \varepsilon E_>$, provided $M \geqq 5$, $\alpha > 1$.
Thus the hypotheses of Lemma 5.3 are fulfilled for each factor $\exp[i2\pi\rho_x^1\phi_x]$, $x \varepsilon \Omega_1$ and, by (5.17), keep being fulfilled after each application of Lemma 5.3, no matter how Ω_1 is chosen. We now set

$$r_0\rho_x^* = \rho_x^2 + \frac{1}{4} \sum_{|y-x|=1} \rho_y^1 \tag{5.21}$$

Recalling the definition (5.16) we now obtain, by successive applications of Lemma 5.3,

$$\begin{aligned} &\int R(\phi)U(\phi)e^{i2\pi\phi(\rho^*)}\,d\mu_\beta(\phi) \\ &= e^{-\frac{\beta\pi^2}{2}\sum_x|\rho_x^1|^2} \int R(\phi)U(\phi)\,e^{i2\pi\phi(r_0\rho^*)}\,d\mu_\beta(\phi) \end{aligned} \tag{5.22}$$

Clearly, we can choose Ω_1 such that

$$\sum_x |\rho_x'|^2 \geqq \frac{1}{2}\,||\rho^*||_2^2 \ . \tag{5.23}$$

On scales $n \geqq 1$, our renormalization transformations always follow the same pattern. Thus we assume we have already constructed $r_n\rho^*$, for some finite $n \geqq 0$. We shall construct $r_{n+1}\rho^*$ by performing a transformation t_s on $r_n\rho^*$, corresponding to each square $s \varepsilon S'_{n+1}(\rho^*)$. Then

$$r_{n+1}\rho^* = \Big(\prod_{s\in S'_{n+1}(\rho^*)} t_s\Big)(r_n\rho^*) \tag{5.24}$$

The transformations t_s , $s \in S'_{n+1}(\rho^*)$ will turn out to commute, because different squares in $S'_{n+1}(\rho^*)$ are far separated. It is convenient to order the squares in $S'_{n+1}(\rho^*)$ in an arbitrary way:

$$S'_{n+1}(\rho^*) = \{s_k\}_{k=1}^{K} . \tag{5.25}$$

We now assume that all transformations, $t_k^{(n+1)} \equiv t_{s_k}$, $s_k \in S'_{n+1}(\rho^*)$, $k \leq l$ have already been performed. We shall construct $t_{l+1}^{(n+1)}$. First, we formulate the <u>induction hypothesis</u>.

<u>$I_{n,l}$:</u>

$$\int R(\phi)U(\phi)\, e^{i2\pi\phi(\rho^*)}\, d\mu_\beta(\phi) \tag{5.26}$$

$$= \int R(\phi)U(\phi)\zeta_{n,l}(\beta,\rho^*)\, e^{i2\pi\phi(r_{n,l}\rho^*)}\, d\mu_\beta(\phi) ,$$

<u>where</u>

$$\zeta_{n,l}(\beta,\rho^*) \leq e^{-\frac{1}{4}\beta\pi^2||\rho^*||_2^2} \Big(\prod_{m=1}^{n} e^{-c\beta V'_m(\rho^*)}\Big)\, e^{-c\beta l} , \tag{5.27}$$

<u>and where</u>

$$r_{n,l}\rho^* \equiv \prod_{k=1}^{l} t_k^{(n+1)}\, r_n\rho^* \tag{5.28}$$

<u>has the following properties:</u>

<u>Let</u> s <u>be an arbitrary square in</u> $S'_{n+1}(\rho^*)$, and let $\rho^*_s \equiv \rho^*\chi_s$, <u>where</u> χ_s <u>is the characteristic function of</u> s . <u>Then</u>

$$r_{n,l}\rho^* = \sum_{k=1}^{K} r_{n,l}\rho^*_{s_k} + r_n\Big(\rho^* - \sum_{k=1}^{K}\rho^*_{s_k}\Big) , \tag{5.29}$$

with

$$\sum_x (r_{n,l}\rho^*_{s_k})_x = \sum_x (\rho^*_{s_k})_x \text{ , for all } k , \tag{5.30}$$

$$r_{n,l}\rho^*_{s_k} = r_n\rho^*_{s_k} \text{ , } \operatorname{supp}(r_{n,l}\rho^*_{s_k}) \subseteq D(n,s_k) \text{ , } k > l , \tag{5.31}$$

$$r_{n,l}\rho^*_{s_k} = t_k^{(n+1)} \, r_n\rho^*_{s_k} \text{ , } \operatorname{supp}(r_{n,l}\rho^*_{s_k}) \subseteq D(n+1,s_k) , \tag{5.32}$$

$k \leq l$, where $D(n,s)$ is a disc of radius 2^{n+1} containing $\operatorname{supp} \rho^*_s$. Furthermore,

$$\left.\begin{aligned} &\operatorname{dist}(r_{n,l}\rho^*_{s_k}, r_{n,l}(\rho^*-\rho^*_{s_k})) > M2^{\alpha(n+1)}-3\;2^{n+1} , \\ &\text{for } k > l \text{ , and} \\ &\operatorname{dist}(r_{n,l}\rho^*_{s_k}, r_{n,l}(\rho^*-\rho^*_{s_k})) > M2^{\alpha(n+1)}-4\;2^{n+1} , \\ &\text{for } k \leq l . \quad \square \end{aligned}\right\} \tag{5.33}$$

Remark. We shall choose $M \geq 5$.

To start our induction, we note that $I_{0,l}$, $l = 1,2,\ldots,|\Omega_1|$ follows from (5.22) and (5.23). [Conditions (5.33) have to be checked only for $n \geq 1$.] The induction step,

$$I_{n,l} \longrightarrow I_{n,l+1} \text{ , } l = 0,\ldots,K-1 \text{ , } (I_{n+1,0} \equiv I_{n,K}) \tag{5.34}$$

is carried out with the help of Lemma 5.4, stated below. We set $s^* \equiv s_{l+1}$. We claim that, since $s^* \in S'_{n+1}(\rho^*)$, $\rho^*_{s^*}$ is charged, in the sense that

$$\left|\sum_x (\rho^*_{s^*})_x\right| \geq 1 , \tag{5.35}$$

because

$$\operatorname{dist}(\rho^*_{s^*},\rho^*-\rho^*_{s^*}) \geq M2^{\alpha(n+1)} \geq Md(\rho^*_{s^*})^\alpha ,$$

so that (5.35) follows from Condition D, c) (Sect. 4).

We now set

$$f \equiv 2\pi\, r_n \rho^*_{s^*} \;,\; q \equiv \sum_x f_x \;. \tag{5.36}$$

By (5.30) and (5.35),

$$|q| \geq 2\pi \;. \tag{5.37}$$

We set

$$F(\phi) \equiv U(\phi)\, e^{i2\pi\phi(r_{n,l}(\rho^* - \rho^*_{s^*}))} \;, \tag{5.38}$$

with $U(\phi)$ as in (5.16) and $r_{n,l}$ as in (5.28). By (5.31),

$$\text{supp } f \subseteq D(n,s^*) \;.$$

Without loss of generality, we may assume that $D(n,s^*)$ is centered at the origin. We choose $D(n+1,s^*)$ to be the disc of radius 2^{n+2} which is centered at the origin, too. By (5.33) and (5.20) [and since $D(n+1,s^*) \supset D(n,s^*) \supseteq \text{supp } \rho^*_{s^*}$,] it is clear that

$$F(\phi) \ \underline{\text{is independent of}} \ \left\{\phi_j : j \in D(n+1,s^*)\right\} . \tag{5.39}$$

The main technical result of this section is the following

Lemma 5.4.

For $\alpha > \frac{3}{2}$ and M large enough,

$$\int e^{i\phi(f)} R(\phi) F(\phi) d\mu_\beta(\phi)$$

$$= e^{-c\beta q^2} \int e^{i\phi(\bar{f})} R(\phi) F(\phi) d\mu_\beta(\phi) \;,$$

where R is given by (5.16), F is defined in (5.38), c is a positive constant and $\bar{f}$ is a function depending linearly on f, with

$$\text{supp } \overline{f} \subseteq D(n+1,s^*) \tag{5.40}$$

and $$\sum_x \overline{f}_x = \sum_x f_x . \tag{5.41}$$

Remark. We set $t_{l+1}^{(n+1)} \; r_n \rho^*_{s^*} = \frac{1}{2\pi} \overline{f}$,

and $r_{n,l+1}\rho^* = \frac{1}{2\pi} \overline{f} + r_{n,l}(\rho^* - \rho^*_{s^*})$, $s^* \equiv s_{l+1}$.

Before we sketch the proof of Lemma 5.4 we want to first note that

$I_{n,l}$, (5.36) and (5.37), and Lemma 5.4 imply $I_{n,l+1}$.

This is checked easily. Thus $I_{n,l}$ holds for all l and n ! [Clearly $r_{n,l}\rho^* = r_{n_0}\rho^*$, for all $n \geq n_0 \equiv n_0(\rho^*)$, where $n_0(\rho^*)$ is the smallest integer for which $V_{n_0}(\rho^*) = 1$, so that $V'_{n_0}(\rho^*) = 0$.] We set

$$\overline{\rho}^* = r_{n_0(\rho^*)} \rho^* \; ; \; (\overline{\rho}^* \text{ depends on } \rho^* \text{ and on } N). \tag{5.42}$$

Next, we observe that from (5.26), (5.27) and $I_{n,l}$, for all l and n , it follows that

$$\int R(\phi)U(\phi)e^{i2\pi\phi(\rho^*)} d\mu_\beta(\phi)$$

$$= \int R(\phi)U(\phi)\zeta_N(\beta,\rho^*)e^{i2\pi\phi(\overline{\rho}^*)} d\mu_\beta(\phi) ,$$

with

$$\zeta_N(\beta,\rho^*) \leq e^{-\frac{\beta\pi^2}{4}||\rho^*||_2^2} \prod_m e^{-c\beta V'_m(\rho^*)} ,$$

and by (5.30), or (5.41) (charge conservation under renormalization)

$$\sum_x \overline{\rho^*}_x = \sum_x \rho^*_x .$$

Finally, by (5.29), (5.31) and (5.32),

$$\text{supp } \overline{\rho}^* \subseteq D(\rho^*)$$

where - we recall - $D(\rho^*)$ is a disc of radius $2d(\rho^*)$ containing $\operatorname{supp} \rho^*$.

Thus we have proven $H_{\alpha+1}$ (see (5.15)), assuming H_α . It follows that H_α holds for all α , and this yields Theorem 5.2.

Proof of Lemma 5.4

We define a function b on $D(n+1,s^*)$ by

$$b_j = \begin{cases} 1, & |j| \le 2^{n+1} \\ \log_2[2^{n+2}|j|^{-1}] , & 2^{n+1} \le |j| \le 2^{n+2} \\ 0, & |j| \ge 2^{n+2} . \end{cases} \tag{5.43}$$

Let $\left\{C_i\right\}_{i=1,2,3,\ldots}$ be the connected components of $\bigcup_{\rho\in N} D(\rho)$.

[Recall that $E = \left\{\varepsilon(\rho)\ \rho : \rho \in N , \varepsilon(\rho) \neq 0\right\}$.] It is easy to see that, given C_i , there is a unique vortex density ρ_i , with $D(\rho_i) \subseteq C_i$, and $d(\rho_i) > d(\rho)$, for all ρ for which $D(\rho) \subset C_i$, $\rho \neq \rho_i$.

We now define a function a which is proportional to b , except on $\cup C_i$:

$$a_j = \begin{cases} \gamma\beta q b_j & , \ j \notin C_i , \quad i = 1,2,3,\ldots \\ \\ \gamma\beta q b_{x_i} & , \ j \in C_i , \end{cases} \tag{5.44}$$

where x_i is a point in C_i chosen so that if C_i meets $D(n,s^*)$ then $x_i \in D(n,s^*)$ and hence $a_j = \gamma\beta q$, for $j \in C_i$, if C_i meets $D(n+1,s^*)^c$ then $x_i \notin D(n+1,s^*)$, hence $a_j = 0$, for all $j \in C_i$, and x_i is the center of $D(\rho_i)$, otherwise. Furthermore γ is a constant to be chosen below. We note that a depends on N .

Using Condition D, b) (Sect. 4) one checks easily (see Appendix E of [5]) that

$$\begin{aligned} &\text{a)} \quad \operatorname{diam} C_i \le \tfrac{5}{2}\, d(\rho_i) \\ &\text{b)} \quad |\partial C_i| \le 10\, d(\rho_i) , \end{aligned} \tag{5.45}$$

where ∂C_i is the set of links joining sites in C_i to sites in the complement of C_i. It follows from Condition D, b) that if, for some k, C_k intersects $D(n+1,s^*)$ then $\text{diam } C_k \ll d(\rho^*)$, and hence

$$4 \cdot 2^{n+1} > \text{dist}(\rho^*,\rho_k) \geq M\, d(\rho_k)^{\alpha} \geq M\, (\tfrac{2}{5} \text{ diam } C_k)^{\alpha} .$$

Thus, for $\alpha > 1$ and M large enough,

$$\text{diam } C_k \ll 2^n , \tag{5.46}$$

(and $\text{diam } C_k \to 0$, as $M \to \infty$).

We now make the change of variables

$$\phi_j = \phi_j' + ia_j . \tag{5.47}$$

This yields, using identity (5.18),

$$\begin{aligned} &\int e^{i\phi(f)} R(\phi)F(\phi)d\mu_\beta(\phi) \\ &= e^{-\frac{1}{2}\beta^{-1}(a,\Delta a)}\, e^{-\gamma\beta q^2} \int e^{i\phi'(f+\beta^{-1}\Delta a)} \\ &\qquad R(\phi'+ia)F(\phi'+ia)d\mu_\beta(\phi') . \end{aligned} \tag{5.48}$$

By (5.39) and (5.44) (i.e. $\text{supp } a \subseteq D(n+1,s^*)$),

$$F(\phi'+ia) = F(\phi') .$$

Since each $\rho \varepsilon E$ is neutral, $(\sum_x \bar{\rho}_x = \sum_x \rho_x = 0)$, and since a is constant on the support of $\bar{\rho}$, for all $\rho \varepsilon E_<$, we have that $\exp i2\pi(\phi'+ia)(\bar{\rho}) = \exp i2\pi\phi'(\bar{\rho})$, for each $\rho \varepsilon E_<$, and hence

$$R(\phi'+ia) = R(\phi') .$$

We set

$$\bar{f} \equiv f + \beta^{-1}\Delta a . \tag{5.49}$$

Thus

$$\int e^{i\phi(f)} R(\phi)F(\phi)d\mu_\beta(\phi)$$

$$= e^{-\frac{1}{2}\beta^{-1}(a,\Delta a)} e^{-\gamma\beta q^2} \int e^{i\phi'(\bar{f})} R(\phi')F(\phi')d\mu_\beta(\phi') ,$$

with

$$\sum_x \bar{f}_x = \sum_x f_x + \beta^{-1} \sum_x (\Delta a)_x$$

$$= \sum_x f_x = q .$$

Furthermore

$$\text{supp } \bar{f} \subseteq D(n+1,s^*) .$$

In order to complete the proof of Lemma 5.4, it remains to estimate

$$\exp[-\frac{1}{2}\beta^{-1}(a,\Delta a) - \gamma\beta q^2] .$$

By the definition of a, see (5.44), and of b, see (5.43),

$$-\frac{1}{2}\beta^{-1}(a,\Delta a) = \frac{1}{2}\beta(q\gamma)^2||\nabla b||_2^2 + \tilde{E} , \tag{5.50}$$

where

$$|\tilde{E}| \leq k_1\beta(q\gamma)^2 \sum_{C_i:C_i\cap A(n,s^*)\neq\emptyset} \max_{j\varepsilon C_i} |(\nabla b)(j)|^2 \text{ diam}(C_i)^3 , \tag{5.51}$$

$A(n,s^*)$ is the annulus $D(n+1,s^*) \sim D(n,s^*)$, and by (5.43)

$$||\nabla b||_2^2 \leq k_2 , \tag{5.52}$$

where k_1 and k_2 are finite constants of $O(1)$.

We are left with estimating the r.s. of (5.51): By (5.43)

$$\max_{j\varepsilon C_i} |(\nabla b)(j)|^2 \leq k_3 |x_i|^{-2} \leq k_3\ 2^{-2(n+1)} \tag{5.53}$$

We now estimate the part of the sum on the l.s. of (5.51) extending over all those C_i for which

$$2^k \leq d(\rho_i) < 2^{k+1}\ , \tag{5.54}$$

where ρ_i is the unique vortex density of maximal diameter $d(\rho_i)$ covered by C_i , and

$$C_i \cap A(n,s^*) \neq \emptyset\ .$$

Subsequently we shall sum over $k = 1,2,3,\ldots$.

Now, by Condition D, b) (Sect. 4) there are no more than

$$\begin{aligned} k_4\ |A(n,s^*)|\ M^{-2} & \Big(\min_{2^k \leq d(\rho_i) < 2^{k+1}} d(\rho_i)\Big)^{-2\alpha} \\ & \leq k_5\ 2^{2n}\ M^{-2}\ 2^{-2\alpha k} \end{aligned} \tag{5.55}$$

components C_i meeting $A(n,s^*)$ for which (5.54) holds. Thus, using (5.51), (5.53) and (5.55), we obtain

$$\begin{aligned} |\tilde{E}| & \leq k_1 \beta(q\gamma)^2 \Big\{ \sum_{k=1}^{\infty} \frac{1}{4}\, k_3\ 2^{-2n}\ 2^{3(k+1)} k_5\ 2^{2n}\ M^{-2}\ 2^{-2\alpha k} \Big\} \\ & \leq k_6 \beta(q\gamma)^2\ M^{-2} \sum_{k=1}^{\infty} 2^{k(3-2\alpha)} \\ & \leq k_7(\alpha) \cdot \beta(q\gamma)^2\ M^{-2} \end{aligned} \tag{5.56}$$

where $k_7(\alpha)$ is a finite constant, provided

$$\alpha > \frac{3}{2}\ ! \tag{5.57}$$

By (5.50), (5.52) and (5.56),

$$\exp[-\tfrac{1}{2}\beta^{-1}(a,\Delta a) - \gamma\beta q^2]$$

$$= \exp[\tfrac{1}{2}\ \beta(q\gamma)^2(k_2+k_7(\alpha)M^{-2}) - \gamma\beta q^2]$$

$$= \exp[-\ \tfrac{1}{2}\ (k_2+k_7(\alpha)M^{-2})^{-1}\ \beta q^2]\ ,$$

if we choose $\gamma = (k_2+k_7(\alpha)M^{-2})^{-1}$.

6. The Balance Between Energy and Entropy

In view of our basic identities (5.1), (5.3) and (5.11) we define a renormalized activity $z_N(\beta,\rho)$ for a vortex molecule in an ensemble N, described by a density ρ, by

$$z_N(\beta,\rho) = \zeta_N(\beta,\rho)K(\rho) \tag{6.1}$$

where the bare activity $K(\rho)$ is given by

$$0 < K(\rho) = \zeta(\rho)\, e^{S(\rho)} , \tag{6.2}$$

$$\zeta(\rho) = \prod_j \zeta_{|\rho_j|} , \quad \zeta_q = \text{const. } e^{\delta q^2} , \tag{6.3}$$

$$S(\rho) \leqq C\, V(\rho) , \text{ with} \tag{6.4}$$

$$V(\rho) = \sum_{n=0}^{n(\rho)} V_n(\rho) , \quad V_n(\rho) = |S_n(\rho)| , \tag{6.5}$$

$$n(\rho) = \text{ smallest integer } \geqq \ln_2 (Md(\rho)^{\alpha}) ;$$

see Theorem 4.1. Moreover $\zeta_N(\beta,\rho)$ has been introduced in Theorem 5.2 and is defined, more precisely, through Lemma 5.3, (5.22), (5.27) and Lemma 5.4. In Theorem 5.2 it is stated that

$$\zeta_N(\beta,\rho) = e^{-\beta E_{loc.}(\rho,N)} , \tag{6.6}$$

with

$$E_{loc.}(\rho,N) \geqq \tilde{c}\, V'(\rho) + \tilde{d}||\rho||_2^2 , \tag{6.7}$$

$$V'(\rho) = \sum_{n=0}^{n(\rho)} V_n'(\rho) , \quad V_n'(\rho) = |S_n'(\rho)| , \tag{6.8}$$

and $S_n'(\rho)$ is the family of "far-isolated n-squares" defined in (5.8) which cover <u>constituents of</u> ρ <u>of non-vanishing total vorticity</u>.

The quantity $S(\rho)$ is an entropy, while $E_{loc.}(\rho,N)$ is a selfenergy of ρ. In order to prove our basic result, Theorem 3.2, we need an upper bound on $z_N(\beta,\rho)$, or, in view of (6.1), (6.2) and (6.6), we must balance entropy against energy. More precisely, we shall show that for sufficiently large β

$$0 < z_N(\beta,\rho) \leq \exp[-\beta(c||\rho||_2^2 + d\ \ell n\ d(\rho))] , \tag{6.9}$$

as claimed in (3.19).

The lower bound follows from the positivity of $\zeta_N(\beta,\rho)$ and $K(\rho)$. To prove the upper bound, we combine (6.1) - (6.4) with (6.6) and (6.7) to get

$$z_N(\beta,\rho) \leqq \zeta(\rho)\ e^{-\beta\tilde{d}||\rho||_2^2}\ e^{C\ V(\rho)-\beta\tilde{c}\ V'(\rho)} . \tag{6.10}$$

By (6.2), (6.4) and the trivial fact that $V_0(\rho) \leqq ||\rho||_2^2$, we have

$$\zeta(\rho)\ e^{-\beta\tilde{d}||\rho||_2^2} \leqq e^{K_1 V_0(\rho)}\ e^{(\delta-\beta\tilde{d})||\rho||_2^2}$$

$$\leqq e^{-\beta c||\rho||_2^2 - \beta K_2\ V_0(\rho)} , \tag{6.11}$$

for constants c and K_2 which are positive if e.g.

$$\beta\tilde{d} - 2\delta - 2K_1 \geqq 0 .$$

Let M and α be the constants introduced in Condition D, Theorem 4.1. The main result of this section is

Lemma 6.1. For $\alpha < 2$ and $M < \infty$,

$$V(\rho) \leqq K_3\ V'(\rho) + K_4\ V_0(\rho) ,$$

for some finite positive constants K_3 and K_4, (depending on M and α).

Before proving this lemma we show that, together with (6.10) and (6.11), it yields the desired upper bound (6.9): We have

$$z_N(\beta,\rho) \leq \exp[-\beta c||\rho||_2^2] \exp[-\beta(K_2 V_0(\rho)+\tilde{c}V'(\rho))] \exp[CV(\rho)] .$$

Lemma 6.1, combined with the trivial inequalities $V(\rho) \geqq V_0(\rho)$ and $V'(\rho) \geqq 0$, yields

$$\tilde{c}\, V'(\rho) + K_2\, V_0(\rho) \geqq K_5\, V(\rho) ,$$

for some constant $K_5 > 0$. Hence

$$z_N(\beta,\rho) \leq \exp[-\beta c||\rho||_2^2] \exp[-\beta K_6 V(\rho)] , \tag{6.12}$$

provided β is large enough; ($\beta > 2\,[\frac{C}{K_5} + \frac{\delta}{\tilde{d}} + \frac{K_1}{\tilde{d}}]$ will do).

Now, by the definition of $V(\rho)$ and $n(\rho)$, see (6.5),

$$V(\rho) \geqq n(\rho) \geqq ln_2(Md(\rho)^{\alpha}) \geqq K_7\, ln\, d(\rho) .$$

Therefore (6.9) follows from (6.12).

Next, we turn to our

Proof of Lemma 6.1.

We define $b \equiv ln_2(2M)$ and introduce a function γ on $[0,\infty)$ by setting

$$\gamma(t) \equiv [\alpha^{-1}(t-b-2)] , \tag{6.13}$$

where M and α are as in Condition D, Theorem 4.1, and $[x]$ denotes the integer part of x. Finally, let

$$S_n''(\rho) \equiv S_n(\rho) \sim S_n'(\rho) . \tag{6.14}$$

We now choose some k with $2^{\gamma(k)} < d(\rho)$, so that $S_{\gamma(k)}(\rho)$ contains at least two squares. Let s_1 be some square in $S''_{\gamma(k)}(\rho)$. [It is imagined, here, that $S''_{\gamma(k)}(\rho) \neq \emptyset$, hence $|S''_{\gamma(k)}(\rho)| \geq 2$.] There exists then some square $s_2 \in S''_{\gamma(k)}(\rho)$ such that

$$\mathrm{dist}(s_1,s_2) < 2\,M\,2^{\alpha\gamma(k)} \leqq 2^{\,b}\,2^{k-b-2} = 2^{k-2} ,$$

and, since $d(s_1) = d(s_2) = \sqrt{2}\,2^{\gamma(k)}$,

$$d(s_1) + d(s_2) + \mathrm{dist}(s_1,s_2) < 2^{k-1} ,$$

provided $b > 1$, $\alpha \geqq 1$. Thus s_1 and s_2 can be covered by a single k -square s in $S_k(\rho)$.

More generally, let $s_1,\dots,s_m$, $m \geqq 3$, be squares in $S''_{\gamma(k)}(\rho)$. We join two squares, s_i and s_j , $(i \neq j)$ by a line iff $\mathrm{dist}(s_i,s_j) < 2\,M\,2^{\alpha\gamma(k)}$, $i,j = 1,\dots,m$. Now, fix i . Then s_i and all squares s_{l_i} , $l_i \neq i$, joined to s_i by a line can be covered by a single k -square. Therefore $[\frac{m}{2}]$ k -squares suffice to cover $s_1,\dots,s_m$, for any $m = 2,3,4,\dots$. From this observation follows the inequality

$$\begin{aligned} V_k(\rho) &\leqq \tfrac{1}{2}|S''_{\gamma(k)}(\rho)| + |S'_{\gamma(k)}(\rho)| \\ &\leqq \tfrac{1}{2}\,V_{\gamma(k)}(\rho) + |S'_{\gamma(k)}(\rho)| , \end{aligned} \tag{6.15}$$

provided $S_{\gamma(k)}(\rho)$ contains at least two squares. We shall need estimate (6.15) only, for $k \leqq n(\rho)$. Since $2^{\gamma(n(\rho))} < d(\rho)$, $|S_{\gamma(k)}(\rho)| \geqq 2$, for all such values of k , so that (6.15) holds.

Clearly (6.15) can be used only if $\gamma(k) \geqq 0$. Thus, for each k , we may iterate (6.15) $l(k)$ times, where $l(k)$ is the maximal integer for which $\gamma^{l(k)}(k) \geqq 0$; (γ^m is the m -fold composition of γ with itself). We conclude that

$$\begin{aligned} V_k(\rho) &\leqq \sum_{m=0}^{l-1} 2^{-m}\,|S'_{\gamma^{m+1}(k)}(\rho)| + 2^{-l}\,V_{\gamma^{l}(k)}(\rho) \\ &\leqq \sum_{m=0}^{l-1} 2^{-m}\,|S'_{\gamma^{m+1}(k)}(\rho)| + 2^{-l}\,V_0(\rho) , \end{aligned} \tag{6.16}$$

with $l \equiv l(k)$. (We have used that $V_j(\rho) \leqq V_0(\rho)$, for $j \geqq 0$.) Now, it is not hard to show (see Sect. 3 of [5]) that

$$l(k) \geq \begin{cases} 0 , & 0 \leq k < k_0 , \\ [(1/ln_2\alpha)\ ln_2\ (k/k_0)] , & \text{otherwise,} \end{cases} \tag{6.17}$$

where $k_0 = (\alpha-1)^{-1}(\alpha+b-2)$.

By (6.16), (6.17) and the definition of $V(\rho)$, we have

$$\begin{aligned} V(\rho) &= \sum_{k=0}^{n(\rho)} V_k(\rho) \\ &\leq \sum_{k=0}^{n(\rho)} 2^{-l(k)}\ V_0(\rho) + \sum_{m=0}^{l(k)-1} 2^{-m}|S'_{\gamma^{m+1}(k)}(\rho)| \qquad (6.18) \\ &\leq K_3\ V'(\rho) + K_4\ V_0(\rho) , \end{aligned}$$

where

$$\begin{aligned} K_4 &\leq \sum_{k=0}^{\infty} 2^{-l(k)} \\ &\leq k_0 + 2 \sum_{k=k_0+1}^{\infty} (k_0/k)^{1/ln_2\ \alpha} . \end{aligned}$$

Thus

$$K_4 < \infty , \text{ provided } 1 \leq \alpha < 2 . \tag{6.19}$$

Next, we estimate K_3 . We define families, $N_{m,j}$, of scales by

$$N_{m,j} \equiv \left\{k : \gamma^m(k) = j\right\} . \tag{6.20}$$

By (6.18), we must estimate

$$\sum_{k=0}^{n(\rho)} \sum_{m=0}^{l(k)-1} 2^{-m} |S'_{\gamma^{m+1}(k)}(\rho)|$$

$$= \sum_{j=0}^{\infty} \Big(\sum_{k=0}^{n(\rho)} \sum_{m=0}^{l(k)-1} 2^{-m} \delta_{\gamma^{m+1}(k),j} \Big) |S_j'(\rho)|$$

$$\leq \sum_{j=0}^{\infty} \Big(\sum_{m,k=0}^{\infty} 2^{-m} \delta_{\gamma^{m+1}(k),j} \Big) |S_j'(\rho)|$$

$$\leq \sum_{j=0}^{\infty} \Big(\sum_{m=0}^{\infty} 2^{-m} |N_{m,j}| \Big) |S_j'(\rho)|$$

$$\leq K_3 \sum_{j=0}^{\infty} |S_j'(\rho)| = K_3 V'(\rho) , \qquad (6.21)$$

where

$$|N_{m,j}| \leq \alpha^m \cdot \frac{2\alpha}{\alpha-1} , \text{ so that} \qquad (6.22)$$

$$K_3 \text{ is finite iff } \alpha < 2 . \qquad (6.23)$$

The proof of (6.22) is as follows: Since the function γ is monotone increasing,

$$\alpha^2 t - \alpha^{-2} d - \alpha^{-1} d - \alpha^{-1} - 1 \leq \gamma(\alpha^{-1}(t-d)-1)$$

$$\leq \gamma(\gamma(t)) \leq \gamma(\alpha^{-1}(t-d)) \leq \alpha^{-2} t - \alpha^{-2} d - \alpha^{-1} d ,$$

see (6.13) . Here $d \equiv b-2$. By induction

$$\alpha^{-m} t - d\Big(\sum_1^m \alpha^{-l}\Big) - \sum_0^{m-1} \alpha^{-l} \leq \gamma^m(t) \qquad (6.24)$$

$$\leq \alpha^{-m} t - d\Big(\sum_1^m \alpha^{-l}\Big) .$$

Let k_- be the minimal and k_+ the maximal integer in $N_{m,j}$. By (6.20) and (6.24),

$$\alpha^{-m} k_+ - d\Big(\sum_1^m \alpha^{-l}\Big) - \sum_0^{m-1} \alpha^{-l} \leq \gamma^m(k_+) = j = \gamma^m(k_-)$$

$$\leq \alpha^{-m} k_- - d\Big(\sum_1^m \alpha^{-l}\Big) ,$$

hence

$$|N_{m,j}| = k_+ - k_- + 1 < \alpha^m \cdot \frac{2\alpha}{\alpha-1} .$$

7. Completion of the Proof of Theorem 3.2

We recall that the goal in Sects. 3-6 was to give a proof of existence of the Berezinskii-Kosterlitz-Thouless transition, as described in Sect. 3, Theorem 3.1 and Remark 2). More precisely, we intended to prove that, for β sufficiently large,

$$\langle\vec{S}_x \cdot \vec{S}_y\rangle (\beta) \geqq \text{const.}\ (1+|x-y|)^{-1/2\pi\beta'} , \tag{7.1}$$

where $\beta' - \beta$ tends to 0, exponentially fast. In Sect. 3, this problem was reduced to proving a somewhat technical identity, stated in Theorem 3.2:

$$Z_{\beta,\Lambda}(\sigma) = \int \prod_{x\varepsilon\Lambda} [1+2 \sum_{n=1}^{\infty} \cos(2\pi n\phi_x+n\sigma_x)]d\mu_\beta(\phi)$$

$$= \sum_{N\varepsilon F_\Lambda} c_N \int \prod_{\rho\varepsilon N} [1+z_N(\beta,\rho)\cos(2\pi\phi(\bar{\rho})+\sigma(\rho))]d\mu_\beta(\phi) , \tag{7.2}$$

where $c_N > 0$, for all ensembles $N \varepsilon F_\Lambda$, and

$$0 < z_N(\beta,\rho) \leq \exp[-\beta(c||\rho||_2^2+d\ ln\ d(\rho))] , \tag{7.3}$$

for some constants c and d which are positive if β is large enough. [The lower bound (7.1) was derived from (7.2) in (3.22), (3.26) and (3.27). See also Sect. 5 of [5] for additional details and further, related results.]

In Sect. 4, we began the proof of (7.2) by establishing the combinatorial identity stated in Theorem 4.1, i.e.

$$\prod_{x\in\Lambda} [1+2 \sum_{n=1}^{\infty} \cos(2\pi n\phi_x + n\sigma_x)] \tag{7.4}$$

$$= \sum_{q_\Lambda} \sum_{N\in F_{q_\Lambda}} z_{q_\Lambda} \tilde{c}_N \prod_{\rho\in N} [1+K(\rho)\cos(2\pi\phi(\rho)+\sigma(\rho))] ,$$

up to a factor $[1+K(\rho^c)\cos(2\pi\phi(\rho^c)+\sigma(\rho^c))]$, with $\sum_x \rho_x^c \neq 0$, which is replaced by 1 upon integration with $d\mu_\beta$. The ensembles N satisfy the (scale-covariant) distance condition, Condition D of Theorem 4.1, the vortex densities ρ are neutral $(\sum_x \rho_x = 0$, for all $\rho \in N)$, and

$$K(\rho) \leqq \zeta(\rho)\, e^{C\,V(\rho)} . \tag{7.5}$$

See Theorem 4.1, Remark (1) following Theorem 4.1 and identity (5.1). [The coefficients $\tilde{c}_N$ were, in Sect. 4, denoted by c_N. Here we set $c_N = z_{q_\Lambda} \tilde{c}_N$, for $N \in F_{q_\Lambda}$.]

In Sect. 5, we first noticed that

$$\prod_{\rho\in N} [1+K(\rho)\cos(2\pi\phi(\rho)+\sigma(\rho))]$$

$$= \sum_{E_\varepsilon \subseteq N} \sum_{\rho\in E_\varepsilon} (\tfrac{1}{2}K(\rho)e^{i\sigma(\varepsilon(\rho)\rho)})e^{i2\pi\phi(\varepsilon(\rho)\rho)} , \tag{7.6}$$

where

$$E_\varepsilon = \left\{\rho \in N : \varepsilon(\rho) \neq 0\right\} , \quad \varepsilon(\rho) = 0 , \pm 1 , \text{ for all } \rho .$$

The punchline of Sect. 5 was proving Theorem 5.2, i.e. establishing the identity

$$\int \prod_{\rho\in E} e^{i2\pi\phi(\rho)} d\mu_\beta(\phi) = \int \prod_{\rho\in E} \zeta_N(\beta,\rho) e^{i2\pi\phi(\bar{\rho})} d\mu_\beta(\phi) , \tag{7.7}$$

where $E = \left\{\rho : \varepsilon(\rho)\ \rho \in N , \varepsilon(\rho) = \pm 1\right\}$,

$$0 < \zeta_N(\beta,\rho) \leq \exp[-\beta(\tilde{c}V'(\rho)+\tilde{d}||\rho||_2^2)] , \tag{7.8}$$

and $\bar{\rho} = \bar{\rho}_N$ is a renormalized vortex density which is linear in ρ and neutral. The proof of (7.7) involved performing renormalization transformations on an infinite sequence of distance scales.

Inserting (7.7) into (7.6) and using the linearity of $\bar{\rho}$ in ρ, we obtain

$$\begin{aligned} &\int \prod_{\rho\in N} [1+K(\rho)\cos(2\pi\phi(\rho)+\sigma(\rho))]d\mu_\beta(\phi) \\ &= \int \prod_{\rho\in N} [1+K(\rho)\zeta_N(\beta,\rho)\cos(2\pi\phi(\bar{\rho})+\sigma(\rho))]d\mu_\beta(\phi) \ . \end{aligned} \tag{7.9}$$

By setting

$$z_N(\beta,\rho) = K(\rho)\zeta_N(\beta,\rho) \tag{7.10}$$

we obtain the basic identity (7.2).

Finally, we have just shown in Sect. 6, by a somewhat subtle energy-entropy argument, that the bound (7.3) on $z_N(\beta,\rho)$ follows from (7.10), (7.8) and (7.5).

The proof of Theorem 3.2 is thus complete.

We refer the reader to Sect. 8 and to [5], Sects. 5.1, 6 and 7 for further applications of our techniques and additional results.

8. Further Properties of Two-Dimensional Models With BKT Transitions

In this section we sketch applications and extensions of the methods and results developed in Sects. 3-7. We shall start with remarks on the long distance behaviour of correlations in the Villain model at low temperatures: We argue that it is Gaussian. Remarks on Debye screening, asymptotic enhancement of symmetries and the roughening transition then follow.

8.1 The Scaling Limit of the Villain Model at Low Temperatures *)

In the theory of critical phenomena one is interested in the behaviour of correlation functions

$$< \prod_{j=1}^{2n} S_{x_j}^{\alpha_j} > (\beta) \ , \ \alpha = 1 \ , \text{ or } 2 \ , \tag{8.1}$$

at large distances, $|x_i - x_j| \to \infty$, for $i \neq j$, and at values of β close to some β_0 , where β_0 is a critical point. In order to analyze that behaviour, one introduces scaled correlations: Let θ be a scale parameter with values in $[1, \infty)$. We propose to choose functions $\alpha(\theta) > 0$ and $\beta(\theta)$ in such a way that the scaled correlations,

$$G_\theta(x_1, \alpha_1, \dots, x_{2n}, \alpha_{2n}) \equiv \alpha(\theta)^{2n} < \prod_{j=1}^{2n} S_{\theta x_j}^{\alpha_j} > (\beta(\theta)) \tag{8.2}$$

have a non-trivial limit when $\theta \to \infty$,

$$G^*(x_1, \alpha_1, \dots, x_{2n}, \alpha_{2n}) \equiv \lim_{\theta \to \infty} G_\theta(x_1, \alpha_1, \dots, x_{2n}, \alpha_{2n}) \ . \tag{8.3}$$

More precisely, we require that, for

$$0 < |x-y| < \infty \ ,$$

$$G^*(x,y) \equiv \lim_{\theta \to \infty} \sum_{\alpha=1}^{2} G_\theta(x, \alpha, y, \alpha) \tag{8.4}$$

be positive and finite.
By correlation inequalities, (8.4) suffices to obtain non-trivial limits in (8.3) , as $\theta \to \infty$, for all $n = 2,3,4\dots$. Condition (8.4) implies that $\beta(\theta)$ must be chosen such that

$$m(\beta(\theta)) \leqq \text{const.}\ \theta^{-1} \ , \tag{8.5}$$

where $m(\beta)$ is the inverse correlation length. Thus

---

*) This section is expository and somewhat heuristic; see also [15] .

$$\lim_{\theta\to\infty} \beta(\theta) = \beta_0 \ , \tag{8.6}$$

where β_0 is a critical point, (meaning that $m(\beta)$ is continous in β near β_0 and vanishes at β_0 ; see Sect. 1). By Theorem 3.1 - see also Theorem A, Sect. 1 - all sufficiently large values of β are critical points of the Villain-, or classical rotor model in zero magnetic field. Furthermore, there is a unique value, β_c , of β such that

$$m(\beta) = 0 \ , \quad \text{for all} \quad \beta \geqq \beta_c \ ,$$

$$m(\beta) > 0 \ , \quad \text{for all} \quad \beta < \beta_c \ , \text{ and}$$

$$\lim_{\beta\uparrow\beta_c} m(\beta) = 0 \ ; \text{ (see Sect. 1 and [25]).}$$

The precise way in which $m(\beta)$ vanishes at β_c and the properties of the scaling limit at β_c remain conjectural and cannot be analyzed by the methods developed in this article. We think, however, that the methods of this and other articles in this book are sufficient to analyze the scaling limits of the correlations of the Villain- and related models at large values of β . We then set

$$\beta(\theta) \equiv \beta >> \beta_c \ . \tag{8.7}$$

By the spin wave upper bound (2.43) ,

$$\langle \vec{S}_x \cdot \vec{S}_y \rangle(\beta) \leqq \text{const.} \ (1+|x-y|)^{-1/2\pi\beta} \ , \tag{8.8}$$

it follows, using (8.4) , that

$$\alpha(\theta) \geqq \theta^{1/4\pi\beta} \ , \tag{8.9}$$

and by Theorem 3.1, for β sufficiently large,

$$\alpha(\theta) \leqq \theta^{1/4\pi\beta'} \ , \tag{8.10}$$

where $\beta' = \beta'(\beta)$ is a function with the property that

$\beta'/\beta \to 1$, as $\beta \to \infty$.

The critical exponent $\eta = \eta(\beta)$ is defined by

$$\eta = \lim_{\theta\to\infty} (\log \alpha(\theta)^2/\log \theta) . \tag{8.11}$$

Our main contention is that a combination of the methods in this and in the article of Gawedzki and Kupiainen, and refs. given there, will suffice to construct the scaling limits $G^*(x_1,\alpha_1,\ldots,x_{2n},\alpha_{2n})$ and to devise an algorithm to calculate, in principle, $\eta(\beta)$. [We have, however, not worked out a proof of this contention!] The starting point of our analysis is the identity

$$\begin{aligned} <\vec{S}_x\cdot\vec{S}_y>(\beta) &= \exp[(2\beta)^{-1}(\delta_0-\delta_x,\Delta^{-1}(\delta_0-\delta_x))]\cdot \\ &\cdot \lim_{|\Lambda|\to\infty} [Z_{\beta,\Lambda}(0)^{-1}Z_{\beta,\Lambda}(\sigma)] , \end{aligned} \tag{8.12}$$

with

$$Z_{\beta,\Lambda}(\sigma) = \sum_{N\varepsilon F_\Lambda} \mathbf{c}_N \int \prod_{\rho\varepsilon N} [1+z_N(\beta,\rho)\cos(2\pi\phi(\bar{\rho})+\sigma(\rho))]d\mu_\beta(\phi) , \tag{8.13}$$

where ρ and $\bar{\rho}$ are neutral, $z_N(\beta,\rho) << 1$ for large β , and

$$\sigma \equiv \sigma_{xy} = -2\pi d\Delta^{-1} f_{\gamma_{0x}} , \text{ with}$$

$$f_\gamma(b) = 1 \text{ if } b \in \gamma , \quad f_\gamma(b) = 0 , \text{ otherwise.}$$

See (2.41), (2.42), (3.15) and (3.18). Similar identities are valid for arbitrary correlations, $<\prod_{j=1}^{2n} S_{x_j}^{\alpha_j}>(\beta)$. It thus suffices to analyze quotients

$$Z_{\beta,\Lambda}(0)^{-1} Z_{\beta,\Lambda}(\sigma_{xy}) , \tag{8.14}$$

when $\Lambda \uparrow Z^2$ and for $|x-y|$ large. By (3.18) and (3.22),

$Z_{\beta,\Lambda}(0)^{-1}Z_{\beta,\Lambda}(\sigma)$ can be expressed, approximately, in terms of the expectation value of some (non-local) observable in the state

$$d\mu_{ren.}(\phi) = \lim_{\Lambda\uparrow Z^2} (Z^V_{\beta,\Lambda})^{-1} \sum_{N\varepsilon F_\Lambda} c_N \cdot \prod_{\rho\varepsilon N} [1+z_N(\beta,\rho)\cos 2\pi\phi(\bar{\rho})] \quad d\mu_\beta(\phi) \ . \tag{8.15}$$

Calculating the scaling limit of (8.14) (behaviour of $Z_{\beta,\Lambda}(0)^{-1}\ Z_{\beta,\Lambda}(\sigma_{xy})$, as $|x-y| \to \infty$) can, in principle, be reduced to calculating the scaling limit of ϕ -correlations in the state $d\mu_{ren}(\phi)$. In order to construct such a scaling limit, one can make use of renormalization group <u>(block spin)</u> transformations, as explained in more detail elsewhere in this book; (see the contribution of Gawedzki and Kupiainen, and refs. given there). Vaguely speaking, the main idea of this method is as follows: With each site $x \varepsilon Z^2$ we associate a block variable $(r\phi)(x)$, defined e.g. by

$$(r\phi)(x) = L^{-2} \cdot \left\{ \sum_{\substack{y\ \varepsilon\ Z^2 \\ -1/2 \leq L^{-1}y^\alpha - x^\alpha \leq 1/2}} \phi_y \right\} , \tag{8.16}$$

$\alpha = 1,2$; ($y^\alpha = \alpha^{th}$ component of y). For technical reasons, the definition of $r\phi$ will usually be more complicated than suggested by (8.16), but the above definition captures the right qualitative features.

Note that r increases the distance scale (in units of the original lattice) by a factor of L . Given a state $d\mu$, we define a transformed state $d(R\mu)$ by the equation

$$\int e^{i\phi(h)}\ d(R\mu)(\phi) = \int e^{i(r\phi)(h)}\ d\mu(\phi) \ , \tag{8.17}$$

hence, heuristically,

$$d(R\mu)(\phi) = \left\{\int[\prod_{x\varepsilon Z^2} \delta(\phi_x-(r\phi')(x))]d\mu(\phi')\right\} \prod_{x\varepsilon Z^2} d\phi_x \ . \tag{8.18}$$

The transformation R is called renormalization (block spin) transformation.

Since the transformation R increases the distance scale, the scaling limit, $d\mu^*$, of a state $d\mu$ is obtained as the limit

$$d\mu^*(\phi) = \lim_{m\to\infty} d(\underbrace{R\circ\cdots\circ R}_{m \text{ times}}\mu)(\phi) \tag{8.19}$$

if it exists, i.e. the renormalization transformation has to be repeated infinitely often to yield the scaling limit of a state.

For the construction of this limit it is important to analyze the states which are fixed points of the block spin transformation R. (The measure $d\mu^*$ in (8.19) is clearly a fixed point of R.) It is easy to verify that R has a line of fixed points consisting of zero-mass Gaussian measures, $d\mu^*_\beta$, with

$$\text{mean } 0 \text{ and "covariance" } \beta C . \tag{8.20}$$

Here C is the Green's function of the continuum Laplacian restricted to the lattice Z^2, and the range of values of β is the positive real axis. Given a fixed point, $d\mu^*_\beta$, of R it is interesting to analyze the action of the transformation R in the vicinity of $d\mu^*_\beta$ An idea about the properties of the action of R near $d\mu^*_\beta$ can be obtained by considering the linearization of R at $d\mu^*_\beta$, i.e. the linear operator $(DR)_{\mu^*_\beta}$ acting on an infinite dimensional space of quasi-local functionals of the random field ϕ. It is easy to verify that the spectrum of $(DR)_{\mu^*_\beta}$ contains an eigenvalue 1 which is at least triply degenerate: One eigenvector corresponding to the eigenvalue 1 consists of the tangent vector to the half line $\left\{d\mu^*_{\beta'} : 0 < \beta' < \infty\right\}$ at $\beta' = \beta$ which is given by the "operator" $\int d^2x : (\nabla\phi)^2 : (x)$. Other eigenvectors corresponding to the eigenvalue 1 are given by the operators

$$\Gamma_\varepsilon \equiv \text{"}\int d^2x : \cos\varepsilon\phi : (x)\text{"} , \quad \Sigma_\varepsilon \equiv \text{"}\int d^2x : \sin\varepsilon\phi : (x)\text{"} ,$$

with

$\varepsilon^2\beta = 8\pi$; (the double colons indicate normal ordering).

All these eigenvectors are called "marginal" (operators). When $\varepsilon^2 > 8\pi/\beta$ then Γ_ε and Σ_ε correspond to eigenvalues of $(DR)_{\mu^*_\beta}$ smaller than 1 ("irrelevant operators"), when $\varepsilon^2 < 8\pi/\beta$ they correspond to eigenvalues larger than 1 ("relevant operators"). The functionals

$$"\int d^2x : (\nabla\phi)^{2m} : (x)" \tag{8.21}$$

correspond to eigenvalues $L^{-(2m-2)}$ of $(DR)_{\mu^*_\beta}$, i.e. are irrelevant.

All this is just standard dimensional analysis (power counting).

Now, the subspace of $(DR)_{\mu^*_\beta}$ corresponding to eigenvalues of modulus > 1 is contained in the tangent space to the unstable manifold $M_u^{(\beta)}$ of R at μ^*_β , the subspace of $(DR)_{\mu^*_\beta}$ corresponding to eigenvalues of modulus < 1 is contained in the tangent space to the stable manifold $M_s^{(\beta)}$ of R at μ^*_β - (if such manifolds exist). States on $M_s^{(\beta)}$ are driven towards μ^*_β under the action of R , states on $M_u^{(\beta)}$ are driven away from μ^*_β under the action of R . [States on a manifold transverse to $M_s^{(\beta)}$ and $M_u^{(\beta)}$ are typically other fixed points of R .] States on $\bigcup_{\beta > 0} M_s^{(\beta)}$ are critical states, in the sense that their correlation length **is divergent.** By a Mermin-type argument one can show that the state $d\mu_{ren.}(\phi)$ defined in (8.15) which describes the Villain model at low temperatures is a critical state:

$$K(\beta)\Delta(k)^{-1} \leqq \langle|\hat{\phi}(k)|^2\rangle(\beta) \leqq \beta\, \Delta(k)^{-1} , \tag{8.22}$$

$$\Delta(k) \equiv 4-2\cos k^1 - 2\cos k^2 ,$$

for some constant $K(\beta)$ which is positive for sufficiently large β , and $K(\beta)/\beta \to 1$, as $\beta \to \infty$. This has been proven in [5] by arguments very similar to the ones in Sects. 3-7. Related results for dipole gases were first proven in [32]. The basic fact underlying (8.22) is that each function ρ indexing a factor $1 + z_N(\beta,\rho)\cos 2\pi\phi(\bar{\rho})$ in the product on the r.s. of (8.15) is neutral, i.e.

$$\sum_x \rho_x = \sum_x \bar{\rho}_x = 0 .$$

Therefore the functional

$$\ln \left\{ \sum c_N \prod_{\rho\in N} [1+z_N(\beta,\rho)\cos 2\pi\phi(\bar{\rho})] \right\}$$

is invariant under the transformations $\phi \to \phi + \text{const.}$ and can thus be approximated by sums of "operators"

$$k(x_1,\ldots,x_{2m}) : (\partial^{\alpha_1}\phi)_{x_1} \ldots (\partial^{\alpha_{2m}}\phi)_{x_{2m}} : ,$$

$m = 1,2,3,\ldots,$ with kernels $k(x_1,\ldots,x_{2m})$ which decay rapidly in $|x_i - x_j|$, $i \neq j$. These operators are all irrelevant, for $m > 1$, and marginal for $m = 1$. It is thus very plausible to conjecture that

$$d\mu_{\text{ren.}}(\phi) \in M_s^{(\beta^*)} , \tag{8.23}$$

for some $\beta^* < \beta$; ($|\beta^* - \beta| \leq \text{const.} \exp[-\text{const.}'\beta]$). We believe that such a result could be shown by using the techniques outlined in other articles in this volume; (Gawedzki-Kupiainen, Magnen-Sénéor). Thus we expect that the scaling limit

$$d\mu^*_{\text{ren.}}(\phi) = \lim_{m\to\infty} d(R^m \mu_{\text{ren.}})(\phi) \tag{8.24}$$

exists and is given by a Gaussian measure

$$d\mu^*_{\text{ren.}}(\phi) = d\mu^*_{\beta^*}(\phi) \text{ , with } \beta^* < \beta .$$

Translated back into the original formulation of the Villain model, our arguments suggest that the scaling limits of the multi-spin correlations in the Villain model at sufficiently low temperature are given by standard spin wave theory, i.e. Gaussian integrals (with respect to $d\mu^*_{1/\beta^*}(\psi)$, $\beta^* < \beta$) of products of $\exp i\psi_x$, i.e.

$S_x \sim e^{i\psi_x}$; see Theorem 3.1.

This ends our discussion of the scaling limit of the Villain model. See also [15], (and [47, 11] for general discussions).

8.2 Debye Screening

We recall that in the Villain model correlations decay exponentially at high temperatures. Because of the fact that the basic spin variables of this model are bounded, this can be demonstrated by applying a standard high temperature expansion. In the sine-Gordon representation, see (3.13) - (3.17), that result is far from obvious. The sine-Gordon representation suggests to study perturbations of the zero-mass Gaussian measure $d\mu_\beta(\phi)$ by multiplicative functionals of the form

$$\exp[z \sum_{x\varepsilon\Lambda} \cos(\varepsilon\phi_x)] , \tag{8.25}$$

with $\varepsilon^2\beta < 8\pi$; (as explained in Sect. 8.1). These models are isomorphic to two-dimensional Coulomb gases [43, 32]. The dual of the Villain model corresponds precisely to choosing $\varepsilon = 2\pi$ and letting $z \to \infty$. In this limit the problem simplifies: The model so obtained is called the discrete Gaussian model, see (3.28) - (3.29). Exponential decay of correlations also follows from a standard low temperature (Peierls contour) expansion, provided β is sufficiently small.

Brydges and Federbush [48, 49] have devised an impressive technique, patterned on [50], to study the models in (8.25) for sufficiently small $\varepsilon^2\beta$ and suitably large z . It combines three expansions, a Mayer-, a Peierls contour- and a cluster expansion. The net result of their sophisticated methods is a proof of exponential fall-off of connected correlations in this system. By expanding $\exp[z \sum_{x\varepsilon\Lambda} \cos(\varepsilon\phi_x)]$ in powers of z and carrying out the Gaussian integrals one sees that, as mentioned already, this system really describes a two-dimensional Coulomb lattice gas of particles with electric charge $\pm\varepsilon$ and activity z , at inverse temperature β , as is well known; (see e.g. [43, 32]). Thus the Brydges-Federbush result [49] establishes Debye screening at high temperature.

We now want to describe some heuristic ideas about this result: The Hamilton function of the Coulomb gas in the sine-Gordon representation is given by

$$H = \sum_{x \in Z_a^d} [(a^{d-2}/2\beta)(\nabla\phi)_x^2 - a^d z \cos(\varepsilon\phi_x)] , \qquad (8.26)$$

where ∇ is the finite difference gradient, and a is the lattice spacing. By applying renormalization transformations on a scale of $O(a)$, very similar to the one constructed in Lemma 5.3, one can show that one may replace $a^d z \cos(\varepsilon\phi_x)$ by a term whose leading behaviour is given by

$$a^d z_{ren.}(\beta\varepsilon^2)\cos(\varepsilon\bar{\phi}_x) , \qquad (8.27)$$

where

$$z_{ren.}(\beta\varepsilon^2) = z \exp[-const.(\beta\varepsilon^2/a^{d-2})] . \qquad (8.28)$$

By changing variables, $\phi \to \sqrt{\beta}\phi$, we now obtain an effective Hamiltonian given by

$$H \cong H_{eff.} = \sum_x [\frac{a^{d-2}}{2}(\nabla\phi)_x^2 + a^d \frac{z_{ren.}\beta\varepsilon^2}{2} \phi_x^2 - a^d \frac{z_{ren.}\beta^2\varepsilon^4}{4!} \phi_x^4 + \dots] \qquad (8.29)$$

By scaling, $x \to \theta x$ $(a \to \theta^{-1}a)$ we see that the dimensionless coupling constant, g, which measures the deviation of H from a quadratic Hamiltonian is given by

$$g = \lambda/m_0^2 = \beta\varepsilon^2 , \; d = 2 , \text{ and}$$
$$g = \lambda/m_0 = \sqrt{z_{ren.}(\beta\varepsilon^2)^3} , \; d = 3 , \qquad (8.30)$$

where

$$\lambda = z_{ren.}\beta^2\varepsilon^4 \ , \text{ and } \ m_0^2 = z_{ren.}\beta\varepsilon^2 \ .$$

In two dimensions g is small only at sufficiently high temperature. In contrast, in three dimensions,

$$g \to 0 \ , \text{ as } \ \beta \to +\infty \ , \tag{8.31}$$

by (8.30) and (8.28).

We now suppose that $0 < g << 1$ and that $\phi_x = 0$, for all $x \notin \Lambda$, for some large finite region Λ . By integrating out fluctuations on scales not exceeding the Debye length

$$l_D \approx (z_{ren.}\beta\varepsilon^2)^{-1/2} \tag{8.32}$$

we obtain an effective theory for "block variables" $\phi_{av.}(x)_{x\varepsilon l_D Z^d}$. In [49] this "block" integration is accompolished with the help of a Mayer expansion. (Such expansions are reviewed in Imbrie's article.) Alternatively, one could probably apply a modified version of the method in Sects. 4 and 5 to fluctuations on scales $\leq O(l_D)$, in particular the renormalization transformations of Lemma 5.3, to achieve the same result. The Hamilton function of the effective theory is approximately given by

$$\sum_{x\varepsilon l_D Z^d} [\tfrac{1}{2}(\nabla\phi_{av.})^2(x) - \text{const. } g^{-1}\cos(\sqrt{g}\phi_{av.}(x)) + \ldots] \ , \tag{8.33}$$

with $\phi_{av.}(x) = 0$, $x \notin \Lambda$. The corresponding equilibrium state is approximately given by

$$Z_\Lambda^{-1} \exp\left\{-\sum_x [\tfrac{1}{2}(\nabla\phi_{av.})^2(x) + \text{const. } g^{-1}\cos(\sqrt{g}\phi_{av.}(x))]\right\} \cdot \prod_x d\phi_{av.}(x) \ . \tag{8.34}$$

By applying a Peierls contour expansion to this state one shows that, predominantly,

$$\phi_{av.}(x) \approx 0 \ .$$

Since $g << 1$, the fluctuations around $\phi_{av.} = 0$ are approximately Gaussian with correlation length ≈ 1, in units of l_D. This is proven with the help of a cluster expansion; see [49]. Thus connected correlations of this system will decay exponentially, with decay rate $\approx l_D^{-1}$, provided $g << 1$. Equ. (8.31) suggests that, in $d \geqq 3$ dimensions, this is true for all sufficiently large values of β, hence (by a correlation inequality [51]) for all values of β. This has been proven in [52] and is in contrast to our results for the two-dimensional system. In other words, while Debye screening breaks down at low temperatures in the two-dimensional Coulomb gas, (due to dipole formation, leading to a Berezinskii-Kosterlitz-Thouless transition), it persists to arbitrarily low temperatures in the three- or higher dimensional gas which always remains in the plasma phase.

8.3 Asymptotic Enhancement of Symmetries

The methods of Sects. 3-7 can also be applied to the Z_N models in two dimensions which admit a global Z_N symmetry group; (see Sect. 1). These models have massive high- and low-temperature phases and a massless intermediate phase in which asymptotically, at large distances, the Z_N -symmetry is enhanced to a full $U(1)$ -symmetry. The Z_N models are defined as follows: The spin variable, $\vec{S}_x$, is given by

$$\vec{S}_x = (\cos\theta_x, \sin\theta_x) \ , \qquad \theta_x = \frac{2\pi n_x}{N} \ , \tag{8.35}$$

$n_x = 0,\dots,N-1$, for all x. The equilibrium state assigns to a configuration of spins $\{\theta_x\}_{x\varepsilon\Lambda}$ the statistical weight

$$\mu^{(N)}_{\beta,\Lambda}(\theta) = (Z^{(N)}_{\beta,\Lambda})^{-1} B(\theta_{\partial\Lambda}) \exp[\beta \sum_{\substack{x,y\in\Lambda \\ |x-y|=1}} \cos(\theta_x - \theta_y)] , \tag{8:36}$$

where $B(\theta_{\partial\Lambda})$ only depends on $\{\theta_x\}_{x\in\partial\Lambda}$ and fixes the boundary conditions. This state can be written as a perturbation of the equilibrium state of the classical rotor model:

$$d\mu^{(N)}_{\beta,\Lambda}(\theta) = (Z^{(N)}_{\beta,\Lambda})^{-1} B(\theta_{\partial\Lambda}) \exp[\beta \sum_{\substack{x,y\in\Lambda \\ |x-y|=1}} \cos(\theta_x - \theta_y)] \cdot \tag{8.37}$$

$$\cdot \prod_{x\in\Lambda} [1+2 \sum_{m=1}^{\infty} \cos(mN\theta_x)] d\theta_x .$$

This expression is analogous to formulas (3.15) - (3.17) for the equilibrium state of the Villain model in the sine-Gordon representation (which can be viewed as a perturbation of the zero-mass Gaussian measure). In particular, the expectation value of the disorder operator D^{ξ}_{xy} of the Z_N model, (see (3.3) and [40, 5] for definitions), is given by a formula closely related to (3.15). It involves a shifted partition function

$$Z^{(N)}_{\beta,\Lambda}(\sigma) \equiv \int B(\theta_{\partial\Lambda}) \exp[\beta \sum_{xy\subset\Lambda} \cos(\theta_x - \theta_y)] \cdot \tag{8.38}$$

$$\cdot \prod_{x\in\Lambda} [1+2 \sum_{m=1}^{\infty} \cos(mN(\theta_x + \sigma))] d\theta_x .$$

In the following it is convenient to choose the b.c. $B(\theta_{\partial\Lambda})$ such that

$$Z^{-1}_{\beta,\Lambda} B(\theta_{\partial\Lambda}) \exp[\beta \sum_{xy\subset\Lambda} \cos(\theta_x - \theta_y)] \prod d\theta_x$$

is the infinite volume equilibrium state of the rotor model conditioned on $\{\theta_x\}_{x\in\Lambda}$, (but other b.c. can be handled, too. The choice of b.c. is really quite irrelevant. See also [5]).

Now, if β is fixed then for large enough N (depending on β) the partition function $Z^{(N)}_{\beta,\Lambda}(\sigma)$ can be analyzed with a variant of the methods explained in Sects. 3-7. The intuitive reason behind this circumstance is, apart from representation (8.38), the fact that

$$\exp[\beta \sum_{|x-y|=1} \cos(\theta_x-\theta_y)]\Pi d\theta_x \approx$$

$$\text{const.} \exp.[-\frac{\beta}{2N^2} \sum_{|x-y|=1} (\phi_x-\phi_y)^2]\prod_x d\phi_x \text{ , (for } |\theta_x-\theta_y| << 1)$$

with $\phi_x = N\theta_x$. The r.s. behaves, for large N , like the Gaussian measure with covariance $-(N^2/\beta)\Delta^{-1}$. See [5] for details.

The net result, after applying the transformations of Sects. 4-6, is the identity

$$Z^{(N)}_{\beta,\Lambda}(\sigma) = \sum_{E\varepsilon F_\Lambda} c_E \int \prod_{\rho\varepsilon E} [1+z_E(\beta,\rho)R_\rho(\theta+\sigma)]d\mu_\beta(\theta) , \tag{8.39}$$

where F_Λ is a family of ensembles which are here denoted (for obvious reasons) by E , $c_E > 0$, for all E , each $\rho \varepsilon E$ is a function on Λ with values in $N\cdot Z$, $d\mu_\beta(\theta)$ is the infinite volume equilibrium state of the classical rotor, and, for each $\rho \varepsilon E$, and all $E \varepsilon F_\Lambda$, $R_\rho(\cdot)$ is a real-valued function with the properties:

1) $R_\rho(\theta)$ only depends on the variables $\{\theta_x\}_{x \varepsilon \text{ supp } \bar{\rho}}$, where $\bar{\rho}$ is a renormalized "vortex density".

2) $\max_\theta |R_\rho(\theta)| \leqq 1$.

3) $R_\rho(\theta+\alpha) = R_\rho(\theta)$, where $(\theta+\alpha)_x = \theta_x+\alpha$, for all x .

Each function $R_\rho(\theta)$ is a renormalized version of $\cos\theta(\rho)$, approximately given by $\cos\theta(\bar{\rho})$.

The coefficients $z_E(\beta,\rho)$ tend to 0 , as $N \to \infty$, faster than any inverse power of N . In particular, for each $\beta < \infty$ and $\varepsilon < 1$, there exists a finite integer $N_0 = N_0(\beta,\varepsilon)$ such that for all $N \geqq N_0$

$$z_E(\beta,\rho) \leqq 1 - \varepsilon \text{ , for all } \rho \varepsilon E ,$$

for all ensembles $E \varepsilon F_\Lambda$, uniformly in Λ . Thus, for $N \geqq N_0$, the equilibrium state,

$$d\mu_\beta^{(N)}(\theta) = \lim_{\Lambda \uparrow Z^2} d\mu_{\beta,\Lambda}^{(N)}(\theta) ,$$

of the Z_N - model, and the renormalized state

$$d\mu_{ren.}^{(N)}(\theta) = \lim_{\Lambda \uparrow Z^2} (Z_{\beta,\Lambda}^{(N)})^{-1} \sum_{E \varepsilon F_\Lambda} c_E \cdot \tag{8.40}$$

$$\cdot \prod_{\rho \varepsilon E} [1+z_E(\beta,\rho)R_\rho(\theta)] \quad d\mu_\beta(\theta)$$

have the same scaling limit, because $d\mu_{ren.}^{(N)}$ is obtained from $d\mu_\beta^{(N)}$ by a succession of integrations over fluctuations of θ on bounded length scales; (see Sect. 5, and Sects. 4,7 of [5]). Now, by property 3) of R_ρ , the state $d\mu_{ren.}^{(N)}$ has a full, global $U(1)$ symmetry group. Since a continuous symmetry cannot be broken in two dimensions (for interactions, as the ones described by $d\mu_{ren.}^{(N)}(\theta)$, of short range), we conclude that $d\mu_{ren.}^{(N)}(\theta)$, and hence $d\mu_\beta^{(N)}(\theta)$, are clustering states, i.e. they do not exhibit long range order. See Sect. 1. From Ginibre's inequality [19] it follows that the correlation length of the Z_N -model is bounded below by the one of the rotor model, for each fixed value of β . Choosing $\beta \geqq \beta_c(\text{rotor})$ and $N \geqq N_0(\beta)$, we conclude that the corresponding Z_N -model has divergent correlation length but, as just shown above, no long range order. Thus each $\beta \geqq \beta_c(\text{rotor})$ is a critical point of the Z_N -model, for $N \geqq N_0(\beta)$, and since for this choice of parameters the scaling limits of $d\mu_\beta^{(N)}(\theta)$ and $d\mu_{ren.}^{(N)}(\theta)$ coincide, the scaling limits of the correlation functions exhibit full $U(1)$ symmetry.

This is an instance of what is meant by asymptotic enhancement of symmetries (in the infrared).

We end by describing this concept in a somewhat abstract setting:

Let X be some space (a simplex) of states of a (classical) physical system, and let R be a renormalization (e.g. Block spin) transformation acting on X which preserves the set $E(X)$ of extremal states. Let $\mathbf{M}_{f.p.}(R) \subset E(X)$ be the subspace of fixed points of R . Let ρ^* be such a fixed point. Let $\mathbf{M}_s(R,\rho^*)$ and $\mathbf{M}_u(R,\rho^*)$

be the stable and unstable manifolds passing trough ρ^* . The states on $M_u(R,\rho^*)$ are scaling limits of families of states converging to states on $M_s(R,\rho^*)$. It may happen that all states on $M_u(R,\rho^*)$ admit a symmetry group, G , which is much larger than symmetry groups of states on $M_s(R,\rho^*)$, (or converging to states on $M_s(R,\rho^*)$, as $\beta \to \beta_c$) . More precisely, let H be a proper subgroup of G , and suppose that all H -invariant states on $M_u(R,\rho^*)$ are automatically G -invariant. Then the scaling limit of families of H -invariant states passing through $M_s(R,\rho^*)$ are G -invariant. Under suitable smoothness assumptions on the action of R on $E(X)$, asymptotic enhancement of H - to G -symmetry is stable, in the sense that if the H -invariant and the G -invariant subspaces of the linear space of marginal and relevant perturbations of ρ^* coincide then, in some open neighborhood of ρ^* , H -invariant fixed points, $\tilde{\rho}$, of R and H -invariant states on $M_u(R,\tilde{\rho})$ are G -invariant. This fact is useful in applications. For heuristic ideas about symmetry enhancement see also [53,17,54].

8.4 The Roughening Transition

In this section we sketch what our techniques have to tell us about the so-called roughening transition. This transition is observed in the statistical physics of crystalline surfaces and of interfaces separating different phases of a physical system. It is the name for the phenomenon that below a certain temperature, T_R , crystalline surfaces, or interfaces, are quite rigid and flat, with bounded fluctuations about some mean position, while above T_R they exhibit logarithmically divergent fluctuations. (An abundance of terrasses and craters appear on such surfaces, for $T > T_R$.) Mathematical results on interfaces are summarized in the article by C.E. Pfister.

In the following, we describe only one class of models of surfaces, a model of "self-avoiding random surfaces", the related solid-on-solid model and the discrete Gaussian model dual to the Villain model. These models were defined in Sect. 1, (1.27) and (1.28). For detailed results see [34,55], with proofs appearing in [5].

Consider a lattice Z^ν , $\nu \geq 3$, and a square Λ^* contained in the coordinate plane $x_3 = \dots = x_\nu = 0$ (where x_α is the αth component of $x \in Z^\nu$) with sides of length L . Let $E_{\Lambda*}$ denote the ensemble of all surfaces S built out of plaquettes (2-cells) of Z^ν, with the following properties:

1) S is connected;

2) $\partial S = \partial\Lambda^*$;

3) each surface S is self-avoiding, i.e. each link in Z^ν belongs to at most two plaquettes in S . We shall give each surface S the statistical weight

$$W_{\beta,\Lambda^*}(S) = e^{-\beta|S|}, \tag{8.41}$$

where $|S|$ is the area of S , i.e. the total number of plaquettes belonging to S . This model was first considered in [55] and its mathematical analysis was begun in [34]. Let

$$Z_\beta(\partial\Lambda^*) \equiv \sum_{S \in E_{\Lambda^*}} W_{\beta,\Lambda^*}(S) . \tag{8.42}$$

It can be shown that

$$Z_\beta(\partial\Lambda^*) \text{ diverges for } \beta < \beta_0 , \tag{8.43}$$

where β_0 is a finite positive number independent of the coice of Λ^*

The model introduced here is difficult to analyze mathematically, and only rather preliminary results have been established, so far. Interesting problems are to show that β_0 is a critical point, in the sense that all correlation lengths diverge, as $\beta \downarrow \beta_0$, and to prove the existence of a roughening transition occurring at some temperature $T_R < \beta_0^{-1}$. Both problems remain open, on rigorous grounds.

In order to describe roughening mathematically precisely we need some definitions:

Let π_0 be the plane $\{x : x_1 = x_2 = 0\}$, and let $P_\beta(d|\partial\Lambda^*)$ be the probability that a surface $S \in E_{\Lambda^*}$ have an intersection with π_0 at a distance $\geq d$ from the origin. One can prove the following result.

Theorem 8.1

For β sufficiently large,

$$P_\beta(d|\partial\Lambda^*) < e^{-c(\beta)d}$$, uniformly in Λ^* , with $c(\beta) \to \infty$, as $\beta \to \infty$.

Although we have not checked the details of the proof, it appears as rather obvious that Theorem 8.1 can be proven with the help of Dobrushin's techniques [56], for example. See also Pfister's article, and refs. given there. On the basis of our results for the solid-on-solid model summarized below we conjecture that there exists some $\beta_R > \beta_0$ such that

$$\sum_{d=1}^{\infty} d^2 P_\beta(d|\partial\Lambda^*) \sim \text{const.}\,(\log L)^\kappa , \tag{8.44}$$

with $\kappa \cong 1$, for L large and all $\beta \in (\beta_0,\beta_R)$. Our conjecture could be tested on a computer, but it seems quite hard to find a mathematical proof.

The model discussed above is of interest, because it provides a rather realistic description of the statistical mechanics of crystalline surfaces. Since its mathematical analysis meets difficulties it is natural to simplify it by restricting the ensemble, E_{Λ^*} , of admissible surfaces. For simplicity we assume that $\nu = 3$, but see [34] for more general considerations, (including $\nu = 3$) . Let Λ be the array of sites dual to the plaquettes in Λ^* . Let E^0_Λ be the ensemble of surfaces in E_{Λ^*} which are graphs of functions, h , defined on the plaquettes $p \subset \Lambda^*$ and vanishing outside Λ^* . The number $h(p)$, $p \subset \Lambda^*$, assumed to be an integer, is interpreted as the height of the surface over p . The weight of the surface $S^{(h)}$ corresponding to h is given, according to (8.41), by

$$W_{\beta,\Lambda^*}(S^{(h)}) = e^{-\beta|\Lambda^*|} \prod_{(p,p')\subset\Lambda^*} e^{-\beta|h(p)-h(p')|} , \tag{8.45}$$

with p and p' adjacent plaquettes.

For $x \in \Lambda$ let $*x$ be the plaquette in Λ^* dual to x . We set

$$\phi(x) \equiv h(*x) \ .$$

Since the factor $e^{-\beta|\Lambda^*|}$ is common to all surfaces in E^0_Λ it can be omitted. We define

$$W^{s-o-s}_{\beta,\Lambda}(\phi) = \prod_{xy\subset\Lambda} e^{-\beta|\phi(x)-\phi(y)|} \ . \tag{8.46}$$

The solid-on-solid model so obtained is an exact description of the interface in the three-dimensional anisotropic Ising model with coupling constants $J_x = 1$, $J_y = 1$, $J_z \to \infty$, the interface being the graph of ϕ.

A related model, expected to belong to the same universality class, is the discrete Gaussian model, with surface weights given by

$$W^{DG}_{\beta,\Lambda}(\phi) = \prod_{xy\subset\Lambda} e^{-\frac{\beta}{2}(\phi(x)-\phi(y))^2} \ . \tag{8.47}$$

If one is interested in the behaviour of phase separation <u>lines</u> one studies these models in <u>one</u> dimension, for Bloch walls (interfaces), or crystalline surface, $\Lambda \subset Z^d$, with $d = 2$, but the models are of some interest in arbitrary dimension. (In two dimensions, there is a third model of interfaces which one expects to be in the same universality class as the s-o-s model and which can be solved exactly [57], because it is equivalent to a six vertex model. We shall not study it here.) Our main results on the s-o-s and the discrete Gaussian model are summarized in

<u>Theorem 8.2</u>

<u>Let</u> $<(\cdot)>(\beta)$ <u>denote the equilibrium expectation of the</u> s-o-s, <u>or the discrete Gaussian model in the thermodynamic limit</u>, $\Lambda\uparrow Z^d$, <u>at inverse temperature</u> β. <u>Then:</u>

1) <u>For</u> $d = 1$,

$$<(\phi_0-\phi_x)^2>(\beta) \cong c_1(\beta)|x| \ , \ \underline{\text{as}} \quad |x| \to \infty \ .$$

2) <u>For</u> $d = 2$,

$$<(\phi_0-\phi_x)^2>(\beta) \leq c_2(\beta) \ ,$$

uniformly in x, provided β is large enough. When β is small enough.

$$c_3(\beta)\log(|x|+1) \leqq \langle(\phi_0-\phi_x)^2\rangle(\beta) \leqq c_4(\beta)\log(|x|+1) . \tag{8.48}$$

3) For $d \geqq 3$

$$\langle(\phi_0-\phi_x)^2\rangle(\beta) \leqq c_5(\beta) ,$$

for all β.

Here $c_1(\beta),\ldots,c_5(\beta)$ are finite, positive constants.

Remarks.

1) Parts 1) and 2) could be rephrased as follows: For $d = 1$, $\Lambda = [-\frac{L}{2}, \frac{L}{2}]$,

$$\langle\phi_0{}^2\rangle_\Lambda(\beta) \sim L , \text{ for large } L .$$

For $d = 2$, Λ a square with sides of length L,

$$\langle\phi_0{}^2\rangle_\Lambda(\beta) \leqq \text{const.} , \text{ uniformly in } L ,$$

if β is large, and if β is sufficiently small

$$\langle\phi_0{}^2\rangle_\Lambda(\beta) \geqq \text{const.} \log L . \tag{8.49}$$

The proof of (8.49) which is analogous to conjecture (8.44) is somewhat more subtle than the one of (8.48). (See Sect. 7 and Appendix D of [5].)

2) Part 1) of Theorem 8.2 follows immediately from the central limit theorem. A related, but more subtle result concerning the phase separation line in the two-dimensional Ising model has been established by Gallavotti [58]. The first half of part 2) (large β) is a standard consequence of low temperature expansions. The upper bound in (8.48) and part 3) follow from infrared bounds, as sketched below; see also [59], or [34]. The deepest result is the lower bound in (8.48), but that follows readily from our results in Sects. 3-7,

(except for some technical complications arising in the s-o-s model for which we refer the reader to Sect. 7 and Appendix C of [5]): We consider the discrete Gaussian model defined in (8.47). In order to prove the lower bound in (8.48) we first show that for real functions f of finite support with $\Sigma f_x = 0$

$$<e^{\phi(f)}>^{DG}(\beta) \geq e^{-(\beta'/2)(f,\Delta^{-1}f)} , \tag{8.50}$$

for some $\beta' < \beta$, with $|\beta'-\beta| \to 0$, as $\beta \to \infty$. Setting $f = \varepsilon h$ and expanding both sides of (8.50) in powers of ε , subtracting 1, dividing by ε^2 and letting ε tend to 0 , we obtain

$$<\phi(h)^2>^{DG}(\beta) \geq - \beta'(h,\Delta^{-1}h) ,$$

from which one derives the lower bound in (8.48) by setting $h(j) = \delta_{0j}-\delta_{xj}$.

To prove (8.50) we set

$$\phi_x = \phi'_x + \sigma_x , \text{ where}$$

$$\sigma_x = -\beta(\delta_x,\Delta^{-1}f) .$$

Then

$$<e^{\phi(f)}>^{DG}_{\Lambda}(\beta) = e^{-(\beta/2)(f,\Delta^{-1}f)} Z_{\beta,\Lambda}(\sigma)/Z_{\beta,\Lambda}(0) \tag{8.51}$$

where $Z_{\beta,\Lambda}(\sigma)$ has been introduced and studied in (3.16) - (3.18), (3.27). A lower bound on $Z_{\beta,\Lambda}(\sigma)$ is obtained as explained in Sect. 3, (3.22), (3.23), (3.24),..., given Theorem 3.2. Inequality (8.50) follows from that lower bound and (8.51).

3) Next, we comment on the proof of the upper bound in part 2) and of part 3) of Theorem 8.2. We propose to show that

$$<|\hat{\phi}(k)|^2>(\beta) \leq c(\beta)[2d-2\sum_{\alpha=1}^{d} \cos k_\alpha]^{-1} , \tag{8.52}$$

for some finite constant $c(\beta)$. By arguments similar to the ones just sketched above, (8.52) follows from the following inequality, called Gaussian domination [9],

$$<\exp\left\{\varepsilon \sum_j h_j(\partial_\alpha\phi)_j\right\}>(\beta) \leq \exp[\tfrac{1}{2}c(\beta)\varepsilon^2||h||_2^2] , \tag{8.53}$$

with ε small, (e.g. $\varepsilon||h||_\infty \leq \varepsilon_0$, for some $\varepsilon_0 > 0$) . Here ∂_α is the finite difference derivative in the α direction, and h is an arbitrary real-valued function on Z^d . Inequality (8.53) is proven by using a transfer matrix in the α direction, $\alpha = 1,\ldots,d$, of Z^d . As explained in [9], the transfer matrix formalism reduces the problem of proving (8.53) to estimating the integral operator with integral kernel given by

$$e^{\varepsilon h(\phi-\phi')} e^{-F_\beta(\phi-\phi')} , \phi , \phi' \in R ,$$

where

$$F_\beta(\phi) = \begin{cases} \beta|\phi| , \text{ or} \\ (\beta/2)\,\phi^2 , \end{cases}$$

in terms of the integral operator determined by

$$e^{-F_\beta(\phi-\phi')}$$

This is done with the help of Fourier transformation. For $F_\beta(\phi) = \beta|\phi|$, the required bound follows by noticing the inequality

$$|[(k+i\varepsilon h)^2+\beta^2]^{-1}| \leq e^{c(\beta)(\varepsilon h)^2}[k^2+\beta^2]^{-1} ,$$

for $|\varepsilon h| \leq \beta/2$. For $F_\beta(\phi) = (\beta/2)\phi^2$, one observes that

$$|\exp[-(1/2\beta)(k+i\varepsilon h)^2]| \leq e^{(\varepsilon h)^2/2\beta} e^{-(1/2\beta)k^2} ,$$

for arbitrary real ε and h .

From these estimates inequality (8.53) follows by the transfer matrix formalism. Inequality (8.53) yields (8.52), by the arguments sketched after (8.50). Finally, the bound (8.52) yields the upper bound in (8.48) and part 3) of Theorem 8.2, by Fourier transformation. This completes our sketch of proof of Theorem 8.2.

Next, we define the step free energy of the s-o-s and the DG model. Let $Z_\beta^0(\partial\Lambda)$ be the usual partition function of the s-o-s, or DG model with zero boundary conditions, and let $Z_\beta^{m,0}(\partial\Lambda)$ be the partition function of the same model with boundary conditions

$$\left.\begin{aligned} \phi(x) &= m \,, \quad \text{for} \quad x \notin \Lambda \,, \quad x^1 > 0 \,; \\ \phi(x) &= 0 \,, \quad \text{for} \quad x \notin \Lambda \,, \quad x^1 < 0 \,. \end{aligned}\right\} \tag{8.54}$$

We define the step free energy, τ_d, by

$$\beta\, \tau_d(m;\beta) \equiv \lim_{L\to\infty} L^{1-d} \log(Z_\beta^0(\partial\Lambda)/Z_\beta^{m,0}(\partial\Lambda)) \,. \tag{8.55}$$

We note that $\tau_2(1;\beta)$ is closely related to the step free energy of the three-dimensional Ising model. By using the methods of [5,49,52] we can prove

Theorem 8.3.

1) $\tau_1(m;\beta) = 0$, for all β and all m.

2) [5] $\tau_2(m;\beta) > 0$, for β large enough,

$\tau_2(m;\beta) = 0$, for small values of β,

for all $m = \pm 1, \pm 2, \ldots$.

3) [52] For the discrete Gaussian model, in dimension $d \geqq 3$,

$\tau_d(m;\beta) > 0$, for all $\beta > 0$

and $m = \pm 1, \pm 2, \ldots$.

Part 1) is trivial. The first half of 2) is proven by a standard low temperature expansion, while the proof of the second half of 2) is closely related to the proof of the main result of this article, Theorem 3.1. (There are some additional, analytical subtleties arising in the s-o-s model for which we refer to [5].)

The proof of part 3) for the discrete Gaussian model in three dimensions is identical to the Göpfert-Mack proof of permanent confinement in the three-dimensional $U(1)$ (-Villain) lattice gauge theory. For $d > 3$ the result follows from the three-dimensional result and correlation inequalities [51].

8.5 Energy-Entropy Arguments in Higher-Dimensional Models with Long Range Interactions, Conclusion.

(1) The classical XY (-Villain) model, $d \geq 3$.

We have shown in Sect. 2 that the $d \geq 3$ dimensional Villain model can be expressed as a gas of $(d-2)$-dimensional, closed vortex "networks" which have Coulomb interactions. As sketched there, one can understand the main features of the transition in the $d \geq 3$ dimensional XY model, (long range order, spontaneous symmetry breaking and Goldstone mode at low temperature) with the help of an energy-entropy, or generalized Peierls argument. This has been made rigorous in [17]. Here we sketch a variant of the arguments in that paper.

The equilibrium state of the $d \geq 3$ dimensional, classical XY model in the sine-Gordon representation is given by

$$Z_\beta^{-1} \sum_n \prod_{c \subset Z^d} e^{2\pi i n(c)\alpha(c)} d\mu_\beta(\alpha) , \tag{8.56}$$

where c denotes a $(d-2)$-cell in Z^d, α is a real-valued and n an integer-valued $(d-2)$-chain, and $d\mu_\beta(\alpha)$ is the Gaussian measure on the equivalence classes $[\alpha]$ of $(d-2)$-chains α labelled by the curl, $d\alpha$, with mean 0 and covariance $\beta(\delta d)^{-1}$, i.e.

$$\int d\mu_\beta(\alpha) e^{i(k,\alpha)} \equiv \begin{cases} 0 & \text{if } \delta k \neq 0 \\ \exp[(\beta/2)(k,\Delta^{-1}k)] & \text{if } \delta k = 0 . \end{cases} \tag{8.57}$$

The proof of (8.56) follows by Fourier transformation in the angular variables, θ_j, of the Villain model, or by the arguments in Sects. 2

and 3. For $d \geq 3$, the measure defined in (8.56) has a gauge invariance: The configurations α and $\alpha + d\gamma$, where γ is an (integer-valued) d-3 chain, have the same weight.

Equation (8.57) shows that the sum in (8.56) extends over (d-2)-chains, n , with vanishing divergence, i.e.

$$\delta n = 0 \; .$$

Thus, the (d-2) chains n form closed, (d-2) -dimensional networks. If we carry out the integration over $[\alpha]$ we see that these networks have Coulomb interactions. They are thus the vortices in the spin configurations of the Villain model.

As in Sects. 2, 3 one may show that the two point function of the XY model is given, in the sine-Gordon representation, by

$$\begin{aligned} \vec{S}_x \cdot \vec{S}_y \; (\beta) &= \langle e^{i(\theta_x - \theta_y)} \rangle (\beta) \\ &= \langle \prod_{b \varepsilon \gamma_{xy}} e^{(2\pi/\beta)(d\alpha)(*b)} e^{-(\pi/\beta)} \rangle^{\alpha}(\beta) \; , \end{aligned} \tag{8.58}$$

where γ_{xy} is a path connecting x to y , and $*b$ is the (d-1)-cell dual to the bond b . Moreover, $\langle(\cdot)\rangle^{\alpha}(\beta)$ is the expectation in the measure defined in (8.56). The expression on the r.s. of (8.58) is not identical to the one obtained from the calculations in Sect. 2, but is related to it by a change of variables:

$$\alpha \to \alpha - \delta \, \Delta^{-1} f_{xy} \; ,$$

where

$$f_{xy}\,(c_{d-1}) = \begin{cases} 2\pi & \text{if } \; *c_{d-1} \,\varepsilon\, \gamma_{xy} \\ 0 & , \text{ otherwise} \; . \end{cases} \tag{8.59}$$

After this change of variables, the r.s. of (8.58) factorizes into a spin wave- and a vortex contribution: We first note that

$$(2\pi)^{-1}*(df_{xy})(j) = \delta_{jx} - \delta_{jy} . \tag{8.60}$$

Hence, using the well-known transformation properties of the Gaussian measure $d\mu_\beta(\alpha)$ under an affine change of variables, we obtain

$$\langle \vec{S}_x \cdot \vec{S}_y \rangle(\beta) = \exp[2\pi^2\beta^{-1}(\delta_x-\delta_y,\Delta^{-1}(\delta_x-\delta_y)] \cdot \cdot Z_\beta^{-1} Z_\beta(\sigma_{xy}) , \tag{8.61}$$

where

$$\sigma_{xy} \equiv - \delta \Delta^{-1} f_{xy} , \bar{\sigma} = (2\pi)^{-1}\sigma , \text{ and}$$

$$Z_\beta(\sigma) \equiv \sum_{n:\delta n=0} \int \prod_{c\in Z^d} e^{2\pi i n(c)(\alpha+\bar{\sigma})(c)} d\mu_\beta(\alpha) . \tag{8.62}$$

The first factor on the r.s. of (8.61) is the spin wave contribution, the second factor the vortex contribution to the two-point function. In order to exhibit long range order at low temperature, it now suffices to establish a lower bound on $Z_\beta(\sigma_{xy})$. [For large β, it will merely lead to a renormalization of the spin wave contribution, $\beta \to \beta_{ren.} > \beta$.] Such a lower bound can be found, for large β, by adapting the techniques developed in Sects. 4-6. However, the analysis in the present case $(d \geq 3)$ is actually much simpler, thanks to the gauge invariance of $d\mu_\beta(\alpha)$ which enforces strict local "neutrality", $\delta n = 0$, (rather than only approximate local neutrality, as in two dimensions.) The simplifications arising in $d \geq 3$ dimensions are comparable to the simplifying features of the ordinary Peierls argument for the Ising model in ≥ 2 dimensions with nearest neighbor interactions, as compared to the generalized Peierls argument which is used to analyze the one-dimensional Ising model with $1/r^2$ interaction energy at low temperatures [35].

The analysis of the three- or higher dimensional, classical XY model proceeds by first partitioning each vertex configuration, n, into (not necessarily connected) closed vortex networks $\{\rho_\alpha\}_{\alpha = 1,2,3,...}$, which are separated from each other by a distance $\geq$ 2 . More precisely,

$$Z_\beta(\sigma) = \sum_{n:\delta n=0} \int e^{i(2\pi\alpha(n)+\sigma(n))} d\mu_\beta(\alpha)$$
$$= \sum_E B_E \int \prod_{\rho\varepsilon E} e^{i(2\pi\alpha(\rho)+\sigma(\rho))} d\mu_\beta(\alpha) , \tag{8.63}$$

where $\alpha(f) \equiv (\alpha,f) = \sum_c \alpha(c)f(c)$, E ranges over all possible ensembles of closed vortex networks given by integer-valued functions, ρ , defined on $d - 2$ cells, with $\delta\rho = 0$, and B_E vanishes if, for two different vortex networks ρ and ρ' in E , $\text{dist}(\text{supp }\rho , \text{supp }\rho') \leq 1$, and $B_E = 1$, otherwise.

Next, we renormalize all vortex networks on the shortest distance scale. A straight forward variant of Lemma 5.3, (see [17]), permits us to show that one can extract a selfenergy factor $\exp[-\beta E_{\text{loc.}}(\rho)]$, with

$$E_{\text{loc.}}(\rho) \geq \text{const.}||\rho||_2^2 , \quad ||\rho||_2^2 \equiv \sum_c \rho(c)^2 , \tag{8.64}$$

for each $\rho \varepsilon E$. In the process, ρ is replaced by a renormalized vortex network $\bar{\rho}$ which depends linearly on ρ . As a result, we obtain the identity

$$Z_\beta(\sigma) = \sum_E B_E \int \prod_{\rho\varepsilon E} \zeta(\beta,\rho) e^{i(2\pi\alpha(\bar{\rho})+\sigma(\rho))} d\mu_\beta(\alpha) ,$$

with

$$\zeta(\beta,\rho) = \exp[-\beta E_{\text{loc.}}(\rho)] . \tag{8.65}$$

In contrast to the renormalizations performed in Sect. 5, $[\rho \to \bar{\rho}_N , K(\rho) \to z_N(\beta,\rho) , \rho \varepsilon N \varepsilon F_\Lambda]$, the renormalizations made here have the pleasant property that if $B_E = 1$

$\zeta(\beta,\rho)$ <u>and</u> $\bar{\rho}$ <u>depend only on</u> ρ , <u>but are independent of all</u> $\rho' \varepsilon E , \rho' \neq \rho$. (8.66)

[This is because the complex translations in α -space performed to renormalize $\exp 2\pi i\alpha(\rho)$, see Lemma 5.3, can be chosen to be independent of all $\rho' \in E$, $\rho' \neq \rho$, because $\mathrm{dist}(\rho,\rho') \geq 2$, so that there is no plaquette bordering supp ρ and supp ρ' .] By means of a standard Mayer expansion (i.e. a convergent form of the linked cluster theorem; see [42,45]) we can exponentiate the expansion for the integrand in (8.65), i.e.

$$\sum_E B_E \prod_{\rho\in E} \zeta(\beta,\rho) \exp i(2\pi(\alpha(\bar{\rho})+\sigma(\rho)))$$
$$= \exp\left\{\sum_Q Z(\beta,Q)\cos[2\pi(\alpha(\bar{Q})+\sigma(Q))]\right\} , \tag{8.67}$$

where

$$Q = \sum_\rho \nu(Q,\rho)\rho \ , \text{ with } \ \nu(Q,\rho) \in Z \ ,$$

and

$$\left\{\rho : \nu(Q,\rho) \neq 0\right\}$$

is a set with the property that the graph obtained by joining by a line any two networks ρ and ρ' in that set which are at a distance ≤ 1 from each other is connected. Furthermore

$$\left.\begin{aligned} \bar{Q} &= \sum_\rho \nu(Q,\rho)\bar{\rho} \\ Z(\beta,Q) &= \phi^T(Q) \prod_\rho z(\beta,\rho)^{\nu(Q,\rho)} \ , \end{aligned}\right\} \tag{8.68}$$

where $\phi^T(Q)$ is some combinatorial coefficient; see [42,45]. Now, it follows from (8.64), (8.65) and (8.68) that

$$|Z(\beta,Q)| \leq \exp[-\mathrm{const.}\beta||Q||_2^2] \ , \tag{8.69}$$

and that the expansion in the exponent on the r.s. of (8.67) converges, provided

$$\beta > b_d \ , \tag{8.70}$$

where b_d is a purely geometrical constant; $(b_d \sim O(1))$. See also [17,42].

By the Poincaré Lemma (see Lemma 2.1) and the fact that $\delta Q = \sum_\rho \nu(Q,\rho)\delta\rho = 0$,

$$\sigma(Q) = (d\sigma)(M) \ , \tag{8.71}$$

where M is an integer-valued solution of the equation

$$\delta M = Q$$

which one can choose in such a way that

$$||M||_\infty \leq ||Q||_1 \ ,$$

and that $\text{supp}\, M$ is contained in the smallest rectangle containing $\text{supp}\, Q$; (see Lemma 2; and [16,17,42]). Now

$$(d\sigma)(M)|\text{mod}.2\pi = (\delta\Delta^{-1}df_{xy})(M)|\text{mod}.2\pi \tag{8.72}$$

if $\sigma \equiv \sigma_{xy}$ is given by (8.62) with f_{xy} as in (8.59). Inserting this equation into (8.71) and the result into (8.67) we obtain

$$Z_\beta(\sigma) = \int \exp \sum_Q Z(\beta,Q)\cos(2\pi\alpha(\overline{Q})) \prod_Q \exp O(\alpha,Q)\exp[-R(\alpha,Q)]d\mu_\beta(\alpha) \ , \tag{8.73}$$

where $O(\alpha,Q)$ is <u>odd</u> in α , and

$$\exp(\sum_Q |R(\alpha,Q)|) \leq \exp[-c(\beta)(\delta_x-\delta_y,\Delta^{-1}(\delta_x-\delta_y))] \ ,$$

with $c(\beta) \to 0$, as $\beta \to \infty$, exponentially fast. Here we have used (8.72), (8.67) and (8.60). See also [17]. By Jensen's inequality we obtain

$$Z_\beta(\sigma) \geq Z_\beta \exp[c(\beta)(\delta_x-\delta_y,\Delta^{-1}(\delta_x-\delta_y))] ,$$

hence, using (8.61),

$$<\vec{S}_x\cdot\vec{S}_y>(\beta) \geq \exp[(2\pi^2\beta^{-1}+c(\beta))(\delta_x-\delta_y,\Delta^{-1}(\delta_x-\delta_y))] ,$$

provided $\beta > b_d$; (see (8.70)).

This proves that, at sufficiently low temperatures, the classical XY- and the Villain model have long range order, and the $O(2)$ - symmetry is spontaneously broken, in dimension $d \geq 3$.

For an analysis of the structure of translation-invariant equilibrium states and of the breaking of translation invariance in the XY- and the Villain model see [34].

(2) Abelian lattice gauge theories.

By the arguments developed in Sect. 2, or by using a duality (Fourier) transformation, the compact $U(1)$ lattice gauge theory can be mapped onto a theory of magnetic monopoles with Coulomb interactions. In the continuum limit, monopoles are topologically stable defects of co-dimension three, labelled by the elements of the first homotopy group of the gauge group, $U(1)$, i.e. by an integer which is interpreted as the magnetic charge. The magnetic charge is actually given by the first Chern class. Thanks to the abelianness of the gauge group, these continuum notions make sense for the lattice theory as well.

In the three (space-imaginary time) dimensional, compact $U(1)$ lattice gauge theory, the magnetic monopoles are pointlike; (they are localized in unit cubes of the lattice Z^3). They have Coulomb two-body interactions among each other. As sketched in Subsect. 8.2, the three-dimensional Coulomb gas exhibits exponential Debye screening, i.e. a finite correlation length, $m(\beta)^{-1}$, at all temperatures. This is intimately related to the fact that the three-dimensional, compact $U(1)$ theory has permanent confinement by a linear potential. Both facts have been established by Göpfert and Mack [52], using techniques of [48,49]. The string tension, $\tau(\beta)$, of the $U(1)$ theory is the surface tension of the discrete Gaussian which is dual to the $U(1)$ theory and which is identical to a

Coulomb gas in the sine-Gordon representation. Thus

$$\tau(\beta) = -\lim_{L\to\infty} \frac{1}{L^2} \ln < \prod_{*<jj'>\varepsilon S_L} e^{\beta^{-1}(\phi_j - \phi_{j'}) - (2\beta)^{-1}} > (\beta) , \qquad (8.74)$$

where S_L is a square with sides of length L in a coordinate plane, and $* <jj'>$ is the plaquette in S_L dual to the bond $<jj'>$ of the dual lattice. As argued in Subsect. 8.2, the state $<(\cdot)> (\beta)$ of the discrete Gaussian is, for large values of β, approximately given by a standard Gaussian measure with inverse correlation length (mass)

$$m(\beta) \approx \text{const. } \beta\, e^{-\text{const.}' \beta} \qquad (8.75)$$

(This is a consequence of Debye screening. See also [52].) In this approximation, we find for the string tension $\tau(\beta)$ the formula

$$\tau(\beta) \simeq \text{const.}'' \beta^{-1} m(\beta) , \qquad (8.76)$$

as follows easily from (8.74). Thus, the dimensionless quantity $m(\beta)^2/\tau(\beta)$ vanishes exponentially fast, as $\beta \to \infty$, i.e. in the continuum limit. See [52] for rigorous arguments and an analysis of the continuum limit.

In four or more dimensions, the magnetic monopoles form closed networks of dimension $\geqq 1$, (analogous to the vortices in the $d \geqq 3$ dimensional, classical XY model). An energy-entropy argument of the type reviewed in Sect. 2 predicts that there is a deconfining transition and a massless phase at large values of β (i.e. at weak coupling). This was first proven rigorously by Guth for the Villain action [16]. A somewhat simpler (and perhaps slightly more natural) proof appeared subsequently in [17]. The methods just sketched above for the $d \geqq 3$ dimensional XY model can be adapted, without difficulties, to the $d \geqq 4$ dimensional compact U(1) lattice gauge theory and yield simple, rigorous proofs of the expected results. When combined with the methods developed in the article by Gawedzki and Kupiainen they could actually be used to show that, at large values of β, the scaling (= continuum) limit is the free (non-compact) Gaussian theory. This result is related to the conjectures discussed in Subsect. 8.1, but its proof is presumably considerably easier.

(3) Non-abelian spin systems, non-abelian gauge theories.

So far, the defect gas approach described and explained in this article has, in a mathematically rigorous form, been limited to abelian spin systems and abelian lattice gauge theories (without Fermions). In such systems it provides us, however, with a very detailed understanding of the low temperature phases, even when these phases are characterized by a divergent correlation length. The obstruction which prevents us from extending our approach to non-abelian lattice systems lies in the circumstance that there are no known, mathematically precise and useful reformulations of non-abelian lattice systems as gases of localized defects with fairly precisely known mutual interactions. Some of the physically interesting non-abelian lattice systems, like the classical Heisenberg model, or the pure SU(2) lattice gauge theory, have formal continuum limits with field configurations exhibiting "topologically stable defects". These defects have become known under the name of instantons. The problems with instantons are that they have a variable (classically, an arbitrary) scale size (which corresponds to conformal invariance properties of the continuum theories in the classical limit), that we have some trouble with defining a useful notion of instantons for the lattice theories and that in cases where such a notion can be introduced we cannot calculate the interactions beween different instantons.

On very heuristic grounds, the instanton picture does, however, lead to predictions that appear to be qualitatively correct. Consider, for example, the pure SU(2) Yang-Mills theory in dimension $d \geqq 4$ This theory admits topologically non-trivial field configurations, corresponding to defects of dimension $d-4$ with a variable scale size, which have a finite action density (= action per cm^{d-4}), the instantons. These configurations drive the vacuum state of the theory away from the perturbative vacuum, (i.e. make "spin-wave theory" inapplicable , in statistical mechanics jargon). In four dimensions, the instantons are point-like, more precisely ball-like, and their mutual interactions appear to tend to 0 , as the separation increases to ∞ . We therefore expect that the "instanton gas" is always in a dense plasma phase, even at large values of β , and that therefore the perturbative ("spin-wave") regime is never reached. This is analogous to the rigorously established situation in the three-dimensional U(1) lattice gauge theory, reformulated as a Coulomb gas of magnetic monopoles, which we have described above, and is consistent with the idea of permanent quark confinement. In five or more dimensions, instantons are defects associated with closed "networks" of dimension $\geqq 1$, and if a naïve energy-entropy argument were applicable to such instantons it would predict a transition to a low temperature phase in which instantons form a very dilute gas, and "spin-wave theory", i.e. perturbation theory, is qualitatively correct in the infrared. This is consistent with the idea that there exists a deconfining transition in the $d \geqq 5$ dimensional SU(2) lattice gauge theory.

It would be most interesting to find a rigorous counterpart for the above heuristic speculations. Thus, much work remains to be done. May this be our conclusion.

References.

1. R. Peierls, Proc. Cambridge Philos. Soc. 32, 477 (1936).

2. L. Onsager, Phys. Rev. 65, 117 (1944).

3. R.B. Griffiths, Phys. Rev. 136A, 437 (1964).
R.L. Dobrushin, Teorija Verojatn, i ee Prim. 10, 209 (1965).

4. D. Ruelle, Phys. Rev. Letters 27, 1040 (1971).

5. J. Fröhlich and T. Spencer, Phys. Rev. Letters 46, 1006 (1981); Commun. Math. Phys. 81, 527 (1981).

6. E.H. Lieb and F.Y. Wu, in "Phase Transistions and Critical Phenomena", vol. 1, C. Domb and M.S. Green (eds.), New York and London: Academic Press 1972.
E.H. Lieb, Physica 73, 226 (1974).
R.J. Baxter, "Exactly Solved Models in Statistical Mechanics", New York and London: Academic Press 1982.

7. R.A. Minlos and Ya. G. Sinai, Trudy Moskov. Math. Obsc. 19, 121 (1968);
S. Pirogov and Ya. G. Sinai, Funct. Anal. Appl. 8, 21 (1974); Theor. Math. Phys. 25, 1185 (1975), and 26, 39 (1976); Ann. Phys. 109, 393 (1977).

8. J. Glimm, A. Jaffe and T. Spencer, Commun. Math. Phys. 45, 203 (1975).
[J. Fröhlich and E.H. Lieb, Commun. Math. Phys. 60, 233 (1978).
J. Fröhlich, R. Israel, E.H. Lieb and B. Simon, J. Stat. Phys. 22, 297 (1980).]

9. J. Fröhlich, B. Simon and T. Spencer, Commun. Math. Phys. 50, 79 (1976).
J. Fröhlich, R. Israel, E.H. Lieb and B. Simon, Commun. Math. Phys. 62, 1 (1978).

10. F.J. Dyson, E.H. Lieb and B. Simon, J. Stat. Phys. 18, 335 (1978).

11. J. Fröhlich and T. Spencer, Séminaire Bourbaki No. 586, Astérisque 92-93, 159 (1982).

12. J. Fröhlich and T. Spencer, in "New Developments in Quantum Field Theory and Statistical Mechanics", M. Lévy and P. Mitter (eds.), New York and London: Plenum Press 1977.

13. J. Fröhlich, Bull. Amer. Math. Soc. 84, 165 (1978).
E.H. Lieb, in "Mathematical Problems in Theorectical Physics", G.-F. Dell'Antonio, S. Doplicher and G. Jona-Lasinio (eds.), Lecture Notes in Physics 80, Berlin, Heidelberg and New York: Springer-Verlag 1978.
J. Fröhlich, Acta Physica Austriaca, Suppl. XV, 133 (1976).

14. P. Collet and J.-P. Eckmann, "A Renormalization Group Analysis of the Hierarchical Model in Statistical Mechanics", Lecture Notes in Physics 74, Springer-Verlag 1978. See also references to the original work of P.M. Bleher and Ya. G. Sinai quoted therein; and P.M. Bleher, Commun. Math. Phys. 84, 557 (1982).

15. V. Berezinskii, Soviet Phys. JETP 32, 493 (1971).
J.M. Kosterlitz and D.J. Thouless, J. Phys. C 6, 1181 (1973).
J.M. Kosterlitz, J. Phys. C 7, 1046 (1974).
J. Villain, J. Phys. (Paris) 32, 581 (1975).
J.V. José, L.P. Kadanoff, S. Kirkpatrick and D.R. Nelson, Phys. Rev. B 16, 1217 (1977).
For a brief review of rigorous results see also:
J. Fröhlich and T. Spencer, in "Mathematical Problems in Theoretical Physics", R. Schrader, R. Seiler and D.A. Uhlenbrock (eds.), Lecture Notes in Physics 153, Springer-Verlag 1982.

16. A. Guth, Phys. Rev. D 21, 2291 (1980).

17. J. Fröhlich and T. Spencer, Commun. Math. Phys. 83, 411 (1982).

18. D. Brydges, J. Fröhlich and E. Seiler, Nucl. Phys. B 152, 521 (1979).

19. J. Ginibre, Commun. Math. Phys. 16, 310 (1970).

20. J. Bricmont, J.L. Lebowitz and C.E. Pfister, Appendix in "Periodic Gibbs States...", J. Stat. Phys. 24, 269 (1981).

21. A. Messager, S. Miracle-Sole and C.E. Pfister, Commun. Math. Phys. 58, 19 (1978).

22. J. Fröhlich and C.E. Pfister, "Spin-Waves, Vortices and the Structure of Equilibrium States in the Classical XY Model", subm. to Commun. Math. Phys. (spring 1983).

23. R.L.Dobrushin and S.B. Shlosman, Commun. Math. Phys. 42, 31 (1975).

24. C.E. Pfister, Commun. Math. Phys. 79, 181 (1981).
J. Fröhlich and C.E. Pfister, Commun. Math. Phys. 81, 277 (1981).
A. Klein, L. Landau and D. Shucker, manuscript.

25. J. Glimm and A. Jaffe, "Quantum Physics", Berlin, Heidelberg and New York: Springer-Verlag 1981.

26. J. Rosen, Adv. Appl. Math. 1, 37 (1980).
J. Glimm and A. Jaffe, Commun. Math. Phys. 52, 203 (1977).
See also:
J. Glimm and A. Jaffe, Phys. Rev. D 10, 536 (1974).

27. D. Brydges, J. Fröhlich and A. Sokal, "A New Proof of the Existence and Nontriviality of the Continuum $\phi 4_2$ and $\phi 4_3$ Quantum Field Theories", Preprint, Spring 1983.

28. J. Bricmont, J.-R. Fontaine, J.L. Lebowitz, E.H. Lieb and T. Spencer, Commun. Math. Phys. 78, 545 (1980); and preceeding articles in the same volume.

29. O. Mc Bryan and T. Spencer, Commun. Math. Phys. 53, 299 (1977).

30. M. Aizenman and B. Simon, Phys. Letts. A 76, 281 (1980).

31. B. Simon, Commun. Math. Phys. 77, 111 (1980).

32. J. Fröhlich and T. Spencer, J. Stat. Phys. 24, 617 (1981).

33. A.M. Polyakov, Phys. Letts. B 59, 79 (1975).

34. J. Fröhlich, C.E. Pfister and T. Spencer, in "Stochastic Processes in Quantum Theory and Statistical Physics", S. Albeverio, Ph. Combe, and M. Sirugue-Collin, Lecture Notes in Physics 173, Springer-Verlag 1982.

35. J. Fröhlich and T. Spencer, Commun. Math. Phys. 84, 87 (1982).

36. J. Fröhlich and T. Spencer "Absence of Diffusion in the Anderson Tight Binding Model for Large Disorder or Low Energy", Commun. Math. Phys. (1983).

37. N.D. Mermin, Rev. Mod. Phys. 51, 591 (1979).

38. L. Michel, Rev. Mod. Phys. 52, 617 (1980).

39. F. Wegner, J. Math. Phys. 12, 2259 (1971).

40. L.P. Kadanoff and H. Ceva, Phys. Rev. B 3, 3918 (1971).

41. J. Glimm and A. Jaffe, Commun. Math. Phys. 56, 195 (1977).

42. E. Seiler, "Gauge Theories as a Problem of Constructive Field Theory and Statistical Mechanics", Lecture Notes in Physics 159, Springer-Verlag 1982.

43. J. Fröhlich, Commun. Math. Phys. 47, 233, (1976).

44. J.Z. Imbrie, Commun. Math. Phys. 85, 491 (1982).

45. C. Cammarota, Commun. Math. Phys. 85, 517, (1982).

46. J. Fröhlich, in "Mathematical Problems..."; see ref. 13.

47. S.K. Ma, "Modern Theory of Critical Phenomena", New York: Benjamin 1976.

48. D.C. Brydges, Commun. Math. Phys. 58, 313 (1978).

49. D. Brydges and P. Federbush, Commun. Math. Phys. 73, 197 (1980).

50. J. Glimm, A. Jaffe and T. Spencer, Ann. Phys. (N.Y.) 101, 610 (1976) and 101, 631 (1976).

51. Y.M. Park, J. Math. Phys. 18, 2423, (1977).
J. Fröhlich and Y.M. Park, Commun. Math. Phys. 59, 235 (1978).

52. M. Göpfert and G. Mack, Commun. Math. Phys. 82, 545 (1982); see also: Commun. Math. Phys. 81, 97 (1981).

53. D. Foerster, H.B. Nielson and M. Ninomiya, Phys. Letts. B 94, 135 (1980); and refs. given there.
J. Iliopoulos, D.V. Nanopoulos and T.N. Tomaras, Phys. Letts. B 94, 141 (1980).
K. Cahill and P. Denes, Preprint, University of New Mexico, 1981.

54. C. Newman and L. Schulman, "Asymptotic Symmetry: Enhancement and Stability", Preprint 1981.

55. A. Maritan and C. Omero, Phys. Letts. B 109, 51 (1982).

56. R.L. Dobrushin, Theor. Prob. and Appl. 17, 582 (1972) and 18, 253 (1973).

57. H. van Beijeren, Phys. Rev. Letts. 38, 993 (1977).

58. G. Gallavotti, Commun. Math. Phys. 27, 103 (1972).

59. J. Bricmont, J.-R. Fontaine and J.L. Lebowitz, J. Stat. Phys. 29, 193 (1982).

INTERFACE AND SURFACE TENSION IN ISING MODEL

Charles-Edouard Pfister

1. Introduction.

The Ising model is one of the simplest non-trivial systems in Statistical Mechanics, for which many physically interesting properties have been established in a rigorous way. It is hoped, that the results obtained in this particular model could help in clarifying some of the questions, which cannot be discussed in a completely satisfactory manner in more complicated situations. I shall come back to this point in the next section. There is an excellent review by Gallavotti [28] on rigorous results for the Ising model. Therefore I concentrate on the properties of the interface between two coexisting phases and on the surface tension, or interfacial free energy. I do not consider the very important papers by Minlos and Sinai on coexistence and separation of phases [45]. (See also Lectures by Minlos [43] and by Gallavotti, Martin-Löf and Miracle-Sole [30].) Very different techniques are used to establish the results below. In this brief report it is impossible to give any detailed information on the proofs. I want only to make some remarks on the main papers, which are considered here. One can distinguish roughly three different methods : exact computations, perturbation expansions and correlation inequalities. Exact computations are very difficult. One has only obtained so far some results for the two-dimensional case, like the computations of the surface tension by Onsager [46], and the magnetization profile by Abraham and Reed [4].

The method of perturbation expansion is a standard tool in Statistical Mechanics, although the calculations are sometimes rather complicated (for details see [44], [45], [30], [48]). Its main advantages are its wide range of applicability and the detailed information obtained. The main disadvantage is that the method works only away from the critical point. The papers of Gallavotti on the two-dimensional case [27]; Bricmont, Lebowitz and Pfister [13], [15]; Higuchi [33] are based on this technique. Dobrushin used another technique in his work on the three-dimensional case [19], [20], but the main results can also be proved by the method used in [27] and [13]. The third method is based on symmetry properties and the ferromagnetic character of the model (see e.g. the review [51]). The proofs are very simple and the results can often be extended to other ferromagnetic models. The papers of van Beijeren [7]; Bricmont, Lebowitz and Pfister [14], and Lebowitz and Pfister [38] are based on this technique.

2. Physical Motivation. Simple Fluid.

There is an important physical system, the one-component fluid, for which there remain questions open. In two reviews, [57] and [58], Widom gives a very clear exposition of the subject. I should like to point out one problem, which motivated a study of the two-dimensional Ising model [15]. Consider a one-component fluid in a vessel V, in the coexistence region. The gravitational field produces a spatial separation of the liquid and vapor phases. The notion of interface is intuitively clear in that particular situation. The problem is to study the intrinsic properties of the interface. This means that one takes the thermodynamic limit and the limit of a vanishing gravitational field, since one wants properties which are independent of the actual area of the interface and of the external field. Suppose that an intrinsic structure for the interface exists, and in particular the thickness of the interface is defined. The scaling and homogeneity hypotheses, or the related hypothesis that there is only one length of fundamental significance in critical phenomena, imply that the thickness is of the order of the correlation length. In particular it diverges at the critical point according to the exponent ν. Moreover one would have (see e.g.[57])

$$\mu+\nu = 2-\alpha \tag{2.1}$$

and

$$\mu = (d-1)\nu \tag{2.2}$$

Here μ, resp. α, are the critical exponents of the surface tension, resp. the heat capacity, and d is the dimensionality of the system. For the two-dimensional lattice gas (Ising model) both relations (2.1) and (2.2) are satisfied, since $\mu = \nu = 1$ and $\alpha = 0$. The interface has not yet been defined in a precise way. Van der Waals proposed a description of the interface by the density profile [52], and Fisk and Widom improved this theory, by adding to it the scaling and homogeneity hypotheses [23]. In such theories one finds a density profile, which varies on a scale determined by the correlation length, and the surface tension is directly related to this profile. However, Buff, Lovett and Stillinger observed that the density profile is quite different, if one takes into account the transverse fluctuations of the interface [16]. Such long-wavelength fluctuations, called capillary waves, produce macroscopic displacements of the interface, which is assimilated to a membrane in their argument. This implies that the density profile, computed according to the rules of Statistical Mechanics, is trivial, i.e constant, unless one does *not* take the thermodynamic limit or the limit of a vanishing gravitational field. Therefore the profile obtained in the theory of van der Waals is not the true equilibrium density profile. It is however believed that the interface has a well-defined local structure, which is not destroyed by the capillary waves, as the local structure of a membrane is not destroyed by its macroscopic displacements. The theory of van der Waals, modified by Fisk and Widom, should give a qualitatively correct description of the intrinsic local structure of the interface. The problem is to describe such a structure, starting from first principles of Statistical Mechanics. This question can be investigated in great details in the two-dimensional lattice gas. In this model, like in the one-component fluid, there is no way to find an equilibrium state in the thermo-dynamic limit with a non-trivial profile, which would describe the interface between the two phases of the model.

However, one can show rigorously that the interface has a well-defined intrinsic structure, in particular a finite thickness, although it undergoes macroscopic fluctuations. The main point is that one can use a random separation line[1] (surface in the three-dimensional case) in order to investigate the properties of the interface (see section 6).

3. Ising Model. Notation.

The model is defined on the cubic lattice $\mathbb{Z}^d$, whose points are $x = (x^1,\ldots,x^d)$. The dimensionality will be 2 or 3, but several results hold for $d \geqslant 2$. At each x there is a spin variable $\sigma(x) = \pm 1$, and a pair of nearest neighbour points on the lattice is denoted by $<xy>$. The system is first confined to a finite box $\Lambda = \Lambda(L,M) = \{x: -M \leqslant x^1 < M, |x^i| \leqslant L, i \geqslant 2\}$. x^1 denotes the vertical direction. The hamiltonian is

$$H^b_\Lambda(\underline{\sigma}) = -\sum_{<xy>\subset\Lambda} \sigma(x)\sigma(y) - \sum_{x\in\Lambda} h(x)\sigma(x) - \sum_{\substack{<xy>\\ x\in\Lambda,y\notin\Lambda}} b(y)\sigma(x) \tag{3.1}$$

The function $\underline{h}$ describes an inhomogeneous magnetic field. The last term is a boundary condition, which is given by the boundary field b. The Gibbs state at inverse temperature β, magnetic field $\underline{h}$ and boundary condition b is the probability measure $<\cdot>^b_\Lambda(\underline{h},\beta)$ given by the normalized density

$$Z(\Lambda,b)^{-1} \exp(-\beta H(\underline{\sigma})) \tag{3.2}$$

where $Z(\Lambda,b)$ is the partition function.

It is well known how to construct two different phases[2], $<\cdot>^+(\beta)$ and $<\cdot>^-(\beta)$, at zero magnetic field and low temperatures [28]. In particular

$$<\sigma(0)>^+ = -<\sigma(0)>^- = m^*(\beta)$$

where $m^*(\beta)$ is the spontaneous magnetization, which is equal to the right derivative of the free energy with respect to an homogeneous magnetic field h at h = 0 [37]. It is true[3], as can be seen in [3] and

[9], but not obvious, that in the two-dimensional case m^* coincides with the famous result of Onsager [47] and Yang [60]. The two states $\langle\cdot\rangle^+$ and $\langle\cdot\rangle^-$ are extremal translation invariant states, they have exponential clustering properties and analyticity properties in $e^{-2\beta}$.

4. Existence of Non-Translation Invariant States.

If one wants to describe the interface between the two phases of the model using a magnetization profile, one must construct a non-translation invariant state. The most natural way to achieve this goal is to try the following construction. One chooses an inhomogeneous magnetic field $\underline{h}$

$$h(x) = h>0,\ x^1 \geqslant 0 \quad \text{and} \quad h(x) = -h,\ x^1<0 \tag{4.1}$$

For technical reasons [4] one chooses also the boundary condition $\underline{+}$ defined by

$$b(x) = +1,\ x^1 \geqslant 0 \quad \text{and} \quad b(x) = -1,\ x^1<0 \tag{4.2}$$

Using correlation inequalities one proves that the limit $\Lambda \uparrow \mathbb{Z}^d$ exists and one obtains in this way a state $\langle\cdot\rangle^{\pm}(\underline{h},\beta)$, which is invariant under horizontal translations of the lattice. Let $m(z,\underline{h}) = \langle\sigma(z,0,0)\rangle^{\pm}(\underline{h},\beta)$. The local magnetization $m(z,\underline{h})$ has the following properties [7]

a) $m(z,\underline{h}) = -m(-z-1,\underline{h})$ for $z \geqslant 0$

b) $m(z+1,\underline{h}) \geqslant m(z,\underline{h})$

c) $\lim\limits_{z\to\infty} m(z,\underline{h}) = \langle\sigma(0)\rangle^+(h,\beta)$

where $\langle\cdot\rangle^+(h,\beta)$ is the unique equilibrium state with homogenous magnetic field $h(x) \equiv h$. All of these properties are expected. A less trivial one is

d) $m(z+1,\underline{h})-m(z,\underline{h}) \leqslant m(z,\underline{h})-m(z-1,\underline{h}),\ z\geqslant 0$

There is a well localized interface in the state $\langle\cdot\rangle^{\pm}(\underline{h},\beta)$, describedby $m(z,\underline{h})$. In particular the last property of $m(z,\underline{h})$ implies that the gradient of the magnetization profile is maximal at $z = 0$. The last step in the construction is to take the limit of a vanishing magnetic field[5]. The existence of the limit follows from correlation inequalities. The limiting state is denoted by $\langle\cdot\rangle^{\pm}(\beta)$. If the interface in the state $\langle\cdot\rangle^{\pm}(\underline{h},\beta)$ remains stable, when h tends to zero, then $\langle\cdot\rangle^{\pm}(\beta)$ is a non-translation invariant state with a non-trivial magnetization profile. Moreover one expects that

$$m^* = \lim_{h\downarrow 0}\lim_{z\to\infty} m(z,\underline{h}) = \lim_{z\to\infty}\lim_{h\downarrow 0} m(z,\underline{h}) \tag{4.3}$$

On the other hand, if the effect of the stabilizing field $\underline{h}$ disappears in the limit, i.e the interface becomes unstable, then one expects a situation analogous to that of the one-component fluid in section 2. The results depend on the dimensionality of the system. They are very different for $d = 2$ and $d = 3$.

Three-dimensional Ising model at low temperatures. The state $\langle\cdot\rangle^{\pm}$ is non-translation invariant and (4.3) is verified. Moreover, $m(z,0) = \langle\sigma(z,0,0)\rangle^{\pm}(\beta)$ tends to $m^*(\beta)$ exponentially fast when z tends to infinity. This is true for any local observables. Consequently the thickness of the interface is finite and one proves that it vanishes exponentially fast when β tends to infinity. The properties a,b,c and d of the magnetization profile $m(z,\underline{h})$ are of course still true for $m(z,0)$. The state $\langle\cdot\rangle^{\pm}$ has exponential clustering properties in all directions [6], for all local observables. For example

$$|\langle\sigma(0)\sigma(x)\rangle^{\pm}-\langle\sigma(0)\rangle^{\pm}\langle\sigma(x)\rangle^{\pm}| \leqslant C\exp(-\alpha|x|)$$

It is also possible to write an expansion for any local correlation functions in powers of $\eta = e^{-2\beta}$, for small absolute values of the complex variable η. The state $\langle\cdot\rangle^{\pm}$ is an extremal Gibbs state. All these properties are proved by Dobrushin in [19], [20], except the analyticity which is proved in [13]. In [7] van Beijeren greatly improved the lower bound on β, above which one *can prove* that the state $\langle\cdot\rangle^{\pm}$ is non-translation invariant. Using correlation inequalities, he showed that the magnetization $\langle\sigma(z,0,0)\rangle^{\pm}$ in the horizontal two-dimensional layer $x^1 = z$, $z \geq 0$, is always greater than the spontaneous magnetization of a two-dimensional Ising model. Therefore for all β larger than the inverse critical temperature $\beta_c(2)$ of the two-dimensional model, the state $\langle\cdot\rangle^{\pm}$ has an interface. It is expected that there exists a β_R, $\beta_c(2) > \beta_R > \beta_c(3)$, below which the state $\langle\cdot\rangle^{\pm}$ is translation invariant and equal to $\frac{1}{2}(\langle\cdot\rangle^{+} + \langle\cdot\rangle^{-})$. This phenomenon is called the roughening transition[7].

Two-dimensional Ising model Gallavotti proved that the state $\langle\cdot\rangle^{\pm}$ is translation invariant[8] and equal to $\frac{1}{2}(\langle\cdot\rangle^{+} + \langle\cdot\rangle^{-})$ [27]. In particular (4.3) is no longer true, since the right-hand side is zero. Intuitively one expects that *all* equilibrium states are translation invariant if the state $\langle\cdot\rangle^{\pm}$ is translation invariant. This natural conjecture was proved independently by Aizenman [5] and Higuchi [34]. However their proofs do not work for the three-dimensional Ising model above the roughening temperature.

The state $\langle\cdot\rangle^{\pm}$ can be constructed in a different way, which is the one used in the work of Dobrushin and Gallavotti. One can first take the limit of a vanishing magnetic field and then the thermodynamic limit. This means that

$$\lim_{h\downarrow 0}\lim_{\Lambda\uparrow\mathbb{Z}^d}\langle\cdot\rangle^{\pm}_{\Lambda}(\underline{h},\beta) = \lim_{\Lambda\uparrow\mathbb{Z}^d}\lim_{h\downarrow 0}\langle\cdot\rangle^{\pm}_{\Lambda}(\underline{h},\beta)$$

The proof of that statement uses correlation inequalities. Let $\underline{h} \equiv 0$. The thermodynamic limit of $\langle\cdot\rangle^{\pm}_{\Lambda}(0,\beta) \equiv \langle\cdot\rangle^{\pm}_{\Lambda}(\beta)$ is done in two steps. (However this is not essential). First one takes the limit M going to infinity, and one obtains a state $\langle\cdot\rangle^{\pm}_{L}$ for the system confined to the infinite parallelepiped $\Lambda(L)$ of finite cross-section. Since the cross-section is finite there is a non-trivial magnetization profile in the state $\langle\cdot\rangle^{\pm}_{L}$, even in the two-dimensional case. In particular, properties a,b,c, and d are verified. The main problem in this approach is to control the interface as the cross-section of $\Lambda(L)$ becomes infinite. This is done using the random separation line if $d = 2$, or random separation surface if d = 3, which are mentionned at the end of section 2. This analysis is presented in section 6.

5. Surface Tension.

The surface tension is the contribution to the free energy resulting from the coexistence of the two phases. There are in principle various possibilities for defining the surface tension. Abraham, Gallavotti and Martin-Löf discussed several definitions and proved their equivalence for the two-dimensional Ising model[9] [2]. The equivalence among these different definitions is not an obvious question. Gallavotti and Martin-Löf investigated in detail the surface tension in an ensemble with fixed magnetization[10] [29]. Since I use in this paper another ensemble[11], the surface tension is defined[12] by

$$\tau(\beta) = -\frac{1}{\beta} \lim_{L\to\infty} \left(\frac{1}{2L+1}\right)^{d-1} \log\left(\lim_{M\to\infty} \frac{Z_{L,M}^{\pm}}{Z_{L,M}^{+}}\right) \qquad (5.1)$$

In this formula $Z_{L,M}^{\pm}$ resp. $Z_{L,M}^{+}$ are the partition functions of the system confined to $\Lambda(L,M)$ with zero magnetic field and boundary conditions $\pm$ resp. + (see note 4 and (4.2)). In note 12 I give an heuristic motivation for (5.1). It is however not obvious that $\tau(\beta)$ is non-zero if and only if $\beta > \beta_c(d)$, where $\beta_c(d)$ is the inverse critical temperature. The following bounds imply that desired property.

$$\tau(\beta) \leqslant 2(m^*(\beta))^2 \qquad (5.2)$$

and

$$\frac{d(\beta\tau)}{d\beta} \geqslant 2(m^*(\beta))^2 \qquad (5.3)$$

The upper bound is proved in [14], and the lower bound[13] in [38]. I sketch the main steps of the proof of (5.3) in section 7.

In two dimensions Abraham and Martin-Löf computed (5.1) [3]. Their result coincides with the result of Onsager [46] :

$$\beta\tau(\beta) = 2(\beta-\beta^*), \quad \beta > \beta_c(2)$$

$$\tau(\beta) = 0 \quad , \quad \beta \leqslant \beta_c(2)$$

where β^* is the dual temperature of the model, and is defined by

$$\exp(-2\beta^*) = \tanh \beta$$

In three dimensions one can prove that $\beta\tau(\beta) - 2\beta$ is an analytic function[14] of the variable $\eta = e^{-2\beta}$ with $|\eta|$ small enough [13].

Using duality [55], [9] one gets new results. In two dimensions the model is self-dual and the surface tension is related to the quantity

$$\kappa(\beta) = \lim_{L\to\infty} -\frac{1}{L} \log\langle\sigma(0,0)\sigma(0,L)\rangle(\beta)$$

Above $\beta_c(2)$, $\kappa(\beta)$ is the inverse of the correlation length $\xi(\beta)$. The exact relation[15] is, [14]

$$\beta\tau(\beta) = \kappa(\beta^*) \tag{5.5}$$

In three dimensions the dual of the Ising model is the $\mathbb{Z}_2$ gauge theory. The corresponding quantity to $\kappa(\beta)$ is the coefficient of the area-decay of the Wilson loop [14]. The same relation as (5.5) holds[16].

6. Random Separation Surface.

In this section I explain how one can study the interface using a random separation surface[17]. I consider first the three-dimensional case.

Each configuration in $\Lambda(L,M)$, with $\pm$ boundary condition, is described geometrically by *contours* (see e.g. [28]). The crucial observation is that each configuration $\underline{\sigma}$ is specified in a unique way by the pairs of points <xy> such that $\sigma(x)\sigma(y) = -1$ or $\sigma(x)b(y) = -1$ (see (3.1)). Instead of specifying the pairs <xy>, one can specify the common face of the unit cubes centered at x and y. The set of these faces is decomposed into connected components and each component is a surface, which is called a contour. Each of these surfaces may intersect itself, but in such a way that each edge of a face belonging to the surface is common to an even number of faces of the surface. It is easy to see that each configuration $\underline{\sigma}$ is described by a family of contours $(\lambda,\gamma_1,\ldots,\gamma_n)$, where each γ_i is closed and there is exactly one surface λ which is not closed. This is the *separation surface* λ. The energy $H(\underline{\sigma})$ is, up to a constant, equal to

$$2|\lambda| + 2\sum_{i=1}^{n} |\gamma_i| \tag{6.1}$$

where $|\lambda|$ is the area of the surface λ. One computes easily the probability of λ. It is given by

$$Pr_{L,M}(\lambda) = \frac{\exp(-2\beta|\lambda|)\, Z^{+}_{L,M}(\lambda)}{Z^{\pm}_{L,M}} \tag{6.2}$$

The presence of the factor $\exp(-2\beta|\lambda|)$ is obvious and $Z^{+}_{L,M}(\lambda)$ is equal[18] to the partition function of a system in the box $\Lambda(L,M|\lambda)$ with + boundary condition, where $\Lambda(L,M|\lambda)$ is the set of points of $\Lambda(L,M)$ which are at a distance greater than one from λ. The probability of the separation surface λ in the infinite parallelepiped $\Lambda(L)$ is given by the limit of $Pr_{L,M}(\lambda)$, as M tends to infinity. This limit is equal to

$$Pr_{L}(\lambda) = \frac{\exp(-2\beta|\lambda|-\beta U_L(\lambda))}{Z_L} \tag{6.3}$$

where

$$Z_L = \lim_{M\to\infty} Z^{\pm}_{L,M} \,/\, Z^{+}_{L,M} \tag{6.4}$$

and

$$\exp(-\beta U_L(\lambda)) = \lim_{M\to\infty} Z^{+}_{L,M}(\lambda)/Z^{+}_{L,M} \tag{6.5}$$

The quantity $U_L(\lambda)$ is a surface free energy associated with the surface λ (see note 12). Moreover, the surface tension τ, defined by (5.1) is equal to

$$\tau = -\frac{1}{\beta} \lim_{L\to\infty} \frac{1}{(2L+1)^2} \log Z_L$$

Let λ be fixed, and let x be a point of $\Lambda(L)$ above λ at a distance δ from λ. The average of $\sigma(x)$ in the state $\langle\cdot\rangle_L^{\pm}$, under the condition that the separation surface is λ, is given by $\langle\sigma(0)\rangle_L^{+}$ plus a correction term which vanishes exponentially fast when δ tends to infinity. The same property is true for any local observable [15]. In other words, the local properties of the interface are related to the local properties of the separation surface.

The study of this ensemble of random surfaces is difficult. I consider an approximate ensemble[19], for which the main ideas and results can be explained in a simple way. Moreover, the results in this approximation become exact when β tends to infinity. The first simplification is to omit $U_L(\lambda)$ in (6.3). The second simplification is to consider only surfaces which are the graphs of functions. Let λ be such a surface. If λ contains the horizontal face with center $(x^1-\frac{1}{2},x^2,x^3)$, then the value at (x^2,x^3) of the associated function, denoted again by λ, is $\lambda(x^2,x^3) = x^1$. Since the area of λ is $(2L+1)^2 + ||\lambda||$, where $(2L+1)^2$ is the number of horizontal faces and $||\lambda||$ is the number of vertical faces of λ, the probability of λ in this new ensemble is proportional to $\exp(-2\beta||\lambda||)$. In order to study the local properties of the surface λ, one describes λ by *local* variables. The support of the function λ is decomposed into maximal connected[20] components. Then λ is written as the sum $d_1 + \ldots + d_n$, where d_i is the restriction of λ on the ith component of its support. Moreover $||\lambda|| = \sum_{i=1}^{n} ||d_i||$. At zero temperature the separation surface λ is a plane with probability one ($\lambda = 0$). The functions d_i describe the local deviations of λ with respect to the situation at zero temperature. By taking the limit of $Pr_L(d_1,\ldots,d_k)$ as L tends to infinity one obtains the probability distribution of any finite family $(d_1,\ldots,d_k)$ in the thermodynamic limit. One can therefore investigate the local structure of the separation surface in the thermodynamic limit, and hence the interface. Since $Pr_L(d) \leqslant \exp(-2\beta||d||)$, one concludes, exactly as in the usual Peierls argument (see e.g.[28]), that the probability $Pr_L(\lambda(0,0) \neq 0)$ tends to zero, uniformly in L, as β tends to infinity. At high β the surface λ is essentially flat and

$$<(\lambda(0,0) - \lambda(x^2,x^3))^2>_L \leq C(\beta)$$

uniformly in L and (x^2,x^3), where $<\cdot>_L$ is the average in the approximate ensemble. These results are valid for the ensemble given by (6.3), i.e. for the Ising model at low temperatures [19], [20].

At high temperatures the properties of the separation surface are totally different and the above description is not very useful. Even in the approximate ensemble, the analysis is highly non-trivial. Fröhlich and Spencer proved, that there exist two constants $C_1(\beta)$ and $C_2(\beta)$ such that for all L

$$C_1(\beta) \log|x| \leq <(\lambda(0,0) - \lambda(x^2,x^3))^2>_L \leq C_2(\beta) \log|x|$$

where $|x|$ is the distance of (x^2,x^3) from (0,0) [25]. The same behaviour for the separation surface in the Ising model is expected to occur above the roughening temperature.

In the two-dimensional case the analysis of the interface is related to the analysis of an ensemble of random separation lines, defined in a completely similar way. In the approximate ensemble the situation is very simple. Let L be fixed. Each line is the graph of a function λ, such that $\lambda(x^2) = 0$ if $|x^2| > L$. The line λ is described by local variables, indexed by x^2, $d(x^2) = \lambda(x^2) - \lambda(x^2-1)$. In particular the probability of λ is proportional to $\exp(-2\beta \sum_{x^2} |d(x^2)|)$ and the height of λ above $x^2 = 0$ is given by

$$\lambda(0) = \sum_{x^2 \leq 0} d(x^2) \tag{6.6}$$

As in the three-dimensional case, $Pr_L(d(x^2) \neq 0)$ tends zero, uniformly in L, as β tends to infinity. At high β the line λ is locally almost horizontal. However, since $\lambda(0)$ is given by (6.6), one has global fluctuations described by the central limit theorem

$$\Pr_L(\lambda(0) < \frac{\alpha}{\sqrt{2}}\sqrt{L}) = (2\pi\gamma^2)^{-\frac{1}{2}} \int_{\infty}^{\alpha} \exp\left(-\frac{u^2}{2\gamma^2}\right) du + O\left(\frac{1}{\sqrt{L}}\right) \qquad (6.7)$$

where $\gamma = (\sqrt{2}\,\mathrm{sh}\beta)^{-1}$. Gallavotti proved the same results for the Ising model at low temperatures [27]. The local random variables, corresponding to the variables $d(x^2)$ of the above case, are non-independent in a non-trivial way, since the term $U_L(\lambda)$ is taken into account in (6.3). The average of the distance between the highest and lowest intersection points of λ with a vertical line is finite, and tends to zero exponentially fast, as β tends to infinity. The distributions of these intersection points with the vertical line $x^2 = 0$ are given by the formula (6.7) (with of course another γ). The results imply that the thickness of the interface is finite, and that there are global fluctuations of order $O(\sqrt{L})$. More details on the local structure of the interface are given in [15]. The *global* fluctuations can be detected by computing a rescaled magnetization profile on the scale $\sqrt{L}$. From (6.7) one finds immediately that

$$M(\alpha) = \lim_{L\to\infty} \left\langle\sigma\left(\frac{\alpha}{\sqrt{2}}\sqrt{L},0\right)\right\rangle_L^{\pm} \qquad (6.8)$$

$$= 2m^*\left(2\pi\gamma^2\right)^{-\frac{1}{2}} \int_0^{\alpha} \exp\left(-\frac{u^2}{2\gamma^2}\right) du$$

This profile has been computed by Abraham and Reed for *all* temperatures [4]. The result is (6.8) with

$$\gamma^2(\beta) = (\mathrm{sh}(2\beta - 2\beta^*))^{-1} = (\mathrm{sh}(\beta\tau(\beta)))^{-1} \qquad (6.9)$$

One can identify γ of (6.7) in the Ising model with the above value (6.9). At high β, $\gamma^2 \sim 2e^{-2\beta}$, like in the approximate ensemble, and near the critical temperature, $\gamma^2 \sim kT/\tau(\beta)$. There is an equivalent way of detecting the global fluctuations. One modifies the boundary conditions for the parallelepiped $\Lambda(L)$, as L tends to infinity, as follows :

$$b(x) = +1,\ x^1 \geq -\frac{\alpha}{\sqrt{2}} L \text{ and } b(x) = -1,\ x^1 < -\frac{\alpha}{\sqrt{2}}\sqrt{L}$$

The modified state in $\Lambda(L)$ is denoted by $\langle\cdot\rangle^{\pm}_{\alpha\sqrt{L}}$. It is equal to the state $\langle\cdot\rangle^{\pm}_{L}$ translated vertically by $(\frac{\alpha}{\sqrt{2}}\sqrt{L},0)$. From the result of Abraham and Reed one obtains immediately[21] that

$$\lim_{L\to\infty} \langle\sigma(x^1,x^2)\rangle^{\pm} = M(\alpha),\ \forall\ (x^1,x^2) \tag{6.10}$$

Since all states are translation invariant one has in fact the stronger result[22]

$$\lim_{L\to\infty} \langle\cdot\rangle^{\pm}_{\alpha\sqrt{L}} = \mu(\alpha)\langle\cdot\rangle^{+} + (1-\mu(\alpha))\langle\cdot\rangle^{-} \tag{6.11}$$

where

$$\mu(\alpha) = (2\pi\gamma^2)^{-\frac{1}{2}} \int_{-\infty}^{\alpha} \exp\left\{-\frac{u^2}{2\gamma^2}\right\} du$$

The interpretation of (6.11) is simple. The interface has zero probability of passing through *any* finite fixed region Ω in the thermodynamic limit. With probability $\mu(\alpha)$ it is below Ω, and with probability $1 - \mu(\alpha)$ it is above Ω. The result (6.11) is valid for all temperatures.

Fluctuations of the separation line have been investigated further by Cammarota [18] and Higuchi [33] at low temperatures. Let $0 < t_1 < t_2 < \ldots < t_k < 1$ be given and let $h(t_i)$ be the height of the highest intersection point[23] of λ with the vertical line $x^2 = 2Lt_i - L$. The main result of Cammarota and Higuchi is the generalization of (6.7), wich corresponds to the case $k = 1$ and $t_1 = \frac{1}{2}$:

$$\lim_{L\to\infty} \Pr_L(a_i\sqrt{2L} \leq h(t_i) \leq b_i\sqrt{2L},\ i = 1,\ldots,k) =$$

$$\left(2\pi\gamma^2\right)^{\frac{1}{2}} \int_{a_1}^{b_1} P(t_1,u_1)\, du_1 \int_{a_2}^{b_2} P(t_2-t_1,u_2-u_1)\, du_2 \ldots \tag{6.12}$$

$$\ldots \int_{a_k}^{b_k} P(t_k-t_{k-1},u_k-u_{k-1})\ P(1-t_k,-u_k)\, du_k$$

where

$$P(t,u) = (2\pi\gamma^2 t)^{-\frac{1}{2}} \exp\left(\frac{-u^2}{2\gamma^2 t}\right)$$

The result (6.12) suggests, that the random separation line λ, suitably normalized as indicated in (6.12), converges to a Brownian bridge (see e.g. [10]) in the thermodynamic limit. This was proved by Higuchi [33].

The results above give a lot of information on the global fluctuations of the interface up to the critical temperature and on the local structure of the interface at low temperatures. Using ideas of Weeks [53], Abraham studied the intrinsic thickness of the interface in a modified Ising model near the critical temperature [1]. The modification consists in suppressing the fluctuations which are coherent over horizontal distances greater than the bulk correlation length. The question of the intrinsic thickness near the critical point is discussed thoroughly by Widom in [58] (see also section 6 in [15]).

7. Surface Tension and Interface.

In the three-dimensional case, at low temperatures, the interface is stable. It is possible to find a connection between the surface tension and the local structure of the interface. Let $x = (z,0,0)$ and

$$e(z) = -\tfrac{1}{2} \sum_y \sigma(x)\sigma(y)$$

where the sum is over all y such that $\langle xy \rangle$. The quantity $e(z)$ is the local energy. In [12] it is shown that

$$-\frac{d(\beta\tau)}{d\beta} = \sum_{z\in\mathbb{Z}} \left(\langle e(z)\rangle^+ - \langle e(z)\rangle^{\pm} \right) \tag{7.1}$$

By symmetry $\langle e(z)\rangle^+ = \langle e(z)\rangle^-$, and therefore $\langle e(z)\rangle^+ - \langle e(z)\rangle^{\pm}$ tends to zero exponentially fast, as $|z|$ tends to infinity. Therefore the main contribution to $\frac{d(\beta\tau)}{d\beta}$ is given by the terms $\langle e(z)\rangle^+ - \langle e(z)\rangle^{\pm}$, when z is in the interface. In the two-dimensional case the formula (7.1) does not make sense, since $\langle e(z)\rangle^+ = \langle e(z)\rangle^{\pm}$ for all z. However, since the interface has a local structure, one can prove that ([15])

$$-\frac{d(\beta\tau)}{d\beta} = \lim_{L\to\infty} \sum_{z\in\mathbb{Z}} \left(\langle e(z)\rangle_L^{+} - \langle e(z)\rangle_L^{\pm} \right) \tag{7.2}$$

The interpretation of (7.2) is the same as the interpretation of (7.1). For fixed L, the interface in the state $\langle\cdot\rangle_L^{\pm}$ does not go to infinity. Hence the main contribution to the sum (7.2) is given by z, when z is in the interface. Indeed

$$\sum_{z\in\mathbb{Z}} \left(\langle e(z)\rangle_L^{+} - \langle e(z)\rangle_L^{\pm} \right) =$$

$$\sum_{\lambda} Pr_L(\lambda) \left(\sum_{z\in\mathbb{Z}} \langle e(z)\rangle_L^{+} - \langle e(z)|\lambda\rangle_L^{\pm} \right) \tag{7.3}$$

where $\langle e(z)|\lambda\rangle_L^{\pm}$ is the conditional expectation value of e(z), given the separation line λ. The main contribution to the last sum in (7.3) comes from the terms $\langle e(z)\rangle_L^{+} - \langle e(z)|\lambda\rangle_L^{\pm}$ with z near the separation line λ. Since the local structure remains well-defined, as L tends to infinity, one obtains the result (7.2). At this point, it is very easy to sketch the proof of the lower bound (5.3). By correlation inequalities $\langle\sigma(x)\sigma(y)\rangle_L^{+}$ is larger than $\langle\sigma(x)\sigma(y)\rangle_L^{\pm}$ for all $\langle xy\rangle$. Therefore (7.2) is bounded below by

$$\sum_{z\in\mathbb{Z}} \left(\langle\sigma(z+1,0)\sigma(z,0)\rangle_L^{+} - \langle\sigma(z+1,0)\sigma(z,0)\rangle_L^{\pm} \right) \tag{7.4}$$

Again, by correlation inequalities,

$$\langle\sigma(z+1,0)\sigma(z,0)\rangle_L^{+} - \langle\sigma(z+1,0)\sigma(z,0)\rangle_L^{\pm} \geq$$

$$\langle\sigma(0)\rangle_L^{+} \left(\langle\sigma(z+1,0)\rangle_L^{\pm} - \langle\sigma(z,0)\rangle_L^{\pm} \right)$$

Therefore (7.4) is bounded below by the gradient of the magnetization profile times the spontaneous magnetization in the state $\langle\cdot\rangle_L^{+}$. The sum over the gradient is easy. Since the interface is localized for fixed L, one obtains $2(\langle\sigma(0)\rangle_L^{+})^2$, which tends to $2(m^*)^2$, as L tends to infinity.

The expression (7.2) has been proved only at low temperatures. Very likely, it holds up to the critical temperature. The lower bound (5.3) is valid for all β.

8. Endnotes.

1. There is only one model of a fluid, the Widom-Rowlinson model [59], for which one can prove rigorously the existence of a phase transition [49]. Unfortunately there is no good analogue of the random separation surface. On the other hand, the lattice versions of the model, [36], can be analyzed in the same way as the lattice gas [11], [12].

2. The state $\langle\cdot\rangle^{+}(\beta)$, which is obtained by taking the thermodynamic limit of $\langle\cdot\rangle_{\Lambda}^{+}(0,\beta)$, where + stands for the boundary condition $b(x) \equiv +1$, can also be constructed as the limit, when h tends to zero, of the unique equilibrium state $\langle\cdot\rangle(h,\beta)$ with homogenous positive magnetic field [37].

3. This result was first proved at low temperatures by Martin-Löf [40] and by Gallavotti and Miracle-Sole [31]. The main result of [31] is however that all translation invariant equilibrium states are convex combinations of $\langle\cdot\rangle^{+}$ and $\langle\cdot\rangle^{-}$ at low temperatures. Messager and Miracle-Sole extended this last result for all temperatures in the two-dimensional case [41]. Later on Lebowitz proved that this result holds if and only if the free energy is differentiable with respect to the temperature [35].

4. As long as the magnetic field is non-zero, the thermodynamic limit should be independent of the boundary condition b. I do not know a proof of this conjecture. The choice of $\underline{h}$ is not unique. One could take any function such that $0 < h(z) = -h(-z-1)$, $z \geq 0$.

5. The inhomogeneous magnetic field plays the rôle of a stabilizing field for the interface, as the gravitational field in the one-component fluid. There is however a difference. The magnetic field produces always an inhomogeneous state with an interface, even at very high temperatures. This is not the case for the gravitational field.

6. This last property is typical for a *lattice* system, because the group of translations is *discrete*. In a continuous system, if the translation invariance is broken, then one has long-ranged correlations, as Wertheim showed in [56]. This is the "Goldstone theorem". (See e.g. [39] and the references of this paper.)

7. There are very few rigorous results on the roughening transition. Let β_R be defined as the largest β, below which $\langle\cdot\rangle^{\pm} = \frac{1}{2}(\langle\cdot\rangle^{+} + \langle\cdot\rangle^{-})$. Weeks, Gilmer and Leamy gave an estimation of β_R by considering the slope S of the magnetization profile at z = 0 [54]. By symmetry $S = 2\langle\sigma(0)\rangle^{\pm}$. Using the concavity of the profile (property d) and an argument of Messager and Miracle-Sole [42], one proves that $S^{-1} < \infty$ if and only if $\beta > \beta_R$. One can also use the quantity $D = (\langle\sigma(0,0,0)\sigma(-1,0,0)\rangle^{+} - \langle\sigma(-1,0,0)\sigma(0,0,0)\rangle^{\pm})$ for estimating the roughening temperature. Indeed, using correlation inequalities one shows that $D^{-1} < \infty$ if and only if $S^{-1} < \infty$. In the SOS approximation of the Ising model the roughening transition has been proved by Fröhlich and Spencer [25]. See section 6.

8. Strictly speaking Gallavotti proved the result at low temperatures. Using the exact computation of Abraham and Reed on the magnetization profile, [4], Messager and Miracle-Sole proved the result for all $\beta > \beta_c(2)$ [42]. Later on Russo gave a completely different proof of this result, without using exact computation [50]. His paper is fundamental for the proofs of Aizenman and Higuchi on the absence of non-translation invariant states [5], [34].

9. Fontaine and Gruber studied the surface tension for general lattice models at low and high temperatures [24].

10. This ensemble corresponds to the canonical ensemble in the lattice gas interpretation of the model.

11. This corresponds to the grand canonical ensemble in the lattice gas interpretation.

12. In (5.1) one can take $M = L$ and do the two limits at one time. Let Λ_L be the cubic box $\Lambda(L,L)$. The free energy of the finite system in Λ_L, with + boundary condition, is $F_+(L) = -\frac{1}{\beta}\log Z^+(\Lambda_L)$. For large L, $F_+(L) = fL^d + g_+L^{d-1} + o(L^{d-1})$, where f is the bulk free energy per unit volume, and g_+ a surface free energy. (Surface free energies are studied in [21], [22] and [17].) By symmetry $F_-(L) = F_+(L)$, and in particular $g_+ = g_-$. For the ± boundary condition $F_\pm(L)$ has a bulk term fL^d and a surface term g_+L^{d-1} coming from the boundary of $\Lambda(L)$. Below the critical temperature one expects another contribution of order $O(L^{d-1})$. Intuitively this contribution comes from the interface between the two phases and is interpreted as the surface tension.

13. Both bounds can be proved for other lattice models. In [14] other upper bounds are derived. Using the results of [14] and the bound (5.3), one proves that the critical temperature of the Ising model coincides with that of the semi-infinite Ising model [38]. Strictly speaking the existence of the derivative in (5.3) is proved only almost everywhere.

14. It is not expected that $\beta\tau(\beta) - 2\beta$ is analytic up to $\beta_c(3)$. One expects a (weak) singularity at the roughening transition. Van Beijeren proved the existence of such a singularity for a model of the interface similar to the SOS model [8].

15. The proof of (5.5) is not difficult. It is however not trivial because one must justify an interchange of limits. If $\beta > \beta_c(2)$, then $\beta^* < \beta_c(2)$ and one can write (5.5) as $\tau(\beta)\,\xi(\beta^*) = kT$. It is interesting to notice that $\tau(\beta)\,\xi(\beta) = \frac{1}{2}kT$, where $\beta > \beta_c(2)$ and $\xi(\beta)$ is the correlation length at inverse temperature β. This implies the scaling relation (2.2) in a strong sense.

16. Relation (5.5) for the three-dimensional Ising model and (5.3) imply that the coefficient of the area-decay of the Wilson loop, $\kappa(\beta)$, is nonzero at $\beta = \beta_R^*$ and non-analytic at this point, if the surface tension is non-analytic at β_R.

17. Aizenman has developped another approach to this problem [6].

Stastical mechanics of surfaces is reviewed in [26].

18. One uses the symmetry property $Z(\Lambda,+) = Z(\Lambda,-)$ (see (3.2)).

19. This corresponds to the so-called solid-on-solid model (SOS-model).

20. A set E is *connected*, if any pair of points of E can be joined by a finite path of adjacent points in E. Two points are adjacent if they are at distance one.

21. If $x = (x^1,x^2)$ depends on L like $x(L) = (x^1+aL^{\frac{1}{2}-\varepsilon}, x^2+bL^{1-\varepsilon})$, for $\varepsilon > 0$, then the result (6.10) is still true.

22. The result (6.11) indicates that *any* Gibbs state can be obtained as the limit of finite volume Gibbs states with boundary conditions, chosen in such way that the ratio of the number of values +1 with the number of values -1 of the boundary field (see (3.1)) tends to 1 as the volume tends to infinity. See also [32].

23. One could take the lowest intersection point or the mean value between the highest and lowest intersection points.

9. References.

[1] Abraham D.B. Capillary waves and surface tension : an exactly solvable model. Phys. Rev. Lett. 47, 545-548 (1981).

[2] Abraham D.B., Gallavotti G., Martin-Löf A. Surface tension in the two-dimensional Ising model. Physica 65, 73-88. (1973).

[3] Abraham D.B., Martin-Löf A. The transfer matrix for a pure phase in the two-dimensional Ising model. Commun. Math. Phys. 32, 245-268 (1973).

[4] Abraham D.B., Reed P. Interface profile of the Ising ferromagnet in two dimensions. Commun. Math. Phys. 49, 35-46 (1976).

[5] Aizenman M. Translation invariance and instability of phase coexistence in two dimensional Ising model. Commun. Math. Phys. 73, 83-94 (1980).

[6] Aizenman M. Geometric analysis of ϕ^4 fields and Ising models I and II. To appear in Commun. Math. Phys.

[7] van Beijeren H. Interface sharpness in the Ising model.

Commun. Math. Phys. 40, 1-6 (1975).

[8] van Beijeren H. Exactly solvable model for the roughening transition of a crystal surface. Phys. Rev. Lett. 38, 993-996 (1977).

[9] Benettin G., Gallavotti G., Jona-Lasinio G., Stella A.L. On the Onsager-Yang value of the spontaneous magnetization. Commun. Math. Phys. 30, 45-54 (1973).

[10] Billingsley P. Convergence of probability measures John Wiley (1968

[11] Bricmont J., Lebowitz J.L., Pfister C.E., Olivieri E. Non-translation invariant Gibbs states with coexisting phases I. Commun. Math. Phys. 66, 1-20 (1979).

[12] Bricmont J., Lebowitz J.L., Pfister C.E. Non-translation invariant Gibbs states with coexisting phases II. Commun. Math. Phys. 66, 21-36 (1979).

[13] Bricmont J., Lebowitz J.L., Pfister C.E. Non-translation invariant Gibbs states with coexisting phases III. Commun. Math. Phys. 69, 267-291 (1979).

[14] Bricmont J., Lebowitz J.L., Pfister C.E. On the surface tension of lattice systems. Annals of the New York Academy of Sciences 337, 214-223 (1980).

[15] Bricmont J., Lebowitz J.L., Pfister C.E. On the local structure of the phase separation line in the two-dimensional Ising system. J. Stat. Phys. 26, 313-332 (1981).

[16] Buff F.P., LovettR.A., Stillinger F.H. Interfacial density profile for fluids in the critical region. Phys. Rev. Lett. 15, 621-623 (1965).

[17] Caginalp G., Fisher M.E. Wall and boundary free energies II. Commun. Math. Phys. 65, 247-280 (1979).

[18] Cammarota C. : private communication.

[19] Dobrushin R.L. Gibbs state describing coexistence of phases for a three-dimensional Ising model. Theory Probab. Appl. 17, 582-600 (1972).

[20] Dobrushin R.L. Investigation of Gibbsian states for three-dimensional lattice systems. Theory Probab. Appl. 18, 253-271 (1973).

[21] Dobrushin R.L. Asymptotic behaviour of Gibbsian distributions for lattice systems and its dependence on the form of the volume. Teoret. and Math. Phys. 12, 115-134 (1972).

[22] Fisher M.E., Caginalp G. Wall and boundary free energies I. Commun. Math. Phys. 56, 11-56 (1977).

[23] Fisk S., Widom B. Structure and free energy of the interface between fluid phases in equilibrium near the critical point. J. Chem. Phys. 50, 3219-3227 (1969).

[24] Fontaine J.R., Gruber C. Surface tension and phase transition for lattice systems. Commun. Math. Phys. 70, 243-269 (1979).

[25] Fröhlich J., Spencer T. The Kosterlitz-Thouless transition in two-dimensional abelian spin systems and the Coulomb gas.

Commun. Math. Phys. 81, 527-602 (1981).

[26] Fröhlich J., Pfister C.E., Spencer T. On the statistical mechanics of surfaces. IHES/P/82/20.

[27] Gallavotti G. The phase separation line in the two-dimensional Ising model. Commun. Math. Phys. 27, 103-136 (1972).

[28] Gallavotti G. Instabilites and phase transitions in the Ising model. A review. Riv. Nuovo Cimento 2, 133-169 (1972).

[29] Gallavotti G., Martin-Löf A. Surface tension in the Ising model. Commun. Math. Phys. 25, 87-126 (1972).

[30] Gallavotti G., Martin-Löf A., Miracle-Sole S. Some problems connected with the description of coexisting phases at low temperatures in the Ising model. Battelle Seattle 1971 Rencontres Ed. A. Lenard. Lectures Notes in Physics 20, 162-204. Springer Berlin Heidelberg New York (1973).

[31] Gallavotti G., Miracle-Sole S. Equilibrium states of the Ising model in the two-phase region. Phys. Rev. B 5, 2555-2559 (1972).

[32] Higuchi Y. On limiting Gibbs states of the two-dimensional Ising models. Publ. R.I.M.S. Kyoto Univ. 14, 53-69 (1978).

[33] Higuchi Y. On some limit theorems related to the phase separation line in the two-dimensional Ising model. Z. Wahrsch. verw. Geb. 50, 287-315 (1979). Fluctuation of the interface of the two-dimensional Ising model. In Quantum Fields-Algebras, Processes Ed. L. Streit, 397-405, Springer Wien, New-York (1980).

[34] Higuchi Y. On the absence of non-translation invariant Gibbs states for the two-dimensional Ising model. Colloquia Mathematica Societatis Jãnos Bolyai 27, Random Fields, Esztergom (1979).

[35] Lebowitz J.L. Coexistence of phases in Ising ferromagnets. J. Stat. Phys. 16, 463-476 (1977).

[36] Lebowitz J.L., Gallavotti G. Phase transitions in binary lattice gases. J. Math. Phys. 12, 1129-1133 (1971).

[37] Lebowitz J.L, Martin-Löf A. On the uniqueness of the equilibrium state for the Ising systems. Commun. Math. Phys. 25, 276-282 (1972)

[38] Lebowitz J.L., Pfister C.E. Surface tension and phase coexistence. Phys. Rev. Lett. 46, 1031-1033 (1981).

[39] Martin P.A. A remark on the Goldstone theorem in statistical mechanics. Il Nuovo Cimento 68 B, 302-314 (1982).

[40] Martin-Löf A. On the spontaneous magnetization in the Ising model. Commun. Math. Phys. 24, 253-259 (1972).

[41] Messager A., Miracle-Sole S. Equilibrium States of the two-dimensional Ising model in the two-phase region. Commun. Math. Phys. 40, 187-196 (1975).

[42] Messager A., Miracle-Sole S. Correlation functions and boundary conditions in the Ising ferromagnet. J. Stat. Phys. 17, 245-263 (1977).

[43] Minlos R.A. Lectures on statistical physics. Russian Math. Surveys

23, 137-196 (1968).

[44] Minlos R.A., Sinai Y. Some new results on first order phase transitions in lattice gas models. Trans. Moscow Math. Soc. 17, 237-267 (1967).

[45] Minlos R.A., Sinai Y. The phenomenon of "phase separation" at low temperatures in some lattice models of a gas I and II. Math. USSR-Sb. 2, 335-395 (1967) and Trans. Moscow Math. Soc. 19, 121-196 (1968).

[46] Onsager L. Crystal Statistics I. A two-dimensional model with an order-disorder transition. Phys. Rev. 65, 117-149 (1944).

[47] Onsager L. Suppl. Nuovo Cimento 6, 261 (1949).

[48] Ruelle D. Statistical Mechanics : Rigorous Results, Benjamin, New York (1969).

[49] Ruelle D. Existence of a phase transition in a continuous classical system. Phys. Rev. Lett. 27, 1040-1041 (1971).

[50] Russo L. The infinite cluster method in two-dimensional Ising model. Commun. Math. Phys. 67, 251-266 (1979).

[51] Shlosman S.B. Correlation inequalities and their applications. J. of Soviet Math. 15, 79-101 (1981).

[52] van der Waals J.D. The thermodynamic theory of capillarity under the hypothesis of a continuous variation of density. English translation J. Stat. Phys. 20, 197-244 (1979).

[53] Weeks J.D. Structure and thermodynamics of the liquid-vapor interface. J. Chem. Phys. 67, 3106-3121 (1977).

[54] Weeks J.D., Gilmer G.H., Leamy H.J. Structural transition in the Ising model interface. Phys. Rev. Lett. 31, 549-551 (1973).

[55] Wegner F.J. Duality in generalized Ising models and phase transitions without local order parameters. J. Math. Phys. 12, 2259-2272 (1971).

[56] Wertheim M.S. Correlations in the liquid-vapor interface : J. Chem. Phys. 65, 2377-2381 (1976).

[57] Widom B. Surface tension of fluids in phase transitions and critical phenomena, in Phase transitions and critical phenomena II, 79-100 Ed. C. Domb, M.S. Green. Academic Press New York (1972).

[58] Widom B. Structure of the interface between fluid phases. Lecture presented at the Faraday Symposium no 16, "Structure of the Interfacial Region", Oxford, December 16-17, 1981.

[59] Widom B., Rowlinson J.S. New model for the study of the liquid-vapor phase transitions. J. Chem. Phys. 52, 1670-1684 (1970).

[60] Yang C.N. The spontaneous magnetization of a two-dimensional Ising model. Phys. Rev. 85, 809-816 (1952).

ITERATED MAYER EXPANSIONS AND THEIR APPLICATION TO COULOMB GASES

John Z. Imbrie

. Introduction

There are many problems in statistical mechanics and field theory where systems of interacting particles (or particle-like defects, contours, molecules, clusters, etc.) arise. Understanding these systems would be simpler if the particles were noninteracting. If the particles are dilute and weakly interacting, then the corrections to the independent particle approximation are small. These can be isolated and regarded as new kinds of (nonlocal) particles by a procedure known as the Mayer expansion. The nonlocal particles (or clusters) carry all the information about the long-distance behavior of the system.

There are many variants of Mayer expansions and cluster expansions that can be used depending on the type of system under consideration. The following grand canonical partition function is a representative example:

$$Z = \sum_{N=0}^{\infty} \frac{z^N}{N!} \int dx_1 \ldots dx_N \exp(-\beta \sum_{i<j} v(x_i - x_j)). \tag{1.1}$$

Here N is the number of particles, z is the activity of each particle, β is the inverse temperature, $x_i \in \Lambda \subseteq \mathbb{R}^d$ is the position of the ith particle, and $v(x_i - x_j)$ is a two-body interaction potential. The basic expansion step is to use the fundamental theorem of calculus to write

$$ze^{-\beta v(x_i - x_j)} = z + z\int_0^1 ds(-\beta)v(x_i - x_j)e^{-s\beta v(x_i - x_j)} \tag{1.2}$$

The first term is an isolated particle, the second is a two-particle cluster. When identities like (1.2) have been applied many times, we find (after some combinatorics) an expansion for $|\Lambda|^{-1}\log Z$ roughly speaking in powers of $z\beta \int |v(x)| dx e^{-\beta E}$, where E is a lower bound for the energy per particle in any N-particle configuration. Thus we obtain convergence for small activities or high temperatures, depending on the strength of the interaction (in the L^1-sense) and on the stability bound E.

A simple Mayer expansion as outlined above is not well suited to interactions that are strong ($\beta E << -1$) only in a region that contributes little to the integral $\int |v(x)| dx$. In this situation we should split the interaction into two or more parts $v^\ell, \ell = 0,1,\ldots$, and expand in the part with the worst stability estimate E^ℓ and shortest range first. The clusters from this expansion are the particles of the next expansion. One can proceed in this way through all the parts of v without ever encountering a very large product $\int |v^\ell(x)| dx\ e^{-\beta E^\ell}$. Then one obtains an expansion that converges with relatively mild conditions on activities.

This procedure of iterated Mayer expansions was formalized and applied to the lattice Yukawa gas by Göpfert and Mack [5]. It fits into the renormalization group idea that the behavior of systems is best understood by considering what happens at successively larger length scales. In its current form, however, the procedure works only when there are just a finite number of length scales to be considered. This limitation doesn't matter for Coulomb systems where screening (exponential clustering) sets in at some length scale to break the scale invariance of the interaction. For truly critical systems, one needs more refined techniques.

To illustrate the technique, we shall consider a soft core continuum Yukawa gas in three dimensions. For the Coulomb gas application we have in mind, it is sufficient to split the interaction into two parts. Let $e_i = \pm 1$ be the charge of the ith particle. Write $\xi_i = (e_i, x_i)$, $\int d\xi_i = \sum_{e_i} \int_\Lambda dx_i$, and put $\ell_D = (2z\beta)^{-1/2}$. Define a two-body interaction

$$v(\xi_i,\xi_j) = e_i \left[\frac{e^{-|x_i - x_j|/(\lambda \ell_D)} - e^{-|x_i - x_j|/R}}{4\pi |x_i - x_j|}\right] e_j. \qquad (1.3)$$

Here λ parametrizes the range of the Yukawa potential, and R is a short distance cutoff, necessary for stability. For simplicity we assume $R \le \beta \le \lambda \ell_D$. The partition function is

$$Z = \sum_{N=0}^{\infty} \frac{z^N}{N!} \int d\xi_1 \ldots d\xi_N \exp(-\beta \sum_{i<j} v(\xi_i, \xi_j)). \tag{1.4}$$

f we put $\lambda = \infty$ in (1.3), then we have a Coulomb potential with soft ore and ℓ_D, the Debye length, is the length for exponential decay of orrelations in the Debye-Hückel theory. Our goal here is to write an xpansion valid for λ small, but without putting restrictions on λ that epend on the other parameters β, R, z. This will ultimately lead to a arge region for screening for the corresponding Coulomb gas [6,7]. In ect. 4 we discuss applications of these expansions to Coulomb gases.

A convenient splitup of v is

$$v = v^0 + v^1 \tag{1.5}$$

$$v^0(\xi_i, \xi_j) = e_i \left[\frac{e^{-|x_i - x_j|/\beta} - e^{-|x_i - x_j|/R}}{4\pi |x_i - x_j|} \right] e_j \tag{1.6}$$

$$v^1(\xi_i, \xi_j) = e_i \left[\frac{e^{-|x_i - x_j|/(\lambda \ell_D)} - e^{-|x_i - x_j|/\beta}}{4\pi |x_i - x_j|} \right] e_j . \tag{1.7}$$

otice that v^0 can be written as

$$v^0(\xi_i, \xi_j) = e_i [(-\Delta + \beta^{-2})^{-1} - (-\Delta + R^{-2})^{-1}](x_i, x_j) e_j . \tag{1.8}$$

his form is useful for stability and two-body estimates. We have

$$\sum_{i,j} v^0(\xi_i, \xi_j) = \langle \rho, [(-\Delta + \beta^{-2})^{-1} - (-\Delta + R^{-2})^{-1}] \rho \rangle \geq 0, \tag{1.9}$$

here $\rho(x) = \sum_i e_i \delta(x - x_i)$. Thus

$$\sum_{1 \leq i < j \leq N} v^0(\xi_i, \xi_j) \geq -\frac{1}{2} \sum_i v^0(\xi_i, \xi_i) = -cN/R, \tag{1.10}$$

nd E^0, the lower bound for the energy per particle in a system with interaction v^0, is equal to $-c/R$ for some constant c. Similarly $E^1 = -c/\beta$; and E, the lower bound for the full system, is equal to $-c/R$. ote that $\int |v^0(\xi_i, \xi_j)| dx_j$ is just the Fourier transform of $(-\Delta + \beta^{-2})^{-1} - (-\Delta + R^{-2})^{-1}$ at $p = 0$. Thus we have

$$\int |v^0(\xi_i, \xi_j)| dx_j = \beta^2 - R^2 \leq \beta^2 \tag{1.11}$$

$$\int |v^1(\xi_i, \xi_j)| dx_j \leq \lambda^2 \ell_D^2 \tag{1.12}$$

$$\int|v(\xi_i,\xi_j)|dx_j \le \lambda^2\ell_D^2. \tag{1.13}$$

We can now apply the conditions $z\beta\int|v^*(\xi_i,\xi_j)|dx_j e^{-\beta E^*} << 1$ to see for what ranges of R,β,z the expansions should converge. For the v^0 expansion we require $z\beta^3 e^{c\beta/R} << 1$. For the v^1 expansion we need $z\beta\lambda^2\ell_D^2 e^c = c\lambda^2 << 1$. Thus expanding in v^0 and v^1 separately put conditions on λ independent of z,β,R as desired. In contrast, expanding in v all at once would require $z\beta\lambda^2\ell_D^2 e^{c\beta/R} = c\lambda^2 e^{c\beta/R} << 1$, or $\lambda^2 << e^{-c\beta/R}$. One is forced to apply the weak stability estimate E over the whole range of v -- a much less efficient procedure than the iterated expansion we describe below.

2. An Iterated Mayer Expansion

We want to develop an expansion for

$$Z^N(\xi_1,\ldots,\xi_N) = \frac{z^N}{N!}\exp(-\beta\sum_{i<j}v(\xi_i,\xi_j)). \tag{2.1}$$

At first we consider only v^0 and expand

$$Q_0^N(\xi_1,\ldots,\xi_N) = z^N\exp(-\beta\sum_{i<j}v^0(\xi_i,\xi_j)). \tag{2.2}$$

One could write

$$e^{\sum v_{ij}} = \prod_{i<j}[1 + (e^{v_{ij}} - 1)] = \sum_G \prod_{\{i,j\}\in G}(e^{v_{ij}} - 1), \tag{2.3}$$

where G runs over all sets of unordered pairs {i,j}, but for estimates it is better to have an interpolation formula. The formula we give is based on [2]. For α a subset of {1,...,N} we write

$$V^0(\alpha) = \sum_{i<j;i,j\in\alpha} v^0(\xi_i,\xi_j). \tag{2.4}$$

The interpolated interaction for the first step is

$$V_{s_1}^0(1;\alpha) = V^0(\{1\}\cup\alpha)s_1 + V^0(\alpha)(1 - s_1), \tag{2.5}$$

where we take $\alpha = \{2,\ldots,N\}$. This yields

$$Q_0^N(\xi_1,\ldots,\xi_N) = zQ_0^{N-1}(\xi_2,\ldots,\xi_N) + z^N \sum_{1<i} \int_0^1 ds_1(-\beta)v^0(\xi_1,\xi_i)\cdot$$

$$\cdot\exp(-\beta V^0_{s_1}(1,\alpha)). \tag{2.6}$$

In the first term particle 1 is isolated from the other particles, and in the second it is connected to particle i through a "line" $v^0(\xi_1,\xi_i)$. The next interpolation depends on i. In an attempt to remove interactions between $\{1,i\}$ and the other particles, we put

$$V^0_{s_1s_2}(1,i;\alpha) = V^0_{s_1}(1;\{i\} \cup \alpha)s_2 + (V^0_{s_1}(1;\{i\}) + V^0(\alpha))(1 - s_2)$$

$$= s_1v^0(\xi_1,\xi_i) + \sum_{j\in\alpha}(s_1s_2v^0(\xi_1,\xi_j) + s_2v^0(\xi_i,\xi_j)) + \sum_{j<k;j,k\in\alpha} v^0(\xi_j,\xi_k), \tag{2.7}$$

where α contains all particles but 1,i. Thus

$$\frac{d}{ds_2} V^0_{s_1s_2}(1,i;\alpha) = \sum_{j\in\alpha}(s_1v^0(\xi_1,\xi_j) + v^0(\xi_i,\xi_j)), \tag{2.8}$$

and the ith term in (2.6) interpolates to

$$z^2 \int ds_1(-\beta)v^0(\xi_1,\xi_i)\exp(-\beta V^0_{s_1}(1,\{i\}))Q^{N-2}(\alpha) +$$

$$+ z^N \sum_{1<j\neq i} \int ds_1ds_2(-\beta)^2v^0(\xi_1,\xi_i)(s_1v^0(\xi_1,\xi_j) + v^0(\xi_i,\xi_j))\cdot$$

$$\cdot\exp(-\beta V^0_{s_1s_2}(1,i;\alpha)). \tag{2.9}$$

We continue in this fashion, always attempting to isolate the group of particles connected to particle 1 in the remainder terms [the second group in (2.9)].

To describe the general step, let $i_1,i_2,\ldots$ be the sequence of particles differentiated down in some term of the expansion ($i_1 = 1, i_2 = i, i_3 = j$ above). We define inductively

$$V^0_{s_1\ldots s_n}(i_1,\ldots,i_n;\alpha) = V^0_{s_1\ldots s_{n-1}}(i_1,\ldots,i_{n-1};\{i_n\} \cup \alpha)s_n +$$

$$+ (V^0_{s_1\ldots s_{n-1}}(i_1,\ldots,i_{n-1};\{i_n\}) + V^0(\alpha))(1 - s_n). \tag{2.10}$$

It can easily be checked that

$$V^0_{s_1\ldots s_n}(i_1,\ldots,i_n;\{i_{n+1},\ldots,i_m\}) = \sum_{1\le\mu<\nu\le m} s_\mu\cdots s_{\min\{\nu-1,n\}}\cdot$$

$$\cdot v^0(\xi_{i_\mu},\xi_{i_\nu})$$

satisfies the recursion (2.10). Thus the general formula analogous to (2.8) is

$$\frac{d}{ds_n} V^0_{s_1\ldots s_n}(i_1,\ldots,i_n;\alpha) = \sum_{i_{n+1}\in\alpha}\ \sum_{\eta(n+1)=1}^{n} s_{\eta(n+1)}\cdots s_{n-1}\cdot$$

$$\cdot v^0(\xi_{i_{n+1}},\xi_{i_{\eta(n+1)}}). \tag{2.11}$$

We use the convention that $s_n\ldots s_{n-1} = 1$. We see that i_{n+1} has been determined for the next interpolation. The procedure also generates a tree graph η, which is a function from $\{1,\ldots,k\}$ to itself satisfying $\eta(n) < n$. The interpolations stop when all N particles have been used, so that all terms have a form analogous to that of the first term in (2.9). The expansion then reads

$$Q^N(\xi_1,\ldots,\xi_N) = \sum_{k=1}^{N}\ \sum_{(i_2,\ldots,i_k)} z^k(-\beta)^{k-1}\int ds_1\ldots ds_{k-1}\sum_\eta$$

$$\prod_{\ell=2}^{k}[s_{\eta(\ell)}\cdots s_{\ell-2}v^0(\xi_{\eta(\ell)},\xi_\ell)]\ \exp(-\beta V^0_{s_1\ldots s_{k-1}}(1,i_2,\ldots,i_{k-1};\{i_k\}))\cdot$$

$$\cdot Q^{N-k}(\xi_{\alpha^c}). \tag{2.12}$$

Here $(i_2,\ldots,i_k)$ is any ordered subset of $\{2,\ldots,N\}$, α^c is the complementary subset, $\xi_\alpha = (\xi_i)_{i\in\alpha}$, and η is a tree on k vertices.

We have isolated connected parts of Q^N which will become the particles for the second expansion. In place of activities z, these nonlocal particles (or 1-vertices, or subsets α of $\{1,\ldots,N\}$) have vertex functions

$$\sigma^1_k(\xi_\alpha) = \mathbb{S}\ z^k\,\frac{(-\beta)^{k-1}}{k}\int ds_1\ldots ds_{k-1}\sum_\eta\prod_{\ell=2}^{k}[s_{\eta(\ell)}\cdots s_{\ell-2}v^0(\xi_{\eta(\ell)},\xi_\ell)]$$

$$\exp(-\beta V^0_{s_1\ldots s_{k-1}}(1,\ldots,k-1;\{k\})). \tag{2.13}$$

We have labeled the particles in α as $1,\ldots,k$. The operator $\mathbb{S}$ symmetrizes the expression following in $\xi_1,\ldots,\xi_k$. Permutations involving

ξ_1 are not present in (2.12), hence the factor 1/k. With these vertex functions (2.12) can be written as

$$Q_0^N(\xi_1,\ldots,\xi_N) = \sum_{k=1}^{N} \sum_{\alpha\subseteq\{1,\ldots,N\},1\in\alpha,|\alpha|=k} k!\sigma_k^1(\xi_\alpha)Q_0^{N-k}(\xi_{\alpha^c}), \quad (2.14)$$

where now we sum over unordered subsets $\alpha \subseteq \{1,\ldots,N\}$.

Equation (2.14) can be inserted into itself repeatedly to yield a complete decomposition of Q_0^N into products of 1-vertex functions:

$$Q_0^N(\xi_1,\ldots,\xi_N) = \sum_{\underline{\pi}\in P(1,\ldots,N)} \prod_{\alpha\in\underline{\pi}} [|\alpha|!\sigma^1_{|\alpha|}(\xi_\alpha)]. \quad (2.15)$$

Here $P(1,\ldots,N)$ is the set of partitions of $\{1,\ldots,N\}$. We would like to write (2.15) in a form more like a grand canonical partition function. To this end we change from summing over partitions to summing over multiplicities N_k for 1-vertices of size $k = |\alpha|$. Given a set of multiplicities N_1, $N_2,\ldots$ and some corresponding partition of $\{1,\ldots,N\}$, a new partition can be obtained by permuting $1,\ldots,N$. However, permutations of particles within any element of the partition do not change the partition, and permutations of elements of the partition of a given size do not change the partition. Thus a combinatoric factor $N!(\prod_{\alpha\in\underline{\pi}} |\alpha|!)^{-1}(\prod_k N_k!)^{-1}$ should be included, and we obtain

$$Q_0^N(\xi_1,\ldots,\xi_N) = \sum_{N_1,N_2,\ldots;\Sigma kN_k=N} N!(\prod_k N_k!)^{-1} \mathbb{S} \prod_{\alpha\in\underline{\pi}} \sigma^1_{|\alpha|}(\xi_\alpha). \quad (2.16)$$

Here $\underline{\pi}$ is an arbitrary partition corresponding to N_1, $N_2,\ldots$, and $\mathbb{S}$ symmetrizes in $\xi_1,\ldots,\xi_N$.

Finally we can apply this expansion to $Z^N(\xi_1,\ldots,\xi_N)$ in (2.1) by reintroducing the v^1-interactions as interactions between 1-vertices. Let us write for two 1-vertices α,β

$$v^1(\alpha,\beta) = \sum_{i\in\alpha,j\in\beta,i\neq j} v^1(\xi_i,\xi_j). \quad (2.17)$$

Then we have

$$Z^N(\xi_1,\ldots,\xi_N) = \sum_{N_1,N_2,\ldots;\Sigma kN_k=N} (\prod_k N_k!)^{-1} \mathbb{S} \prod_{\alpha\in\underline{\pi}} \sigma^1_{|\alpha|}(\xi_\alpha)\cdot \cdot\exp(-\frac{\beta}{2}\sum_{\alpha,\beta} v^1(\alpha,\beta)). \quad (2.18)$$

This expression is similar in structure to (2.1), only here there are more kinds of particles and the particles have more structure in their activities $\sigma^1_{|\alpha|}(\xi_\alpha)$. Thus it is possible to formulate these expansions inductively (see [5]). This would be especially useful if v had been decomposed into many pieces. Since we have only one expansion more to do, we will simply write it with equations (2.10) - (2.16) as a guide.

Interpolating interactions are defined as before by taking convex sums. Letting $\underline{\pi}$ denote a collection of 1-vertices $\{\alpha_1,\ldots,\alpha_M\}$ we define

$$V^1(\underline{\pi}) = \frac{1}{2}\sum_{\alpha,\beta\in\underline{\pi}} v^1(\alpha,\beta) \tag{2.19}$$

$$V^1_{s_1\ldots s_n}(\alpha_{i_1},\ldots,\alpha_{i_n};\underline{\pi}) = V^1_{s_1\ldots s_{n-1}}(\alpha_{i_1},\ldots,\alpha_{i_{n-1}};\{\alpha_{i_n}\}\cup\underline{\pi})s_n + $$
$$+ (V^1_{s_1\ldots s_{n-1}}(\alpha_{i_1},\ldots,\alpha_{i_{n-1}};\{\alpha_{i_n}\}) + V^1(\underline{\pi}))(1 - s_n). \tag{2.20}$$

Then the differentiated interaction is

$$\frac{d}{ds_n} V^1_{s_1\ldots s_n}(\alpha_{i_1},\ldots,\alpha_{i_n};\underline{\pi})$$
$$= \sum_{i_{n+1}:\alpha_{i_{n+1}}\in\underline{\pi}} \sum_{\eta(n+1)=1}^{n} s_{\eta(n+1)}\cdots s_{n-1} v^1(\alpha_{i_{n+1}},\alpha_{i_{\eta(n+1)}}). \tag{2.21}$$

With

$$Q^M_1(\underline{\pi}) = \prod_{\alpha\in\underline{\pi}} \sigma^1_{|\alpha|}(\xi_\alpha)\exp(-\beta V^1(\underline{\pi})), \tag{2.22}$$

The first interpolation yields

$$Q^M_1(\underline{\pi}) = \sigma^1_{|\alpha_1|}(\xi_{\alpha_1})Q^{M-1}_1(\underline{\pi}\setminus\{\alpha_1\})$$
$$+ \sum_{j\neq 1}\int ds_1 \prod_{\alpha\in\underline{\pi}} \sigma^1_{|\alpha|}(\xi_\alpha)(-\beta)v^1(\alpha_1,\alpha_j)\exp(-\beta V^1_{s_1}(\alpha_1;\underline{\pi}\setminus\{\alpha_1\})), \tag{2.23}$$

and in general

$$Q_1^M(\underline{\pi}) = \sum_{k=1}^{M} \sum_{(i_2,\ldots,i_k)} \prod_{\nu=1}^{k} \sigma^1_{|\alpha_{i_\nu}|}(\xi_{\alpha_{i_\nu}})(-\beta)^{k-1}\int ds_1\ldots ds_{k-1} \sum_{\eta}$$

$$\prod_{\ell=2}^{k} [s_{\eta(\ell)}\cdots s_{\ell-2} v^1(\alpha_{i_{\eta(\ell)}},\alpha_{i_\ell})]\cdot$$

$$\cdot\exp(-\beta V^1_{s_1\ldots s_{k-1}}(\alpha_{i_1},\ldots,\alpha_{i_{k-1}};\{\alpha_{i_k}\}))Q_1^{M-k}(\underline{\pi}^c). \qquad (2.24)$$

Just as a 1-vertex α is a collection of particles or 0-vertices, we define a 2-vertex $\underline{\alpha}$ to be a collection of 1-vertices $\{\alpha_1,\ldots,\alpha_k\}$, and give it coordinates $\xi_{\underline{\alpha}} = (\xi_i)_{i\in\alpha_j\in\underline{\alpha}}$. Let $N_t^{\underline{\alpha}}$ be the number of 1-vertices in $\underline{\alpha}$ of size t. Then we define 2-vertex functions

$$\sigma^2_{\underline{\alpha}}(\xi_{\underline{\alpha}}) = k!(\prod_t N_t^{\underline{\alpha}}!)^{-1}\,\hat{\sigma}^2_{\underline{\alpha}}(\xi_{\underline{\alpha}})$$

$$\hat{\sigma}^2_{\underline{\alpha}}(\xi_{\underline{\alpha}}) = \mathbb{S} \prod_{j=1}^{k} \sigma^1_{|\alpha_j|}(\xi_{\alpha_j}) \frac{(-\beta)^{k-1}}{k}\int ds_1\ldots ds_{k-1} \sum_{\eta}$$

$$\prod_{\ell=2}^{k} [s_{\eta(\ell)}\cdots s_{\ell-2} v^1(\alpha_{\eta(\ell)},\alpha_\ell)]\cdot$$

$$\cdot\exp(-\beta V^1_{s_1\ldots s_{k-1}}(\alpha_1,\ldots,\alpha_{k-1};\{\alpha_k\})), \qquad (2.25)$$

where $\mathbb{S}$ symmetrizes in $\alpha_1,\ldots,\alpha_k$. In terms of these, (2.24) becomes

$$Q_1^M(\underline{\pi}) = \sum_{k=1}^{M} \sum_{\underline{\alpha}\subseteq\underline{\pi},\alpha_1\in\underline{\alpha},|\underline{\alpha}|=k} k!\hat{\sigma}^2_{\underline{\alpha}}(\xi_{\underline{\alpha}})Q_1^{M-k}(\underline{\pi}\setminus\underline{\alpha}). \qquad (2.26)$$

Just as in (2.15) we insert (2.26) into itself to obtain

$$Q_1^M(\underline{\pi}) = \sum_{p\in P(\underline{\pi})} \prod_{\underline{\alpha}\in p} [|\underline{\alpha}|!\hat{\sigma}^2_{\underline{\alpha}}(\xi_{\underline{\alpha}})]. \qquad (2.27)$$

This expansion can be inserted into (2.18) to yield

$$Z^N(\xi_1,\ldots,\xi_N) = \sum_{N_1,N_2,\ldots;\Sigma kN_k=N} (\prod_k N_k!)^{-1}\,\mathbb{S} \sum_{p\in P(\underline{\pi})} \prod_{\underline{\alpha}\in p} [|\underline{\alpha}|!\hat{\sigma}^2_{\underline{\alpha}}(\xi_{\underline{\alpha}})]. \qquad (2.28)$$

As before, we would rather sum over multiplicities of types of 2-vertices. The type $[\underline{\alpha}]$ of a 2-vertex is just the set of multiplicities $(N_k^{\underline{\alpha}})_{k=1,2,\ldots}$ of 1-vertices of each size in $\underline{\alpha}$. If there are $N_{[\underline{\alpha}]}$ vertices of type $[\underline{\alpha}]$, let p be a corresponding partition in (2.28).

A combinatoric analysis as in the first expansion shows that there are $\prod_k N_k!(\prod_{[\underline{\alpha}]} N_{[\underline{\alpha}]}!)^{-1}(\prod_{\underline{\alpha}\in p}\prod_k N_k^{\underline{\alpha}}!)^{-1}$ terms in (2.28) corresponding to $\{N_{[\underline{\alpha}]}\}$. The third group of factorials and each $|\underline{\alpha}|!$ in (2.28) converts $\hat{\sigma}^2$ to σ^2, and we obtain

$$Z^N(\xi_1,\ldots,\xi_N) = \sum_{\{N_{[\underline{\alpha}]}\},\ \sum_{[\underline{\alpha}]} N_{[\underline{\alpha}]}\sum_k kN_k^{\underline{\alpha}}=N} (\prod_{[\underline{\alpha}]} N_{[\underline{\alpha}]}!)^{-1} \mathbb{S} \prod_{\underline{\alpha}\in p} \sigma^2_{\underline{\alpha}}(\xi_{\underline{\alpha}}). \quad (2.29)$$

We can now obtain the full partition function by summing over N and integrating over $\xi_1,\ldots,\xi_N$. Symmetrization is no longer necessary, and furthermore the sums over $N_{[\underline{\alpha}]}$ factor into a product of sums

$$\sum_{N_{[\underline{\alpha}]}=0}^{\infty} (N_{[\underline{\alpha}]}!)^{-1} \int d\xi_{\underline{\alpha}}\sigma^2_{\underline{\alpha}}(\xi_{\underline{\alpha}}) = \exp(\int d\xi_{\underline{\alpha}}\sigma^2_{\underline{\alpha}}(\xi_{\underline{\alpha}})).$$

The expansion reaches its final form:

$$Z = \sum_{N=0}^{\infty} \int d\xi_1\ldots d\xi_N Z^N(\xi_1,\ldots,\xi_N) = \exp(\sum_{[\underline{\alpha}]} \int d\xi_{\underline{\alpha}}\sigma^2_{\underline{\alpha}}(\xi_{\underline{\alpha}})). \quad (2.30)$$

We have achieved our goal of writing the partition function as an ensemble of noninteracting 2-vertices. Equation (2.30) also gives us an expansion for the pressure:

$$\lim_{\Lambda\to\mathbb{R}^3} |\Lambda|^{-1}\log Z = \sum_{[\underline{\alpha}]} \int_{x_1=0} d\xi_{\underline{\alpha}}\sigma^2_{\underline{\alpha}}(\xi_{\underline{\alpha}}), \quad (2.31)$$

where the x-integrals on the right extend over $\mathbb{R}^3$. Of course the validity and usefulness of (2.30), (2.31) depend on some convergence estimates; these will be discussed in the next section.

3. Convergence of the Expansion

We estimate the vertex functions σ^ℓ in (2.13) and (2.25). The restrictions derived in Sect. 1,

$$z\beta^3 e^{c\beta/R} \ll 1, \quad \lambda \ll 1, \quad (3.1)$$

will guarantee convergence of the first and second expansions, respectively. We claim that

$$\int \prod_{i\in\alpha, i\neq i_0} d\xi_i |\sigma_k^1(\xi_\alpha)| \leq z(z\beta^3 e^{c\beta/R})^{k-1}. \tag{3.2}$$

The definition of $V^0_{s_1 \ldots s_{k-1}}$ through successive convex combinations (2.10) preserves the stability estimate derived in Sect. 1. Thus

$$\exp(-\beta V^0_{s_1 \ldots s_{k-1}}(1,\ldots,k-1;\{k\})) \leq e^{ck\beta/R}. \tag{3.3}$$

The ξ-integrals are handled using (1.11); we integrate out sequentially the positions of extremal vertices of the tree η, leaving in the end only the i_0 vertex, which is fixed. Each integration produces a factor β^2. The remaining sums and integrals in (2.13) are handled using

$$\int ds_1 \ldots ds_{k-1} \sum_\eta \prod_{\ell=2}^{k} [s_{\eta(\ell)} \ldots s_{\ell-2}] \leq e^{k-1}. \tag{3.4}$$

This is a standard estimate that appears in many kinds of cluster expansions. It is quite important as it controls the sum over the $(k-1)!$ trees η. We prove it as a special case of (3.7) below. Putting all these estimates together, we obtain (3.2). Note that when $k = 1$ we have $\sigma_k^1(\xi_\alpha) = \pm z$.

The estimate for σ^2 proceeds similarly, and we can see an inductive structure emerging. Let $\underline{\alpha} = \{\alpha_1,\ldots,\alpha_k\}$ with $\sum_i |\alpha_i| = t$. We prove that

$$\int \prod_{i\neq i_0} d\xi_i |\hat{\sigma}^2_{\underline{\alpha}}(\xi_{\underline{\alpha}})| \leq z(cz\beta^3 e^{c\beta/R})^{t-k}(c\lambda^2)^{k-1}. \tag{3.5}$$

Again the stability estimate

$$\exp(-\beta V^1_{s_1 \ldots s_{k-1}}(\alpha_1,\ldots,\alpha_{k-1};\{\alpha_k\})) \leq c^k \tag{3.6}$$

is preserved in the interpolation process. If we expand out the sums over particles in $\alpha_{\eta(\ell)},\alpha_\ell$ in $v^1(\alpha_{\eta(\ell)},\alpha_\ell)$, then (2.25) is represented as a sum of $\prod_{\ell=2}^{k} [|\alpha_{\eta(\ell)}||\alpha_\ell|]$ terms. In each term, and for each $\ell = 2,\ldots,k$, a particle in α_ℓ is connected to a particle in $\alpha_{\eta(\ell)}$ through a line $v^1(\xi_i,\xi_j)$. We use (3.2) to integrate over the coordinates of all but one particle in each α_ℓ and we use (1.12) to integrate out the remaining particle. Extremal 1-vertices are integrated out first. Estimate (3.4) generalizes to

$$\int ds_1 \ldots ds_{k-1} \sum_{\eta} \prod_{\ell=2}^{k} [s_{\eta(\ell)} \cdots s_{\ell-2} |\alpha_{\eta(\ell)}|] \le \exp\Big(\sum_{\ell=1}^{k-1} |\alpha_\ell|\Big), \tag{3.7}$$

and the remaining $|\alpha_\ell|$ factors are controlled using

$$\prod_{\ell=2}^{k} |\alpha_\ell| \exp\Big(\sum_{\mu=1}^{k-1} |\alpha_\mu|\Big) \le \exp\Big(2 \sum_{\ell=1}^{k} |\alpha_\ell| - 2\Big) = e^{2(t-1)}. \tag{3.8}$$

Altogether we have factors $(cz\beta^3 e^{c\beta/R})^{|\alpha_\ell|-1}$ at each α_ℓ and factors $cz\beta\lambda^2 \ell_D^2 = c\lambda^2$ for each v^1-line. This proves (3.5).

To prove (3.7) write the sum over η as $k - 1$ sums over $\eta(j) \in [1, j - 1]$ for $2 \le j \le k$, and notice that the left-hand side is less than

$$\int ds_1 \ldots ds_{k-1} \sum_{\eta(2)\ldots\eta(k)} \prod_{\ell=2}^{k} [s_{\eta(\ell)} \cdots s_{\ell-2} |\alpha_{\eta(\ell)}|] \cdot$$

$$\cdot \exp\Big(\sum_{\mu=1}^{k-1} s_\mu \cdots s_{k-1} |\alpha_\mu|\Big)$$

$$= \int ds_1 \ldots ds_{k-1} \sum_{\eta(2)\ldots\eta(k-1)} \prod_{\ell=2}^{k-1} [s_{\eta(\ell)} \cdots s_{\ell-2} |\alpha_{\eta(\ell)}|] \frac{d}{ds_{k-1}} \cdot$$

$$\cdot \exp\Big(\sum_{\mu=1}^{k-1} s_\mu \cdots s_{k-1} |\alpha_\mu|\Big)$$

$$\le \int ds_1 \ldots ds_{k-2} \sum_{\eta(2)\ldots\eta(k-1)} \prod_{\ell=2}^{k-1} [s_{\eta(\ell)} \cdots s_{\ell-2} |\alpha_{\eta(\ell)}|] \cdot$$

$$\cdot \exp\Big(\sum_{\mu=1}^{k-2} s_\mu \cdots s_{k-2} |\alpha_\mu| + |\alpha_{k-1}|\Big). \tag{3.9}$$

These two steps can be repeated for each s_ℓ-integration, and in the end we obtain the right-hand side of (3.7).

We now use (3.5) to control the full expansion (2.31) for the pressure. We have

$$\sum_{[\underline{\alpha}]} \int_{x_1=0} d\xi_{\underline{\alpha}} |\sigma^2_{\underline{\alpha}}(\xi_{\underline{\alpha}})| \le 2 \sum_{N_1^{\underline{\alpha}}, N_2^{\underline{\alpha}}, \ldots} z (cz\beta^3 e^{c\beta/R})^{t-k} (c\lambda^2)^{k-1}, \tag{3.10}$$

where the 2 accounts for the sum over e_1, and where

$t = \sum_j jN_j^{\alpha}$, $k = \sum_j N_j^{\alpha} = |\underline{\alpha}|$. Let us write $y = cz\beta^3 e^{c\beta/R}$, $w = c\lambda^2$; then

$$\sum_{N_1^{\alpha}, N_2^{\alpha}, \ldots} zy^{t-k} w^{k-1} = z \sum_{k=1}^{\infty} w^{k-1} \left(\sum_{m=1}^{\infty} y^{m-1} \right)^k$$

$$= zw^{-1} \sum_{k=1}^{\infty} \left(\frac{w}{1 - y}\right)^k = \frac{z}{1 - y - w}. \tag{3.11}$$

Thus the series for the pressure is absolutely convergent and bounded by $2z(1 - cz\beta^3 e^{c\beta/R} - c\lambda^2)^{-1}$, provided the two expansion parameters y and w are small. We can see that after many iterated expansions we would lose control of the sums; if $\lambda_0, \lambda_1, \ldots, \lambda_R$ were the expansion parameters, then a bound of the form $2z(1 - \lambda_0 - \lambda_1 \cdots - \lambda_R)^{-1}$ would result. The λ_ℓ would also have a tendency to grow beyond their naive values $\lambda_\ell = z\beta \int |v^\ell(x)| dx e^{-\beta E^\ell}$ because of the factors arising from estimates like (3.7),(3.8).

4. Coulomb Gas Applications

In this section we describe how iterated Mayer expansions can be used in studying screening in the three dimensional Coulomb gas [6,7]. Let us consider charges ± 1 again, with a soft-core Coulomb potential

$$v_c(\xi_i, \xi_j) = e_i \left[\frac{1 - e^{-|x_i - x_j|/R}}{4\pi |x_i - x_j|}\right] e_j. \tag{4.1}$$

At activity z and inverse temperature β, the partition function is given by (1.4) with v_c replacing v. To see screening, we wish to perform a sine-Gordon transformation whereby the system becomes a scalar field theory with cosine interaction. The curvature at the minimum of the cosine should act like a mass and lead to exponential decay of correlation functions, or screening. The problem with going directly to the sine-Gordon form is that the short distance cutoff length R is too small to control the analysis of the sine-Gordon theory, particularly if we are interested in the region where the expected screening length $\ell_D = (2z\beta)^{-1/2}$ is much larger then R.

The solution to this problem lies in the use of a mixed gas-sine-Gordon representation. The Mayer expansion of Sects. 1 - 3 is well suited to the short distance analysis, but reaches its limit at a length scale $\lambda\ell_D$. Constructive field theory techniques applied to the sine-Gordon representation work well at length scales above $\lambda\ell_D$, but break down if λ is too small. Thus the two approaches complement each other. The splitting of the problem according to length scale fits well with the renormalization group philosophy that one should never attempt to treat at one time a greater range of lengths than can easily be accomodated in a given procedure.

The splitting we take consists in writing

$$v_c(\xi_i,\xi_j) = v(\xi_i,\xi_j) + v^2(\xi_i,\xi_j), \tag{4.2}$$

where $v = v^0 + v^1$ is the interaction considered in Sects. 1 - 3 and

$$v^2(\xi_i,\xi_j) = e_i\left[\frac{1 - e^{-|x_i-x_j|/(\lambda\ell_D)}}{4\pi|x_i - x_j|}\right]e_j. \tag{4.3}$$

Let $\rho(x) = \sum_i e_i\delta(x - x_i)$ be the charge density. The interaction can be written as

$$V = \sum_{i<j} v_c(\xi_i,\xi_j) = \frac{1}{2}\int dxdy\rho(x)u(x,y)\rho(y) - \sum_i(8\pi\lambda\ell_D)^{-1} + \sum_{i<j} v(\xi_i,\xi_j), \tag{4.4}$$

where

$$u(x,y) = \frac{1 - e^{-|x-y|/(\lambda\ell_D)}}{4\pi|x - y|} = [(-\Delta)^{-1} - (-\Delta + \lambda^{-2}\ell_D^{-2})^{-1}](x,y). \tag{4.5}$$

The second term on the right-hand side of (4.4) cancels the self-energies that were included in the first term. It merely changes the effective activity to $\tilde{z} = ze^{\beta/(8\pi\lambda\ell_D)}$, and for $(\beta/\ell_D)^2 = 2z\beta^3$ small this is only a slight modification. The sine-Gordon transformation treats the first term by using the identity

$$e^{-\beta\langle\rho,u\rho\rangle/2} = \int e^{i\beta^{1/2}\langle\phi,\rho\rangle}d\mu_u(\phi), \tag{4.6}$$

where $d\mu_u(\phi)$ is the Gaussian measure with covariance u. We see that every particle i has a phase factor $\exp(i\beta^{1/2}e_i\phi(x_i))$ associated to it. The third term in (4.4) is the interaction we expanded in (2.30). The same expansion applies here, only the phase factors must be carried along. Thus the Coulomb gas partition function can be represented as

$$Z = \int \exp\Big(\sum_{[\underline{\alpha}]} \int d\xi_{\underline{\alpha}} \sigma^2_{\underline{\alpha}}(\xi_{\underline{\alpha}}) \prod_{i\in\alpha\in\underline{\alpha}} e^{i\beta^{1/2} e_i \phi(x_i)}\Big) d\mu_u(\phi). \tag{4.7}$$

(Actually to proceed with the screening proof we modify u near the boundary by replacing Δ in (4.5) with Δ_V, the Laplacian with Dirichlet boundary conditions at ∂V, $V \subset \Lambda$.)

A final transformation of (4.7) writes $e^{i\beta^{1/2} e_i \phi(x_i)} = 1 + \varepsilon(\xi_i)$ and expands the resulting product. This yields

$$Z = \int \exp\Big(\rho_0 + \sum_{s=1}^{\infty} \frac{1}{s!} \int d\xi_1 \ldots d\xi_s \rho_s(\xi_1,\ldots,\xi_s)\varepsilon(\xi_1)\ldots\varepsilon(\xi_s)\Big) d\mu_u(\phi), \tag{4.8}$$

where

$$\rho_s(\xi_1,\ldots,\xi_s) = s! \sum_{t\geq s} \binom{t}{s} \int d\xi_{s+1} \ldots d\xi_t \sum_{[\underline{\alpha}]:\sum_k kN_k^{\underline{\alpha}}=t} \sigma^2_{\underline{\alpha}}(\xi_{\underline{\alpha}}) \tag{4.9}$$

is the s-point truncated correlation function of the Yukawa gas. To understand this action, look at the leading term $s = 1$ in (4.8) and $t = s$ in (4.9). It is

$$\int dx\, \tilde{z}(e^{i\beta^{1/2}\phi(x)} + e^{-i\beta^{1/2}\phi(x)} - 2) = \int dx\, 2\tilde{z}(\cos \beta^{1/2}\phi(x) - 1). \tag{4.10}$$

Thus we have a generalized sine-Gordon theory, with interaction $2\tilde{z} \cos \beta^{1/2}\phi$ plus higher order nonlocal corrections. The nonlocal corrections are the price we pay for having a good short distance cutoff (at length $\lambda\ell_D$). They are well controlled by estimates like the ones in Sect. 3. A strong form of exponential decay of the functions $\rho_s(\xi_1,\ldots,\xi_s)$ results, see [3,7]. The usefulness of the Mayer expansion is now clear. It expressed the partition function of the Yukawa gas in exponential form, with exponentially localized vertex functions. These properties are basic to the subsequent sine-Gordon analysis.

If we expand the cosine about $\phi = 0$ we obtain

$$2\tilde{z}(\cos \beta^{1/2}\phi - 1) = -\tilde{z}\beta\phi^2 + \frac{1}{12}\tilde{z}\beta^2\phi^4 + \ldots . \tag{4.11}$$

The leading term adds $\tilde{\ell}_D^{-2} = 2\tilde{z}\beta \approx \ell_D^{-2}$ to the inverse covariance of $d\mu_u(\phi)$. The new covariance is

$$(u^{-1} + \tilde{\ell}_D^{-2})^{-1}(x,y) = (\lambda^2\ell_D^2(-\Delta_V)^2 - \Delta_V + \tilde{\ell}_D^{-2})^{-1}(x,y), \tag{4.12}$$

which decays exponentially like $e^{-|x-y|/\tilde{\ell}_D}$ and is bounded by $c\lambda^{-1}$ on the diagonal. We can begin to see how the screening arises. To see what is necessary to control the corrections to the Gaussian, choose units where $\tilde{\ell}_D = 1$. Then the quartic term in (4.11) is proportional to β. Thus we need $(\beta/\ell_D)^2 = 2z\beta^3$ small -- it turns out that $z\beta^3 \ll e^{-c/\lambda}$ is small enough. Note that with no Mayer expansion we would have $\lambda = R/\ell_D$ and thus we would need both $z\beta^3$ and β/R small to satisfy $z\beta^3 \ll \exp(-c(z\beta^3)^{-1/2}\beta/R)$.

Space does not allow a discussion of the expansion of this sine-Gordon model about the massive Gaussian; the reader should consult [1,3,4,7]. However we can see what conditions are needed on z,β,R to obtain convergent expansions. The iterated Mayer expansion of Sects. 1 - 3 required $z\beta^3 \ll e^{-c\beta/R}$, $\lambda \ll 1$. The cluster expansion for the sine-Gordon model requires $z\beta^3 < e^{-c/\lambda}$, $\lambda \ll 1$. Altogether $z\beta^3 \ll e^{-c\beta/R}$ is sufficient to prove screening. This is a considerable improvement over the region obtained with a single Mayer expansion with the requirement $\lambda^2 \ll e^{-c\beta/R}$. The β/R-dependent restriction on λ entails that $z\beta^3 \ll \exp(-ce^{c\beta/R})$, which is much worse for $\beta \gg R$.

An even more striking improvement in screening regions was made by Göpfert and Mack [6]. They studied the lattice Coulomb gas dual to the three dimensional U(1) lattice gauge theory. In this model the activity is linked to β according to $z = e^{-\beta v_c(0)/2}$, where v_c is now the lattice Coulomb potential. [In other words, the self-energy terms with $i = j$ are included in V whereas we have omitted them -- see (4.4).] The lattice spacing plays the role of our short distance cutoff R, and we set both to 1 here by a choice of units. Using very precise stability estimates in the first Mayer expansion, they were able to sharpen the requirement $z\beta^3 \ll e^{-c\beta}$ to one like $z\beta^3 \ll e^{-(1-\varepsilon)\beta v_c(0)/2}$ with $\varepsilon > 0$. Thus with z as above, the condition on β becomes $\beta^3 \ll e^{\varepsilon\beta v_c(0)/2}$, and screening can be proven for large β. The expansion also proves confinement in the U(1) model at large β, and confinement then follows at all temperatures by correlation inequalities.

Acknowledgements

The author is a Junior Fellow in the Harvard University Society of Fellows, and is currently at the Department of Physics, Harvard University, Cambridge MA 02138, USA. This work was supported in part by the National Science Foundation under Grant No. PHY79-16812.

References

1. Brydges, D.: A rigorous approach to Debye screening in dilute classical Coulomb systems. Commun. Math. Phys. 58, 313-350 (1978).
2. Brydges, D., Federbush, P.: A new form of the Mayer expansion in classical statistical mechanics. J. Math. Phys. 19, 2064-2067 (1978).
3. Brydges, D., Federbush, P.: Debye Screening. Commun. Math. Phys. 73, 197-246 (1980).
4. Brydges, D., Federbush, P.: Debye screening in classical Coulomb systems. In: Velo, G., Wightman, A. (eds.): Rigorous atomic and molecular physics--Erice, 1980. New York: Plenum 1981.
5. Göpfert, M. and Mack, G.: Iterated Mayer expansion for classical gases at low temperatures. Commun. Math. Phys. 81, 97-126 (1981).
6. Göpfert, M. and Mack, G.: Proof of confinement of static quarks in 3-dimensional U(1) lattice gauge theory for all values of the coupling constant. Commun. Math. Phys. 82, 545-606 (1982).
7. Imbrie, J.Z.: Debye screening for jellium and other Coulomb systems. Commun. Math. Phys., to appear.

RIGOROUS RESULTS ON THE CRITICAL BEHAVIOR IN STATISTICAL MECHANICS

Michael Aizenman[1]

Summary. We review a collection of rigorous results, which delineate the usefulness of mean field type approximations, and shed light on the phenomenon of the existence of an upper critical dimension -- above which the critical behavior is considerably simpler. The results are relevant for the analysis of ferromagnetic spin systems, percolation models, and quantum field theory.

1. Introduction

An early success of statistical-mechanics has been the provision of a solid framework which encompasses both the laws of mechanics and the basic principles of thermodynamics. Among the spectacular results of this approach are the explanations, from basic principles, of the various phase transitions which are observed in nature. A not less challenging goal is to reach a complete explanation of the critical behavior of bulk systems. The well known universality, which has been observed experimentally, is an intrinsically significant effect. It both calls for -- and raises the possibility of a mathematical elucidation of the subject.

Great advances towards the formation of a global picture, and the approximate calculation of critical exponents, have been made by renormalization-group related methods. By their nature, these methods do not offer exact solutions to any specific statistical-mechanical model. (Although they do come with sharp predictions above, and at, the upper-critical dimension.) Instead, the analysis incorporates the notion of universality, which from the point of view of physics may indeed be better founded than any given model. Despite, or rather **because of** this, we still regard the problem of the critical behavior as well worth further attention. More so -- in view of the presently unsatisfactory level of explanation of the critical exponents in low dimensions (as attested to by recent discussions of hyperscaling in $d = 3$ dimensions).

In this article we review a collection of rigorous results, which delineate the usefulness of some approximations, and shed light on the basic phenomenon of an upper critical dimension, above which the critical behavior is considerably simpler. These results demonstrate that the possibility of a rigorous analysis of the critical behavior should not be given up, even in the face of the lack of exact solutions.

The titles of Sections 2 through 5 should provide a clue to the way this article is organized.

Let me insert here a note of gratitude. During the work on these, and related subjects, I have had a number of times the pleasure of being a guest at the I.H.E.S.. Invariably, these were very rewarding visits. I wish to congratulate the Institute for its conducive and stimulating atmosphere, and thank Professor N. Kuiper, as well as J. Fröhlich, D. Ruelle, and the other members, for these experiences.

2. The Mean-Field Approximation for Ferromagnetic Systems

One of the simplest methods of deriving some idea about a phase transition is to apply a "mean-field" approximation. However, a brief look at the procedure may easily produce strong doubts about the reliability of its predictions. Nevertheless, it turns out that there often is a finite upper-critical dimension above which some of the predictions, which are easily obtained this way, are exactly right. In fact, even in lower dimensions some of the predictions are still more useful than one could expect them to be -- since they provide rigorous bounds. We shall see results of this gender for Ising type systems, and percolation models. First, however, let us introduce this approximation.

Standard models of ferromagnets consist of lattice arrays of spins, $\sigma_x, x \in Z^d$, which interact via the Hamiltonian

$$H = -\frac{1}{2} \sum_{x,y} J_{x-y} \sigma_x \sigma_y - h \sum \sigma_x , \quad \text{with } J_z \geq 0 . \tag{2.1}$$

For Ising spins $\sigma_x = \pm 1$. We shall refer also to other (one component) spin variables, described by a probability measure $\rho_0(d\sigma)$ -- assuming always that $\int e^{\mu\sigma} \rho_o(d\sigma) < \infty$ for all $|\mu| < \infty$.

Let $M(\beta,h)$ be the magnetization, i.e. the expectation value

$$M = \langle\sigma_o\rangle \qquad , \tag{2.2}$$

in the Gibbs state at the inverse temperature β and the magnetic field h. In $d > 1$ dimensions, such systems exhibit a phase transition, which is manifested by a discontinuity of $M(\beta,h)$ accross part of the $h = 0$ line in the (β,h) plane. Quantitative characteristics of the phase transition include:

i) the critical value $\beta_c (= 1/T_c)$, such that for $\beta > \beta_c$ the residual magnetization, $M(\beta,0+)$, does not vanish,

ii) critical exponents, an example of which is γ - defined by the expected power law

$$\chi \cong t^{-\gamma} \tag{2.3}$$

where $\chi = \frac{1}{\beta}\frac{\partial M(\beta,h)}{\partial h}$, and $t = (\beta_c - \beta)/\beta_c$.

We shall now describe a derivation of these quantities within a mean-field approximation.

The interaction of σ_o with the other spins results in the conditional distribution:

$$\text{Prob.}(d\sigma_o \mid \{\sigma_x\}_{x\neq 0}) = e^{\sigma_o \beta[\sum_{x\neq o} J_x \sigma_x + h]} \rho_o(d\sigma)/\text{Norm.} \tag{2.4}$$

(an example of the Dobrushin-Lanford-Ruelle equation). Performing the conditional average we find the following <u>exact</u> relation

$$\langle\sigma_o\rangle = \langle f(\beta \sum_{x\neq o} J_x \sigma_x + \beta h)\rangle \qquad , \tag{2.5}$$

where

$$f(\mu) = \int \sigma e^{\sigma\mu} \rho_o(d\sigma) / \int e^{\sigma\mu} \rho_o(\sigma) . \tag{2.6}$$

For example, in the Ising case (with $\rho_o(d\sigma) = \delta(\sigma^2-1)d\sigma$) :

$$f(\mu) = \tanh(\mu) \qquad . \tag{2.7}$$

Replacing the sum over spins, in the argument of f, by a sum of their averages - one obtains the mean-field equation for $M = M(\beta,h)$:

$$M = f(\beta\,|J|\,M + \beta h) \tag{2.8}$$

with $|J| = \Sigma_{x\neq o} J_x$. Needless to say, (2.8) is no longer an exact relation. The law of large numbers suggests that as an approximation it should be better for systems in which σ_o is coupled to a large number of spins.

The solution of (2.8) is obtained by studying the intersection of two graphs. Let $\mathfrak{M}$ be the class of even spin distributions for which $f(\mu)$ is concave in $[0,\infty)$*. For spins in this class, which includes the Ising case, there is a non-vanishing solution at $h = 0$ if and only if

$$\beta\,|J|\,f'(0) > 1 \tag{2.9}$$

i.e.

$$\beta > \beta_c^{M.F.} \equiv [|J| <\sigma^2>_o]^{-1} , \tag{2.10}$$

where $<\sigma^2>_o \equiv \int \sigma^2\, \rho(d\sigma) = f'(0)$. (Actually, the above conclusion holds for an even more general class of functions f.) A simple analysis of thermodynamic stability indicates that the positive solution of (2.8) is the one which corresponds to $M(\beta,0^+)$.

Differentiating (2.8) with respect to h one obtains, for $h = 0$ and $\beta < \beta_c$ (which here is $\beta_c^{M.F.}$)

$$\chi = (\beta\,|J|\,\chi + 1)\, f'(0) ,$$

or

$$\chi = \frac{1}{|J|(\beta_c-\beta)} = \frac{t^{-1}}{|J|\,\beta_c} \tag{2.11}$$

with $t = (\beta_c-\beta)/\beta_c$.

---

* The class $\mathfrak{M}$ includes the measures $\rho(d\sigma) = e^{-V(\sigma)}d\sigma$, with an even function $V \in C^1(\mathbb{R})$ for which V' is convex on $(0,\infty)$; as well as limits of such measures [10].

Therefore, the mean-field approximation leads to the following "predictions":

i) β_c has, approximately, the value given in (2.10),

ii) the critical exponent γ_+ has the value $\gamma = 1$, regardless of any other "finer" details of the coupling (like the dimension d, and the interaction's range).

Of course, one can calculate also other critical exponents from (2.8).

It should be mentioned here that the results derived above are reproduced by the Landau description on a second order phase transition. A more refined analysis is contained in the Ginzburg-Landau theory, which takes into account also fluctuation effects - at the level of a Gaussian approximation. A stability test suggests there that there may be a significant difference between $d > 4$ and $d < 4$ dimensions. Further details of these, and other more sophisticated methods, are contained in references [21,5].

3. Mean Field Bounds

In the mean-field approximation, β_c can be defined either by the onset of the residual magnetization (i.e. symmetry-breaking), or the divergence of χ. However, strictly speaking, there is no proof that the two critical points coincide. In fact, for systems with a continuous symmetry (e.g. two component spins) in two dimensions, they do not. We shall now adopt the latter, i.e. take

$$\beta_c = \sup \{\beta \mid \chi(\beta) < \infty\} \quad . \tag{3.1}$$

as our definition.

Differentiating the Gibbs state one obtains the following statistic-mechanical expression for χ and its derivative at $h = 0$ and $\beta < \beta_c$,

$$\chi = \sum_x \langle\sigma_o \sigma_x\rangle$$

$$\frac{\partial\chi}{\partial\beta} = \frac{1}{2} \sum_{x,u,v} J_{u,v} (\langle\sigma_o \sigma_x \sigma_u \sigma_v\rangle - \langle\sigma_o \sigma_x\rangle \langle\sigma_u \sigma_v\rangle)$$

$$\equiv \sum_{u,v} \langle\sigma_o \sigma_u\rangle J_{u,v} \langle\sigma_v \sigma_x\rangle + \frac{1}{2} \sum_{x,u,v} J_{u,v}\, u_4(0,x,u,v) \tag{3.2}$$

where u_4 is the function

$$u_4(x_1,x_2,x_3,x_4) = \langle\sigma_{x_1} \sigma_{x_2} \sigma_{x_3} \sigma_{x_4}\rangle -$$

$$-[\langle\sigma_{x_1}\sigma_{x_2}\rangle\langle\sigma_{x_3}\sigma_{x_4}\rangle + \langle\sigma_{x_1}\sigma_{x_3}\rangle\langle\sigma_{x_2}\sigma_{x_4}\rangle + \langle\sigma_{x_1}\sigma_{x_4}\rangle\langle\sigma_{x_2}\sigma_{x_3}\rangle] \tag{3.3}$$

For Ising spins, the Lebowitz inequality [18,20] states that, pointwise:

$$u_4(x_1,\ldots,x_4) \leq 0 \quad . \tag{3.4}$$

This inequality is valid also for other spin distributions. Among them is the class described in the footnote in Section 2.**

The Lebowitz inequality (3.4) has a number of consequences which are of interest for us. First - as the basis for one of the versions of Simon's inequality, it is useful in establishing a number of preliminary results. These include the proof of the exponential decay of correlations for $\beta < \beta_c$, of the actual divergence of $\chi(\to \infty)$ as $\beta \nearrow \beta_c$, and of the actual validity of (3.2).

Furthermore, substituting (3.4) in (3.2) and applying translation invariance one obtains:

$$\frac{\partial\chi}{\partial\beta} \leq |J|\, \chi^2 \quad ;$$

which is conveniently rewritten as follows

$$\left| \frac{\partial\chi^{-1}}{\partial\beta} \right| \leq |J| \; . \tag{3.5}$$

** By the multilinearity of (3.3), it is also satisfied by variables which can be represented as weighted sums of Ising spins, which are ferromagnetically coupled. This construction (with its limits) defines the Griffiths-Simon class [17], which is also included in $\mathfrak{M}$ of Section 2. It includes the classical spin with

$$\rho(d\sigma) = \sum_{-n}^{n} \delta(\sigma-k) \; .$$

The function $\chi^{-1}(\beta)$ is continuous and differentiable in the interval $[0,\beta_c]$, with the boundary values $<\sigma^2>_0^{-1}$ and 0. Integrating (3.5) both from 0 and β_c, we learn that

$$\frac{1}{|J|\ |\beta_c^{M.F.} - \beta|_+} \geq \chi \geq \frac{1}{|J|\ (\beta_c-\beta)} \equiv \frac{t^{-1}}{|J|\ \beta_c} \quad , \tag{3.6}$$

where $\beta_c^{M.F.}$ is taken from (2.10), $|\tau|_+ = \max(\tau,0)$, and $t = (\beta_c-\beta)/\beta_c$.

A comparison of (3.6) with (2.11) shows that the mean field law actually forms both an upper and a lower bound - depending on the interpretation of β_c in (2.11).

Corollary 3.1. In a ferromagnetic system where the Lebowitz inequality is satisfied, the critical parameters satisfy:

i) $\beta_c \geq \beta_c^{M.F.}\ [= (|J|\ <\sigma^2>_o)^{-1}]$ (3.7)

ii) $\gamma_+ \geq 1$ (3.8)

(γ_+ being defined by (2.3), for $t \geq 0$).

Remarks

1. The fact that the mean field value, $\beta_c^{M.F.}$, is a lower bound for the Curie point has by now been demonstrated in a variety of ways, and under various (rather general) assumptions. The first proofs were derived independently by Griffiths [16], via his "third inequality", and Fisher [11], who used a random walk expansion. Both arguments refer only to Ising spins, for which their bounds are even slightly better than the upper bound in (3.6). Implicit among Fisher's results is also the exponential decay of correlations. That Griffiths' inequality can be used for a similar end, has been demonstrated in reference [19], where a somewhat improved inequality is used, and in reference [24] - by a simpler argument. In a much greater generality, such results have been obtained in reference [7] for spins in the class $\mathfrak{m}$ - by means of the Dobrushin technique. Mean field bounds for multicomponent spins are presented if reference [9] , where Ward identities are combined

with correlation inequalities. Arguments based on the Lebowitz inequality appear in references [14,24,26].

2. It is also known that in a large class of systems, the mean field value of $M(\beta,h)$ is an upper bound for the actual magnetization (at $h \geq 0$) - references [28,26].

3. The mean field bound on the critical exponent γ, i.e. (3.8), is due to Glimm and Jaffe [14], who used the above argument.

4. Vindication of the Mean Field Approximation above The Upper Critical Dimension

There is an important difference between the two "predictions" mentioned in Section 2. The value of β_c is clearly model dependent. It should not be expected to be, and is not, exactly equal to $\beta_c^{M.F.}$. On the other hand, the critical exponents are expected to be universal. Indeed, the mean field result exhibits universality. (In fact too much of that - since it is false in low dimension.) It may sound remarkable that even though $\beta_c \neq \beta_c^{M.F.}$, the power law of χ as a function of $\beta-\beta_c$ could be predicted correctly by this approximation. That however is the case in the "high dimensions" - $d > 4$.

We shall now describe a rigorous proof of the last assertion for systems of Ising spins, and, more generally, variables in the Griffiths Simon class (defined in the footnote in Section 3). For such systems, the following inequality was derived in reference [2] .

$$\left| \frac{\partial \chi^{-1}}{\partial \beta} \right| \geq \frac{|J|}{1 + (\beta\, |J|)^2 \sum_x \langle \sigma_o \sigma_x \rangle^2} \tag{4.1}$$

(up to a factor which tends to 1 when $t \searrow 0$).

Clearly, (4.1) offers a converse bound to (3.5), from which it differs only by the presence of the (dressed) "bubble diagram" in the denominator.

The inequality (4.1) has a simple origin. In the geometric representation of reference [1] the truncated correlation, $\langle \sigma_o \sigma_x \sigma_u \sigma_v \rangle - \langle \sigma_o \sigma_x \rangle \langle \sigma_u \sigma_v \rangle$ of (3.2), can be represented as the increase in the partition function of a gas of random currents (which resembles a "sea" of Feynman diagrams) due to the insertion of two current lines, one from 0 to u and the other from v to x, taking into account only the configurations where the two corresponding current clusters do not intersect. The simpler inequality (3.5) corresponds to dropping the constraint of non-intersection. The complementary inequality (4.1) is a consequence of an upper bound on the probability of intersection - in the form of the number of sites which have some of the characteristics of a "first hit". A sketch of the key step is provided in figure 1. We would not go here beyond the above loose description of the proof of reference [2].

To draw conclusions from (4.1), further information is needed about the critical behavior of the "bubble diagram":

$$(\beta\,|J|)^2 \sum_x \langle\sigma_o\sigma_x\rangle^2 \;=\; \frac{(\beta\,|J|)^2}{(2\pi)^d}\int_{[-\pi,\pi]^d} dp\,|G(p)|^2 \quad , \tag{4.2}$$

expressed here in terms of the Fourier transform

$$G(p) \;=\; \sum_x e^{ipx}\langle\sigma_o\sigma_x\rangle \quad . \tag{4.3}$$

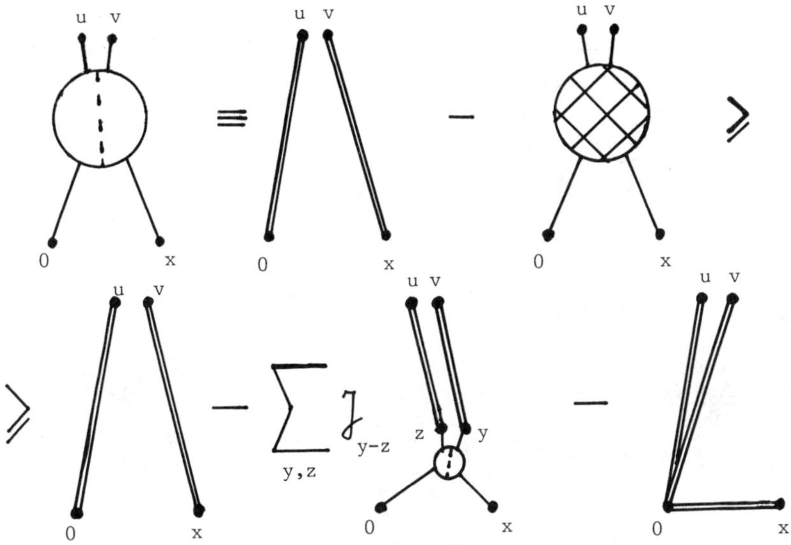

Figure 1 Schematic representation of a key step in the derivation of (4.1) (starting from (3.2)).

A solid line represents $\langle\sigma_x\sigma_y\rangle$. The blobs are the partially truncated, and the fully connected, four point functions. The rightmost term is negligible. To arrive at (4.1) one should sum - and isolate the term which appears on the left.

Under the further restriction to the nearest neighbor interaction (to ensure reflection positively) the above term can be controlled by the "Gaussian (equipartition) bound" of Fröhlich, Simon and Spencer [13]

$$G(p) \leq \frac{1}{\beta J 4 \sum_{i=1}^{d} \sin^2(p_i/2)} + c(\beta)\,\delta(p) \quad , \tag{4.4}$$

with $c(\beta) = 0$ for $\beta < \beta_c$.

The last inequality is a remarkably useful, yet tantalizingly delicate, result. With the Plancherel identity (4.2) it implies that .

$$(\beta\,|J|)^2 \sum_x \langle\sigma_o \sigma_x\rangle^2 \leq \begin{cases} c_d & d > 4 \\ \bar{c}[1 + \ln_+(\beta\chi)] & d = 4 \end{cases} \tag{4.5}$$

where

$$c_d = \frac{(2d)^2}{(2\pi)^d} \int_{[-\pi,\pi]^d} dp\, [4 \sum_{i=1}^{d} \sin^2(p_i(2)]^{-2} \quad (<\infty, \text{ for } d > 4).$$

Substituting (4.5) in (4.1) and integrating from β_c (with an extra argument for the dimension d = 4), we obtain

$$\left((|J|\ \beta_c)^{-1}\ t^{-1} \leq\right)\chi \leq \begin{cases} [(1+c_d)/|J|\ \beta_c]t^{-1} & d > 4 \\ c'\ t^{-1}\ (|\ln t| + 1) & d = 4 \end{cases} \tag{4.6}$$

(The lower bound is just (3.5).)

<u>Corollary 4.1</u>. For the nearest-neighbor model, with variables in the Griffiths-Simon class, in <u>d > 4 dimensions</u> the critical exponent γ_+ takes exactly its mean field value:

$$\gamma_+ = 1 \tag{4.7}$$

With more general interactions, (4.7) holds whenever the "bubble diagram" does not diverge as $\beta \nearrow \beta_c$.

In d = 4 dimensions, under similar assumptions, there can be only logarithmic corrections to the mean field law.

One can see a precursor of the above result in Sokal's [25] proof of the finiteness of the specific heat at $\beta_c - 0$, i.e. the equality $\alpha_+ = 0$, for d > 4 dimensions. There too, the "high dimensions" were characterized by the finiteness of the "bubble diagram" (with the full propagators) - which by just the Lebowitz inequality forms an upper bound on the specific heat.

Presumably, similar results hold also for the other critical exponents. For example, one might expect that

$$\eta = 0, \qquad \text{for} \quad d > 4 \quad , \tag{4.8}$$

which has not yet been proven. Some useful general results about η are: $\eta \geq 0$ (by (4.4), and $\eta \leq 1$ - by Simon's inequality [24], or the very general analysis of Dobrushin and Pecherski [8].

Remark. The equality (4.7) has been first proven in reference [1] strictly for d > 4. For sufficiently soft ϕ^4 fields (and d > 4) it has been also shown in reference [12]. The improved argument described here, which is of relevance also for d = 4, is of reference [2].

5. Scaling Limits in High and Low Dimensions

So far, we have discussed the critical behavior in terms of the critical exponents. The correlation functions entered the discussion merely as calculational tools. However, their structure is of considerable interest.

A characteristic feature of the critical regime is the divergence of the scale of distances over which the spins exhibit significant correlations. Rescaling the distances, and the spin magnitudes, let us denote

$$S_k(x_1,\ldots,x_k) = \lim_{\substack{\beta \nearrow \beta_c \\ (\alpha, \zeta \to \infty)}} \alpha^k \langle \sigma_{[x_1\zeta]} \cdots \sigma_{[x_k\zeta]} \rangle \quad . \tag{5.1}$$

The functions S_k are defined over the continuum, [y] being the lattice site which is closest to y. The variables ζ and α are adjusted with β.

The correlation length is usually defined by the maximal value of the coefficient ξ in the asymptotic bound:

$$\langle\sigma_o\sigma_x\rangle \leq \text{const. } e^{-|x|/\xi} \quad . \tag{5.2}$$

ξ is finite for $\beta < \beta_c$, and diverges as $\beta \nearrow \beta_c$ [22,24].

Two interesting limits in (5.1) are i) when $\eta = O(\xi)$ - in which case the scaling limit should also exhibit exponential decay, and ii) $\zeta << \xi$ - which should be equivalent to fixing $\beta = \beta_c$ and then letting $\zeta \to \infty$. These are refered to as scaling limits near, and at, the critical point. Since $\zeta \to \infty$, the functions described by (5.1) can be also regarded as the correlation functions of a field which describes local averages of σ_x - over regions which are large, yet small on the scale of ζ. A most interesting question is whether in the scaling limit the fluctuations of this field are Gaussian or not.

Gaussian fields are characterized by the Wick identities, which express the k-point functions in terms of just S_2 :

$$S_{2n}(x_1,\ldots,x_{2n}) = \sum_{\substack{\text{pairings}\\ \text{of } (1,\ldots,2n)}} S_2(x_{i_1},x_{j_1}) \ldots S_2(x_{i_n},x_{j_n}) \tag{5.3}$$

The answer to the above question is again dimension dependent. It can now be shown rigorously that in d > 4 dimensions the Wick identities, (5.3), are satisfied in any scaling limit of a system of spin variables in the Griffiths-Simon class, with a nearest-neighbor ferromagnetic interaction. Conversely, it can also be shown that the scaling limits of, say, the Ising system with a finite range interaction are non-Gaussian in d = 2 dimensions.

The high dimensional result is a consequence of an inequality which for Ising systems takes the form:

$$|u_4(x_1,\ldots,x_4)| \leq 2 \sum_y \langle\sigma_{x_1}\sigma_y\rangle\langle\sigma_{x_2}\sigma_y\rangle\langle\sigma_{x_3}\sigma_y\rangle\langle\sigma_{x_4}\sigma_y\rangle \quad . \tag{5.4}$$

The scaling limit of this, and the more general bound, is

$$|U_4(x_1,\ldots,x_4)| \leq G \int dy\; S_2(x,y) \ldots S_2(x_4,y) \tag{5.5}$$

$$G = O\left(\frac{1}{\zeta^{d-4}}\right) = 0 \quad \text{for } d > 4 \;,$$

where U_4 is the correction term for the Wick identity (5.3) with $2n = 2$ (i.e. U_4 is the scaling limit of u_4). The transition from (5.4) to (5.5) requires only the Gaussian bound (4.4), in its position space form of reference [27]:

$$\langle\sigma_o\sigma_x\rangle \leq \frac{\text{const.}}{\beta J\,|x|^{d-2}} \quad , \qquad \text{for} \quad \beta < \beta_c \quad . \tag{5.6}$$

For the system discussed here is is known that U_4 vanishes if and only if the Wick identities (5.3) are satisfied for all n (Newman [23]) This effect is demonstrated in an explicit form in reference [1].

The bound (5.4) is quite suggestive from a number of points of view. Its basic structure is that of the first order term (in a naive perturbation theory) of the fully connected four point function in a ϕ^4 field theory. From a different angle - we note that (5.4) holds as an identity for Ising models on a Cayley tree. What we have already seen does suggest that a tree approximation (in which loop effects are ignored) may be quite effective in $d > 4$ dimensions.

Nevertheless, it is a different aspect of (5.4) which was used for its proof. In the aforementioned representation of reference [1], u_4 corresponds to events in which two currents intersect. The inequalities (5.4) and (5.5) were derived there by bounding the probability of intersection by the expected number of lattice sites at which it occurs. An analogy with random walks suggests that this approach should be very effective in $d > 4$ dimensions - as it really is. This analogy sheds additional light on the criticality of $d = 4$ dimensions - since it is at $d = 4$ that the intersection probability for pairs of Brownian paths (which form the scaling limit of simple random walks) starts to vanish.

The last remarks explain also why in low dimensions the scaling limit is non Gaussian. Indeed, in $d = 2$ dimensions the intersection is most natural. This is reflected in the non-vanishing of U_4.

Another derivation of (5.5), based on a different random walk representation, is presented in reference [12], where the result is extended to two component systems.

A useful measure of the strength of the deviation from the Gaussian law is provided by the renormalized coupling constant:

$$g = \frac{|\overline{u_4}|}{\chi^2\xi^d} \tag{5.7}$$

with $|\overline{u_4}| \equiv \Sigma_{x_2,x_3,x_4} |u_4(0,x_2,x_3,x_4)|$. g is dimensionless, in terms of the scaling parameters α and ζ of (5.1), and has therefore an identical expression in terms of the continuum notation - which is useful in the scaling limit.

The results discussed above are reflected in the following pair of bounds:

$$g \leq \frac{2(2d)^2}{\xi^{d-4}} \left[1 + O\left(\frac{1}{\xi^2}\right)\right] \tag{5.8}$$

$$(\rightarrow 0 \text{ when } \beta \nearrow \beta_c \quad \text{for } d > 4)$$

for a general system in the class described above; and

$$g \geq [2R^2]^{-1} \qquad (\not\rightarrow 0) \tag{5.9}$$

for a two dimensional Ising system with a symmetric interaction of range R.

The last inequality proves also that "hyperscaling" is universally valid in d = 2 dimensions (see reference [1]). Similarly, (5.8) shows also that "hyperscaling" fails in d > 4 dimensions.

The above analysis is still incomplete with regard to the dimensions d = 4 and d = 3. Improvements which should be significant for d = 4 dimensions are discussed in references [2,4] . A particularly explicit bound is derived in reference [2] , where it is shown that:

$$g \leq \frac{4d}{\xi^{d-4}} \left(\left|\frac{\partial\chi^{-1}}{\partial\beta}\right| + O\left(\frac{1}{\xi^2}\right)\right) \tag{5.10}$$

Notice that the discussion of Section 4 allows for logarithmic corrections in d = 4 dimensions, which would cause $\left(\frac{\partial\chi^{-1}}{\partial\beta}\right)$ to vanish at β_c. Such corrections are indeed predicted by the ε-expansion, and are very consistent with our stochastic-geometric analysis. If this is the case, then the relation (5.10) would tie the high-dimensional behavior of g, to that low-dimensional remnant in the behavior of the susceptibility, χ.

Thus we have seen here rigorous arguments which prove the existence of an "upper critical dimension", above which the critical behavior simplifies considerably. The simplification is in terms of:

i) the critical exponents - which take their mean field values,

ii) the structure of the correlation functions - whose scaling limits are Gaussian.

While derived by independent arguments, the above results lend strong support to the widely held picture of the stability and domination of the Gaussian fixed point above $d > d_c$, and, by association, to the more general validity of the renormalization group approach. At the same time, these results demonstrate the possibility of a rigorous analysis which can go into the "heart of the matter" even in the absence of exact solutions, and present us with the challenge of developing a better understanding of the non-trivial behavior in low dimensions.

6. Results on The Critical Behavior in Percolation Models

The methods described in Sections 2 through 5, and the conclusions summarized above, have wider applicability. In order to demonstrate this, we shall briefly discuss here percolation models. While some arguments require non-trivial adaption, and some of the inequalities are considerably different, the picture which emerges is strikingly similar to the one pointed above. Differences are unavoidable. In fact, the interacting systems considered above and the percolation models seem to have different upper-critical dimensions.

6.1 The Model

Our discussion would focus on the Bernoulli (independent) bond percolation model. Its configurations are random arrays of rods, which are placed over the bonds (linking neighboring sites) of the lattice Z^d. The bonds are occupied with the probability p - independently of each other.

Two sites linked by a rod are regarded as connected. The main objective is to study the structure of the connected clusters - as a function of p. The quantities of interest include:

P_∞ = the probability that a site belongs to an infinite connected cluster

$\tau_n(x_1,\ldots,x_n)$ = the probability that the sites $x_1,\ldots,x_n$ all belong to the same cluster

$\chi = \Sigma_x \, \tau_2(0,x)$ = the expected size of the cluster which contains the site 0.

For reasons into which we shall not go here, P_∞ and χ can be regarded as analogous to the spontaneous magnetization $M(\beta,0+)$ and the magnetic susceptibility, which was also denoted by χ. The critical densities are defined by

$$\bar{p}_c = \sup \{p \in [0,1] \mid P_\infty(p) = 0\}$$

and (6.2)

$$p_c = \sup \{p \in [0,1] \mid \chi(p) = 0$$

An example of a critical exponent is γ, defined by the expected power law:

$$\chi(p) \cong c(p_c-p)^{-\gamma} \tag{6.3}$$

6.2 The Tree Approximation

Let us now calculate γ within an analog of the mean field approximation. That approximation is characterized by the absence of "loop effects". Thus, some of its spirit, and in fact the key predictions, are captured by considering an analogous system on a Cayley tree (i.e. a Bethe lattice) - with the same number of neighbors (2d) for each site. On such a tree, the probability that the origin is connected to a given site, at a distance k, is p^k. The number of such sites is $2d(2d-1)^{k-1}$. Hence

$$\chi = 1 + \sum_{k=1}^{\infty} 2d(2d-1)^{k-1} p^k =$$

$$= \frac{1+p}{|1-p(2d-1)|_+} = \frac{1+p}{2d-1} |p_c-p|_+^{-1} , \tag{6.4}$$

which is quite similar to (2.11).

It is also quite easy to derive an equation for P_∞ which is quite analogous to (2.8). It shows that $p_c = \bar{p}_c$.

Thus, within the Cayley tree approximation we get the following values for the critical density and the critical exponent:

i) $p_c = (2d-1)^{-1}$ (6.5)

ii) $\gamma = 1$ (6.6)

We shall now quote the rigorous results, of reference [3] , about the actual values of p_c and γ.

6.3 Mean Field Bounds

In Section 3 we saw that the mean field calculations offer bounds on the exact values of β_c and γ. The same is true here, as a consequence of the following inequality (reference [3])

$$\frac{\partial\chi}{\partial p} \leq (2d-1)\,\chi^2 \quad . \tag{6.7}$$

By the same arguments which led from (3.5) to Corollary 3.1, we obatin:

Corollary 6.1. The actual values satisfy:

i) $p_c \geq (2d-1)^{-1}$ (6.8)

ii) $\gamma \geq 1$ (in any dimension) . (6.9)

6.4 Characterization of The Upper Critical Dimension

For the interacting system considered above, dimensions above the upper critical point were characterized by the finiteness of the "bubble diagram", $\Sigma_x \langle\sigma_o\sigma_x\rangle^2$. The relevant criterion for percolation models which was derived in reference [3] , concerns the "triangle diagram":

$$\nabla = \sum_x \tau_2(0,x)\,\tau_2(x,y)\,\tau_2(y,0) \quad . \tag{6.10}$$

Proposition 6.1. In a percolation model in which $\nabla(p)$ is uniformly bounded in $[0,p_c]$, the critical exponent γ takes the exact value:

$$\gamma = 1 \tag{6.11}$$

(with no logarithmic corrections).

Remark. In terms of the Fourier transform, $\hat{\tau}(q)$, we have:

$$\nabla = \frac{1}{(2\pi)^d} \int_{[-\pi,\pi]^d} dq\, \hat{\tau}(q)^3 \tag{6.12}$$

Had the bound (4.4) been valid also for $\hat{\tau}(q)$, it would readily follow that the criterion of Proposition 6.1 is met in dimensions

$$d > 6 \quad . \tag{6.13}$$

There are, however, indications that the analog of (4.4) is not valid for $\hat{\tau}(q)$, in other words - that the critical exponent η is negative for the nearly critical values of d. (I would like to thank A.B. Harris and Y. Shapir for bringing this to my attention.) Thus, the task of proving that $\gamma = 1$ for $d > 6$, may be somewhat harder.

The reason for the importance of the triangle diagram would not be explained here. Let us just say that it is related to the structure of the connected correlation functions, which are discussed next.

6.5 Structure of The Connected Correlation Functions

General bounds, which hold for the percolation model on every lattice, are provided by the following inequalities

$$\tau_3(x_1,x_2,x_3,x_4) \leq \sum_y \tau_2(x_1,y)\tau_2(x_2,y)\, \tau_2(x_3,y) \tag{6.14}$$

and

$$\tau_n(x_1,\ldots,x_n) \leq T_n(x_1,\ldots,x_n) \quad , \tag{6.15}$$

where T_n is the sum of "tree diagrams" which have $x_1,\ldots,x_n$ as external vertices and n-2 internal vertices of the third order. Each diagram should be interpreted as the corresponding product of two-point functions. (Thus (6.14) is a special case of (6.15).)

The bounds (6.14) are made somewhat intuitive by considering first the low p limit. In this case, the main contribution to τ_n is from the minimal configuration(s) of bonds which interconnects $x_1,\ldots,x_n$. Such configurations correspond to trees which are among the terms which contribute to T_n. For a higher value of p, below p_c, the above description may still be correct, if applied to intermediate size clusters.

One could also point out that the diagrams of T_n have the structure of the tree diagrams in a ϕ^3 field theory. A relation between percolation and ϕ^3 field theory has indeed been expected - on the basis of arguments which we find far less convincing than even the above heuristic discussion.

The analysis of reference [3] suggests that above an upper critical dimension - where the criterion of Proposition 6.1 is satisfied, the long distance behavior of τ_n, in the critical regime, may actually be given by the function T_n - modified by simple vertex factors. Once again we see the simplification of the scaling limit, and the reduction of τ_n to simple expressions in terms of the two point function.

7. Implications for Quantum Field Theory

The two subjects - statistical mechanics and quantum field theory, are (at present) intricately related to each other. In both, the main effects which are studied are observed on a scale which is enormously large compared with the one at which the elementary constituents interact with each other. The subtle propagation of the effects across this large gap of scales is a typical feature of systems of infinitely many degrees of freedom which are at - or very near, a critical point. The observed structure is therefore that of a certain scaling limit.

We shall not describe here the strong relations by which the basic models of statistical mechanics and field theory are intertwined. However, it might be pointed out that the perspectives of the two subjects diverge when it comes to the phenomenon of the upper critical dimension. This is due to the fact that the simple structure described by (5.3) corresponds to a field of non-interacting particles.

When the scaling regime becomes Gaussian (as it does in $d > 4$ dimensions) the main problems of statistical mechanics attain simple solutions. At the same time, the task of formulating a local field theory of interacting particles falls beyond the present reach of the available techniques.

Conversely, the super-renormalizability of the scalar ϕ^4 field theory is coupled with the nontrivial behavior of the scaling limit, and the critical exponents in statistical mechanics. (A simple proof of the "nontriviality" of the scaling limits for sufficiently soft ϕ^4 lattice fields is given in reference [6].)

References

[1] M. Aizenman, Phys. Rev. Lett. 47, 1 (1981); and Commun. Math. Phys. 86, 1 (1982).

[2] M. Aizenman and R. Graham "On the renormalized coupling constant and the susceptibility in ϕ_4^4 field theory and the Ising model in four dimensions", to appear in Nucl. Phys. B [FS].

[3] M. Aizenman and C.M. Newman, "Tree diagram bounds and the critical behavior in percolation models", in preparation.

[4] C. Aragao de Carvalho, S. Caraciolo and J. Fröhlich, Nucl. Phys. B215 [FS7], 209 (1983).

[5] E. Brezin, J.C. Le Guillou and J. Zinn-Justin, in Phase Transitions and Critical Phenomena, C. Domb and M.S. Green (eds.) (Academic Press; London New York San Fransisco, 1976).

[6] D.C. Brydges, J. Fröhlich and A.D. Sokal, "A new proof of the existence and nontriviality of the continuum ϕ_2^4 and ϕ_3^4 quantum field theories." Courant Institute preprint.

[7] M. Cassandro, E. Olivieri, A. Pellegrinotti and E. Presutti, Z. Wahrs. v. Geb. 41, 313 (1978).

[8] R.L. Dobrushin and E.A. Pecherski, in Random Fields, Vol. I, J. Fritz, J.L. Lebowitz and D. Szász (eds.) (North Holland, Amsterdam Oxford New York, 1981).

[9] W. Driesler, L. Landau and J. Fernando-Perez, J. Stat. Phys. 20, 123 (1979).
B. Simon, J. Stat. Phys. 22, 491 (1980).
M. Aizenman and B. Simon, Commun. Math. Phys. 77, 137 (1980).

[10] R.S. Ellis, J.L. Monroe and C.M. Newman, Commun. Math. Phys. 46, 167 (1976).
R.S. Ellis and C.M. Newman, Trans. Am. Math. Soc. 237, 83 (1978).

[11] M.E. Fisher, Phys. Rev. 162, 480 (1967).

[12] J. Fröhlich, Nucl. Phys. B200 [FS4], 281 (1982).

[13] J. Fröhlich, B. Simon and T. Spencer, Commun. Math. Phys. 50, 79 (1976).

[14] J. Glimm and A. Jaffe, Phys. Rev. D10, 536 (1974).

[15] J. Glimm and A. Jaffe, Quantum Physics (Springer-Verlag, New York Heidelberg Berlin, 1981).

[16] R.B. Griffiths, Commun. Math. Phys. 6, 121 (1967).

[17] R.B. Griffiths, J. Math. Phys. 10, 1559 (1969).
B. Simon and R. Griffiths, Commun. Math. Phys. 33, 145 (1973).

[18] R. Griffiths, C. Hurst and S. Sherman, J. Math. Phys. 11, 790 (1970).

[19] S. Krinsky and V.J. Emery, Phys. Lett. 50A, 235 (1974).

[20] J.L. Lebowitz, Commun. Math. Phys. 35, 87 (1974).

[21] S.-K. Ma, Modern Theory of Critical Phenomena (Benjamin; Reading Mass., 1976).

[22] O.A. McBryan and J. Rosen, Commun. Math. Phys. 51, 97 (1976).

[23] C.M. Newman, Commun. Math. Phys. 41, 1 (1975).

[24] B. Simon, Commun. Math. Phys. 77, 111 (1980).

[25] A.D. Sokal, Phys. Lett. 71A, 451 (1979).

[26] A.D. Sokal, J. Stat. Phys. 28, 431 (1982).

[27] A.D. Sokal, Ann. Inst. Henri Poincaré, 37A, 317 (1982).

[28] C.J. Thompson, Commun. Math. Phys. 24, 61 (1971).
P.A. Pearce, J. Stat. Phys. 25, 309 (1981).
C.M. Newman, Univ. of Arizona preprint.
J. Slawny, "On the mean field theory bound on the magnetization", to appear in J. Stat. Phys.

[1] Departments of Mathematics and Physics, Rutgers University, New Brunswick, N.J. 08903, U.S.A.
Sloan Foundation Research Fellow. Supported in part by the National Science Foundation Grant PHY-8301493.

NON-PERTURBATIVE METHODS FOR THE STUDY OF MASSLESS MODELS

Jean-Raymond FONTAINE
Institut de Physique Théorique
Université Catholique de Louvain
chemin du Cyclotron, 2
1348 Louvain-la-Neuve
Belgium

0. Introduction.

In this paper we describe some non-perturbative methods suitable to study massless models of classical statistical mechanics. By massless models, we mean systems that are only polynomially(not exponentially) clustering. Most of the models we shall consider have a clustering like $|x|^{-d}$ in d dimensions. Their correlation functions are therefore even not absolutely integrable, which implies that they are outside the range of powerful techniques like for instance the "cluster expansion" at least in its original form [24,34]. We shall mainly concentrate on three types of systems.

0.1 The Anharmonic Crystal.

It is described by the Hamiltonian

$$\beta H = \frac{1}{2} \sum_{<ij>} (\phi_i - \phi_j)^2 + \lambda/4 \sum_{<ij>} (\phi_i - \phi_j)^4$$

where the sums run over all nearest neighbour pairs of $\mathbb{Z}^d$, ϕ_i is a real random variable uniformly distributed on $\mathbb{R}$ and $\beta = T^{-1}$ is the inverse temperature.

0.2 The Lattice Dipole Gas.

After a sine-Gordon transformation [12,28] it is equivalent to a spin system of the type 0.1 described by the Hamiltonian :

$$\beta H = \frac{1}{2} \sum_{<ij>} (\phi_i - \phi_j)^2 + z \sum_{<ij>} \cos(\phi_i - \phi_j)$$

z is called the fugacity of the gas.

0.3 The Low Temperature Rotator Model.

Its Hamiltonian is given by :

$$\beta H = \beta \sum_{<ij>} \cos(\phi_i - \phi_j)$$

where ϕ_i is now uniformly distributed on *the circle* : $\phi_i \in [-\pi,\pi]$.

The three models are related by two facts.

0.4 They are described by an Hamiltonian which possesses an Abelian continuous symmetry group : translations on $\mathbb{R}$ for models 0.1 and 0.2, and translations on the circle for model 0.3.

0.5 They can be expressed as a perturbation of a massless free field (or harmonic crystal) described by $\beta H_0 = \frac{1}{2}\sum_{<ij>}(\phi_i-\phi_j)^2$. Whereas this is clear for models 0.1 and 0.2, the way one can see it for model 0.3 is to make the change of variable $\phi_i \rightsquigarrow \sqrt{\beta}\,\phi_i$, expand $\beta \cos \sqrt{T}\,(\phi_i-\phi_j)$ into a power series in T and neglect (for the time being) the restriction $\phi \in [-\pi\sqrt{\beta},\pi\sqrt{\beta}\,]$. Of course fact 0.4 is related to the non-exponential clustering of the system, at least when the continuous symmetry is spontaneously broken ($d \geqslant 3$); this is the so-called Goldstone phenomenon [26].

The natural questions to ask about the described systems are :

0.6 What is the relation between the correlation functions and their perturbation theory in λ,z or T ?

0.7 What type of clustering have the correlation functions ?

0.8 What do we know about the scaling limit of these systems ?

For weak coupling (λ,z or T small) these problems are usually solved by doing a cluster expansion. However as already mentioned this technique, as described in [24], does not apply. Nevertheless, several authors [20-23,10,11,33] have been able to combine the "cluster expansion" with renormalization group ideas to solve some of the above questions. For further details we refer to the contributions of Gawedzki and Kupiainen, Magnen and Sénéor, and Federbush to this book.

In this report we shall summarize methods presented in [6,7,8, 12,13,14] which can give at least partial answers to the above questions. Since our arguments are not based on convergent expansions, most of our results will be independent from the strength of the coupling constants λ,z or T (except of course for those relative to question 0.6).

The paper is organized as follows. In section 1 we give a precise definition of the models and for each of them construct the states we shall consider in the rest of the paper. In section 2,3, and 4 we investigate respectively questions 0.6, 0.7, 0.8. In the whole paper we mainly concentrate on models 0.1 and 0.3, and just indicate some remarks about model 0.2 .

1. Description of the Models.

In this section we start by constructing infinite volume states for system 0.1 . As remarked in [3] this can be done using Brascamp-Lieb inequalities (B-L inequalities) [4]. For the rotator model, we define the "+ state" which has a nonzero magnetization at low temperatures ($d \geqslant 3$).

1.1 Unbounded Spin Systems : model 0.1

We first state B-L inequalities which will be a basic tool to investigate these systems. They are similar to the infrared bounds of [17] but can be applied to systems where reflection positivity does not necessarily hold ; it is the case of model 0.1

1.1.1 Definition : A function $F : \mathbb{R}^n \to \mathbb{R}_+$ is *log concave* if $F(\Phi) = \exp f(\Phi)$ and f is concave. (It is understood that f can take the value $-\infty$)

1.1.2 Let A be a strictly positive symmetric n x n matrix, and denote by $\langle\ \rangle_0$ the Gaussian measure associated to it :

$$\langle\cdot\rangle_0 = [\int \exp[-\tfrac{1}{2}(\Phi,A\Phi)]\, d\Phi\,]^{-1} \int \cdot \exp[-\tfrac{1}{2}(\Phi,A\Phi)]\, d\Phi \quad ,$$

consider a perturbation of $\langle\ \rangle_0$ by a log concave function F :

$$\langle\cdot\rangle_F = [\int \exp[-\tfrac{1}{2}(\Phi,A\Phi)]\, F(\Phi)\, d\Phi\,]^{-1} \int \cdot \exp[-\tfrac{1}{2}(\Phi,A\Phi)]\, F(\Phi)\, d\Phi\ ;$$

ϕ_x, $x \in \{1,\dots,n\}$ denotes a component of Φ. We can now state the basic theorem :

1.1.3 Theorem [4] : The covariance matrix $M^F_{xy} = \langle \phi_x \phi_y\rangle_F - \langle\phi_x\rangle_F \langle\phi_y\rangle_F$ satisfies the inequality

$$M^F_{xy} \leqslant (A^{-1})_{xy}$$

in the sense of forms. (This will be denoted by B-L inequalities)

1.1.4 Definition of the system. To each region $\Lambda \subset \mathbb{Z}^d$ (Λ is a parallelipiped) is associated an Hamiltonian H_Λ with periodic boundary conditions on $\partial\Lambda$:

$$H_\Lambda = \frac{1}{2} \sum_{\langle ij\rangle \subset \Lambda} (\phi_i-\phi_j)^2 + \lambda/4 \sum_{\langle ij\rangle \subset \Lambda} (\phi_i-\phi_j)^4$$

$\langle ij\rangle$ denotes a pair of nearest neighbour points or points which are at opposite ends of Λ. We also consider the Hamiltonian

$$H_{\Lambda,m} = H_\Lambda + \frac{m^2}{2} \sum_{i\in\Lambda} \phi_i^2 \quad , \ m \in \mathbb{R}$$

If A is a "subset" of $\mathbb{Z}^d$ (we allow repetitions of the same element in A, which is the reason why we put the word subset in between quotation marks), $\prod_{i\in A} \phi_i$ will be denoted by ϕ^A. It is also useful to consider difference variables : let $\{ e_\alpha \}$ $\alpha = 1,\dots,d$ be a basis of $\mathbb{Z}^d$ given by $e_\alpha = \delta_{i\alpha}$, $\phi_i - \phi_{i+e_\alpha} \equiv \nabla_i^{e_\alpha}\phi$; Σ_{e_α} $\alpha = 1,\dots,d$ will be denoted by $\sum_e$ or $\sum_\xi$. If A is a "subset" of $\Lambda \times \{e_1,\dots,e_d\}$, $\prod_{(i,\xi)\in A} \nabla_i^\xi \phi \equiv \nabla\phi^A$. In the whole paper c will stand for a positive constant which can take different values at different places.

We now want to apply B-L inequalities to our model and in particular deduce the exponential bounds (see also [5]).

1.1.5 Proposition : Assume $d \geq 3$, then there exists $\alpha_0(d) > 0$ such that $\forall\ \alpha \leq \alpha_0$ all limit points of $< \exp \alpha\, \phi_0^2 >_{\Lambda,m}$ as $\Lambda \to \infty$ are uniformly bounded in m by a finite constant.

Proof : Consider the normalized measure $<\cdot>_{\alpha,\Lambda,m} \equiv Z^{-1}_{\alpha,\Lambda,m}$ $\int .\ \exp(-H_{\Lambda,m} + \alpha\phi_0^2)\ \prod_{i\in\Lambda} d\phi_i$. We have :

$$\begin{aligned}
\ln\,(Z_{\alpha,\Lambda,m} / Z_{\alpha=0,\Lambda,m}) &= \int_0^\alpha \frac{d}{d\alpha'}\ Z_{\alpha',\Lambda,m}\ d\alpha' \\
&= \int_o^\alpha d\alpha'\ < \phi^2 >_{\alpha',\Lambda,m} \\
&\leq \int_o^\alpha d\alpha'\ < \phi^2 >_{\alpha',\Lambda,m,\lambda = o}
\end{aligned}$$

where in the last inequality we used B-L inequalities. When $\Lambda \to \infty$ this quantity is uniformly bounded in m by a finite constant if α is sufficiently small.

Similarly one can show :

1.1.6 Proposition : In any dimension d, there exists $\alpha_0(d) > 0$ such that for any $\alpha \leq \alpha_0(d)$ all limit points of $< \exp \alpha(\nabla\phi)^2 >_{\alpha,m}$ as $\Lambda \to \infty$ are uniformly bounded in m.

1.1.7 Definition of the states. Given propositions 1.1.5 and 1.1.6, one can define the state

$$\lim_{m\downarrow 0}\ \lim_{\Lambda\uparrow\infty}\ < \quad >_{\Lambda,m}$$

where the limits may have to be taken through subsequences. Let us define by $<\ \ >$ any of these limit points. $<\ \ >$ is defined on functions of the type ϕ^A for $d \geq 3$ and on functions of the type $\nabla\phi^A$ for any d.

1.1.8 The Pressure. Using B-L inequalities one can easily prove the existence of the pressure :

$$|\Lambda|^{-1}[\log Z_{\Lambda,m} - \log Z_{\Lambda,m'}] = \int_{m'}^{m} \tilde{m}\, |\Lambda|^{-1} \sum_{i\in\Lambda} < \phi_i^2 >_{\Lambda,\tilde{m}}\, d\tilde{m}$$

$$= \int_{m'}^{m} \tilde{m} <\phi_o^2>_{\Lambda,\tilde{m}} \quad d\tilde{m}$$

Using th. 1.1.3 we have the following bounds (in the thermodynamic limit) : $<\phi_o^2>_m \leqslant c$ for $d \geqslant 3$, $<\phi^2>_m \leqslant c \log m$ for $d = 2$ and $<\phi^2>_m \leqslant c\, m^{-1}$ in one dimension. Denote $P_m(\lambda) \equiv \lim_{\Lambda\uparrow\infty} |\Lambda|^{-1} \log Z_{\Lambda,m}$ which exists by standard arguments;we then have $|P_m(\lambda)-P_{m'}(\lambda)| \leqslant c|m-m'|$ which implies that $P(\lambda) \equiv \lim_{m\downarrow o} P_m(\lambda)$exists.

1.2 The Rotator model.

1.2.1 Definition of the model. To each parallelipiped $\Lambda \subset \mathbb{Z}^d$, we associate an Hamiltonian

$$-\beta\, H_\Lambda = \sum_{<ij> \subset \Lambda} \beta \cos(\phi_i - \phi_j)$$

where the sum runs over all nearest neighbour pairs of the torus associated to Λ (periodic boundary conditions), and $\beta = T^{-1}$ is the inverse temperature. We can also put the system in an external field h; the Hamiltonian becomes then :

$$-\beta H_{\Lambda,h} = -\beta H_\Lambda + h \sum_{i\in\Lambda} \cos \phi_i$$

We define a probability measure on $[-\pi,\pi]^{|\Lambda|}$ by :

$$d\mu_{\Lambda,h} = Z^{-1}_{\Lambda,h} \exp -\beta H_{\Lambda,h}$$

For any functions $m : \mathbb{Z}^d \to \mathbb{Z}$ of compact support, we can define the correlation functions

$$<\cos m\, \phi>_{\Lambda,h} = \int \cos m\, \phi \quad d\mu_{\Lambda,h}$$

1.2.2 Definition of the state. We shall consider the Gibbs state :

$$\lim_{h\downarrow o} \lim_{\Lambda \uparrow \mathbb{Z}^d} < \ >_{\Lambda,h} \equiv < \ >_+$$

As usual the pressure is defined by :

$$P(T) = \lim_{\Lambda\uparrow\mathbb{Z}^d} |\Lambda|^{-1} \log Z_{\Lambda,h=o}$$

All the above limits make sense,see [8] for references.

1.2.3 For $d \geqslant 3$ it has been proven that there is a non zero spontaneous magnetization $\bar{m}(\beta) = < \cos\phi >_+$ at low temperature. This was obtained using the infrared bounds [17] :

$$(1) \quad < \exp \ \hat{\sigma}(g)> \leqslant \exp[(g, (-\Delta)^{-1} g)/2\beta]$$

where $\sigma_i = (\cos\phi_i, \sin\phi_i)$, $g : \mathbb{Z}^d \to \mathbb{R}^2$, $\hat{\sigma}_i = \sigma_i < \sigma_i >$,
$\hat{\sigma}(g) = \sum_i g(i). \hat{\sigma}_i$, and Δ is the finite difference Laplacian.

2. Asymptotic Expansion.

In this section we shall developp a method to show that the correlation functions of the massless models considered above have an asymptotic expansion to all orders in λ,z or T. As in the massive case, we generate the perturbation expansion using the integration by parts formula for Gaussian measures (I.P) [24]. However since the expansion is about a massless Gaussian field whose covariance is not summable, we have to introduce a regularization of the I.P. What we do in [6] is to apply the I.P. formula with respect to a massive Gaussian field whose mass $m(\lambda)$ depends upon the coupling constant λ. By a suitable choice of $m(\lambda)$, we can estimate all the terms produced by successive applications of the regularized I.P. formula and get the result.

The same method essentially can be applied to the rotator model, but we have to take a special care of the constraint $\phi \in [-\pi\sqrt{\beta}, \pi\sqrt{\beta}]$ (see 0.5). This represents the main difficulty of the rotator model. Our results not only show that the spin wave approximation is asymptotically correct (as $T \to o$), but also gives a way to get all higher order correction in T for all correlation functions and in particular for the spontaneous magnetization (see [8]).

2.1 Unbounded Spin Systems

2.1.1 Theorem [6] : The free energy $P(\lambda)$ and the correlation functions ($<\nabla \phi^A>$ for all d and $<\phi^A>$ for $d \geq 3$) have an asymptotic expansion to all orders in λ whose coefficients are given by the usual perturbation theory.

2.1.2 Let us first formulate the *integration by parts formula for Gaussian measures*. Let $\{ \varphi_z \}_{z=1}^N$ be a set of Gaussian variables and μ the corresponding Gaussian measure with covariance C_{xy}. If F is a differentiable function and $\phi_y F \in L^1(d\mu)$, then I.P. reads [24] :

$$(2) \quad \int \phi_y \, F(\{\phi_z\}_{z=1}^N) \, d\mu = \sum_{x=1}^N C_{xy} \int \frac{d}{d\phi_x} F(\{\phi_z\}_{z=1}^N) \, d\mu$$

2.1.3 In order to give an idea of the proof of th. 2.1.1. in the simplest case, we restrict ourselves to the zeroth order of $<(\nabla_0^e \phi)^2>$ in $d \geq 3$. We first consider finite volume expectation values and absorb the interaction into the function F of formula (2). This yields :

$$(3) \quad <(\nabla_0^e \phi^2)>_{\Lambda,m} = \nabla_0^e \nabla_0^e C_{00}^{\Lambda,m}$$

$$- \lambda \sum_{x \in \Lambda} \nabla_0^e C_{ox} \sum_{e_\alpha} [< \nabla_0^e \phi(\phi_x - \phi_{x+e_\alpha})^3 > - < \nabla_0^e \phi(\phi_{x-e_\alpha} - \phi_x)^3 >]$$

where $C_{xy}^{\Lambda,m} = (-\Delta_\Lambda + m^2)^{-1}_{xy}$ and $-\Delta_\Lambda$ is the finite difference Laplacian with periodic boundary conditions. Using the usual integration by parts formula, and taking the limit $\Lambda \uparrow \mathbb{Z}^d$ of both sides of (3) we get

$$(4) \quad < (\nabla_0^e \phi)^2 >_m = \nabla_0^e \nabla_0^e C_{00}^m - \lambda \sum_{\substack{x \in \mathbb{Z}^d \\ \xi}} \nabla_0^e \nabla_0^\xi C_{ox}^m < \nabla_0^e \phi \, (\nabla_0^\xi \phi)^3 >_m$$

In the massive case ($m \neq o$) the result follows easily. By B-L inequalities $|< \nabla \phi_0^e (\nabla_0^\xi \phi)^3 >_m|$ is uniformly bounded in λ (and m). On the other hand, a simple Gaussian estimate implies [6] $| \sum_{x,\xi} \nabla_0^e \nabla_x^\xi C_{ox}^m | \leq c \ln m$ (for m small).

For fixed m, the remainder is therefore bounded by $\lambda c \ln m$ which goes to zero with λ. However in the massless case ($m \downarrow o$) the bound we have on the remainder is divergent and this reflects the difficulty of massless theories.

The way we solved the problem in [6] is as follows. Assume here was a small λ-dependent mass in the theory (smaller than any ower of λ as $\lambda \to o$) : $m(\lambda) = \exp - (\ln \lambda)^2$; the second term of (4) ould then be bounded by $c\lambda \ln \exp-(\ln\lambda)^2$ which goes to zero as $\lambda \to o$. ow the first term $\nabla_0^e \nabla_0^e C_{00}^m$ becomes λ-dependent through $m(\lambda)$. owever we have : $|\nabla_0^e \nabla_0^e C_{00}^{m(\lambda)} - \nabla_0^e \nabla_0^e C_{00}| \leqslant c\, m^2$, $C_{xy} = (-\Delta)^{-1}_{xy}$) see [6]. The problem would be completely solved f we could show that expectation values computed with $m = o$ are xponentially close (with λ) to the ones computed in a massive theory f mass $m(\lambda)$. This can be done for the pressure since we showed in .1.8) that $|P_m(\lambda) - P(\lambda)| \leqslant cm$. It is harder to get similar stimates for the correlation functions. Nevertheless we can mplement the ideas explained before using a regularized form of he I.P. formula. Instead of doing the I.P. with respect to a Gaussian heory of mass m as in (3), do it with respect to a theory of mass $m' = m + m(\lambda)$. So add and substract a mass term $\frac{1}{2} m(\lambda) \sum_{i \in \Lambda} \phi_i^2$ to $H_{\Lambda,m}$ see 1.1.4) and absorb the term $\exp \frac{1}{2} m(\lambda) \sum_{i\in\Lambda} \phi_i^2$ into the function of formula (2). This yields, after taking the limits $\Lambda \uparrow \mathbb{Z}^d$ and $m \downarrow o$:

$$(5)\quad \langle (\nabla_0^e \phi)^2 \rangle = \nabla_0^e \nabla_0^e C_{00}^{m(\lambda)} - \lambda \sum_{\substack{x\in\mathbb{Z}^d \\ \xi}} \nabla_0^e \nabla_x^\xi C_{ox}^{m(\lambda)} \langle \nabla_0^e \phi (\nabla_x^\xi \phi)^3 \rangle + m^2(\lambda) \sum_{x\in\mathbb{Z}^d} \nabla_0^e C_{ox}^{m(\lambda)} \langle \nabla_0^e \phi \phi_x \rangle \equiv \nabla_0^e \nabla_0^e C_{00} + R(\lambda)$$

The way we estimate $R(\lambda)$ is in two steps

1) $|\nabla_0^e \nabla_0^e C_{00}^{m(\lambda)} - \nabla_0^e \nabla_0^e C_{00}| \leqslant m(\lambda)$ see [6]

2) i) Using the proposition 1.1; we have the bounds

$|\langle \nabla_0^e \phi (\nabla_x^\xi \phi)^3 \rangle| \leqslant c$, $|\langle \nabla_0^e \phi \phi_x \rangle| \leqslant c$

ii) Using simple Gaussian estimates we also get [6] :

$\sum_{x\in\mathbb{Z}^d} | \nabla_0^e \nabla_x^\xi C_{ox}^{m(\lambda)} | \leqslant c \ln m(\lambda)$

$\sum_{x\in\mathbb{Z}^d} | \nabla_0^e C_{ox}^{m(\lambda)} | \leqslant c\, m^{-1}(\lambda)$.

These estimates imply $R(\lambda) \leqslant c\lambda \ln m(\lambda) + c\, m(\lambda)$. So with our choice of $m(\lambda)$, $R(\lambda)$ goes to zero with λ. Since the third term of the r.h.s. of (5) is exponentially small (with λ), higher orders of the expansion are obtained by repeating the above procedure to the second term of the r.h.s. of (5) (see [6] for details).

2.1.4 For functions like $< \phi^A >$, the terms produced by the regularized I.P. have a slower convergence because there is one derivative less on each covariance. We can however get the result using the Brascamp-Lieb bound on the full two point function (th.1.1.3) and being more flexible on the choice of the function $m(\lambda)$ (see [6])

2.1.5 Up to some technicalities the same ideas can also be used to study the two (and one) dimensional case; for this we refer to the original paper [6].

2.2 The Rotator Model.

2.2.1. Theorem [8] : For any dimension d, the correlation functions $< \cos m\,\phi >_+$ of the rotator model have an asymptotic expansion to all orders in T whose coefficients are given by the usual perturbation theory.

2.2.2 In this paragraph we just want to relate the low temperature expansion of the rotator model to the λ-expansion of section 2.1, and present some of the additional ideas developped in [8] to prove th. 2.2.1 . Let us first restrict ourselves to the case of difference variables ($\Sigma\, m(i)=o$) and consider $< \cos \nabla_0^e \phi>_+$. Using definition 1.2.1 and making a change of variables $\phi_i' = \sqrt{\beta}\,\phi_i$, we have :

$$< \cos \nabla_0^e \phi >_{\Lambda,h} = \int_{-\pi\sqrt{\beta}}^{\pi\sqrt{\beta}} \prod_{i\in\Lambda} d\phi_i \cos \nabla_0^e\phi \exp\{\beta \sum_{\substack{i\in\Lambda\\ \xi}} \cos\sqrt{T}\, \nabla_i^\xi \phi + h \sum_{i\in\Lambda} \cos \sqrt{T}\, \phi_i\}\, Z_{\Lambda,h}^{-1}(\beta) ,$$

where $Z_{\Lambda,h}(\beta) = \int_{-\pi\sqrt{\beta}}^{\pi\sqrt{\beta}} \prod_{i\in\Lambda} d\phi_i \exp\{\beta \sum_{\substack{i\in\Lambda\\ \xi}} \cos \sqrt{T}\, \nabla_i^\xi \phi + h\sum_{i\in\Lambda} \cos\sqrt{T}\, \phi_i\}$

If we forget about the external field (which will be set to zero at the end) and expand the cosines into power series in T, we see that the interaction can be written as

$$-\frac{1}{2} \sum_{i,\xi} (\nabla_i^\xi \phi)^2 + \sum_{i,\xi} V(T,\nabla_i^\xi \phi)$$

where $V(T,\nabla_i^\xi \phi) = \sqrt{\beta} \cos \sqrt{T}\, \nabla_i^\xi \phi - 1/2(\nabla_i^\xi \phi)^2$ is an infinite power eries in T whose first term is $T(\nabla_i^{\xi}\phi)^4/4!$. However the ϕ_i variables re now restricted to belong to the interval $[-\sqrt{\beta}\,\pi, \sqrt{\beta}\,\pi]$. So the ow temperature rotator model can be expressed as a double perturbation the massless Gaussian field. The first perturbation $\sum_{i,\xi} V(T, \nabla_i^{\xi}\phi)$ an be handled as in section 2.1. We now want to explain hat the second perturbation $\prod_{i \in \Lambda} \chi(|\phi_i| \leq \pi\sqrt{\beta}) = V_0$ (where $\chi(A)$ is he characteristic function of the set A) will only produce exponenially small terms (with T) in the expansion.

Using the regularized I.P. formula (5) to generate the expansion f $<(\nabla_0^e\phi)^2>_+$, we see that the contribution of V_2 will be of the orm

$$
\begin{aligned}
(6)\quad & \sum_i \nabla_0^e C_{0i} < \nabla_0^e \phi \frac{d}{d\phi_i} V_0 >_+ \\
&= \sum_i \nabla_0^e C_{0i} < \nabla_0^e\phi \, [\, \delta(\phi_i + \pi\sqrt{\beta}) + \delta(\phi_i - \pi\sqrt{\beta})] >_+ .
\end{aligned}
$$

he basic result of [8] is to show that the infrared bounds in heir exponential form (1) not only imply $< \cos\phi >_+ > 0$, but can lso be used to prove that the equilibrium measure associated to the "+ state" is very much concentrated about $\phi = 0$. This in particular imply the estimate

$$
< \delta\,(\phi \pm \pi\sqrt{\beta}) >_+ \leq c \exp - c\sqrt{\beta}
$$

which together with our choice of $m(T) = \exp -(\ln T)^2$ is enough to prove that (6) is exponentially small with T as $T \to 0$.

2.2.3 The way we prove the result for functions with $\sum_i m(i) \neq 0$, like the spontaneous magnetization, is in two steps :

(i) For x proportional to some power β^n, do the expansion of $< \cos(\phi_0 - \phi_x) >_+$ using the above method.

(ii) Use the decay : $|< \cos(\phi_0 - \phi_x)>_+ - < \cos\phi_0>_+^2 | \leq c\,|x|^{-1}$ (proved in section 3) to get from (i) the expansion of $\bar{m}(\beta) = <\cos\phi>_+$.

2.3 Remarks

2.3.1 Th. 2.1.1 can be easily extended to non nearest neighbour interactions, and to more general polynomial in $(\nabla\phi)^2$ see [6].

2.3.2 In the case of the dipole gas (model 0.2) th. 2.1.1 is also valid but the proof is different : the perturbation theory is generated using the method of complex translations of [9] (see [12] for details).

3. The Decay of Correlations.

3.1 In order to explain the basic ideas of this section, let us first consider the grad-grad two point function of model 0.1 . Its Fourier transform $S^{ee}(p) \equiv < \nabla_0^e \phi \, \nabla_x^e \phi >^{\sim} (p)$ obeys the bounds :

$$(7) \quad [\, 1+3\lambda <(\nabla_0^e \phi)^2>]^{-1} S_0^{ee}(p) \leqslant S^{ee}(p) \leqslant S_0^{ee}(p)$$

where $S_0^{ee}(p) = (2 - 2 \cos p_e)[\, 2 \sum_e (1-\cos p_e)]^{-1}$. The lower bound results from a Mermin-Wagner argument [35]; it shows in particular that $< \phi^2>$ is infinite in $d \leqslant 2$ (the translation group $\phi \rightsquigarrow \phi + c$ is not broken in $d \leqslant 2$). The upper bound is B-L inequality (th. 1.1.3). As remarked by Fröhlich and Spencer [18], and by Park [39], (7)implies that $S^{ee}(p)$ is not continuous at $p = o$ $(d > 1)$. Therefore its Fourier transform $< \nabla_0^e \phi \, \nabla_x^e \phi >$ cannot be absolutely integrable,which is a basic fact of the models considered in this paper (note that the corresponding property for the 2-d rotator at low temperatures is considerably harder to prove,see [19]). Actually (7) gives the exact (up to a multiplicative constant) low momentum singularity of the two point function $S(p) = <\phi_0 \, \phi_x>^{\sim}(p)$ $(d \geqslant 3)$. It is given by $\varepsilon \, [\, 2 \sum_e (1 - \cos p_e)]^{-1}$. So up to a constant ε, the *dielectric constant,* it is the same as in the purely Gaussian case (of course ε obeys the bounds $[\, 1+3\lambda <(\nabla_0^e \phi)^2>]^{-1} \leqslant \varepsilon \leqslant 1$). This indicates that the long distance behaviour of $< \phi_0 \phi_x>$ should be the same as in the purely Gaussian case (i.e. $\sim \varepsilon/(4\pi|x|)$ in $d = 3$ (see th. 3.2.1 and 3.2.4). Moreover as suggested by renormalization group arguments [31], we expect this to be true for general correlation functions. In other words we expect that up to a multiplicative constant, the low momentum behaviour of general correlations is the same as in the theory with $\lambda = o$. This is proven in th.3.3.2, for all λ, when the correlation functions are truncated into two clusters. Using this we are able to get an asymptotic expansion (up to second order in λ) of the dielectric constant ε. This quantity is particularly interesting to study because it is believed that the scaling limit [1] of this theory is Gaussian, and that therefore ε is the only relevant parameter of the theory (see section 4). We finally want to mention that Magnen and Sénéor have been able to compute ε as sum of the convergent series (not in λ)(see their contribution to this book).

3.2 X-space bounds.

3.2.1 Theorem [7] : The two point function of model 0.1 obeys the bounds :

$$(8)\quad 0 \leqslant \langle \phi_0 \phi_x \rangle \leqslant c\,\frac{\ln |x|}{|x|} \qquad (d = 3)$$

$$\leqslant \frac{c}{|x|} \qquad (d \geqslant 4)$$

3.2.2 Let us first remark that up to logarithms, th. 3.2.1 is optimal in d = 3 because, as noticed in the introduction to section 3, it is known that the second derivative of $\langle \phi_0 \phi_x \rangle$ cannot decay faster than $|x|^{-3}$ (at least in some directions).

The positivity in (8) follows from F.K.G. inequalities see (3.2.5). Since B-L inequality has to be understood in the sense of forms, it does not imply directly anything pointwise. However combining it with the following weak nonotonicity property (P) of $\langle \phi_0 \phi_x \rangle$ we could get the bounds (8) in [7].

(P) : $\langle \phi_0 \phi_{2x} \rangle$ is a monotone decreasing function of $|x|$ when x is along a coordinate axis ($x = (x_1, o, o)$). As noticed in section 1.1, since $\exp(-x^2 - \lambda x^4)$ is not of positive type, we do not know wether our model is reflection positive. Nevertheless it is reflection positive with respect to planes crossing the sites because the interaction is nearest neighbour [16]. This is enough to deduce (P).

3.2.3 Property (P) says that for any interval $[-a,a]$ (centered at the origin), if $x \in [-a,a]$ then $\langle \phi_0 \phi_{2x} \rangle$ takes its minimum when x is at the edges. A. Sokal has remarked [41] that if we introduce the stronger property (P') below, then (P') together with the infrared bounds (th.1.1.3 or (1)) imply a bound of the type $\langle \phi_0 \phi_x \rangle \leqslant c|x|^{-(d-2)}$.

(P') : For any d-dimensional hypercube C centered at the origin, if $x \in C$, then $\langle \phi_0 \phi_x \rangle$ takes its minimum at the corners of the cube. Using correlation inequalities [36,40] one can show that (P') holds for the two point function of the Ising or Rotator models [41]. Actually for the rotator model, we could obtain in [8] the stronger result :

3.2.4 Theorem : For any $d \geqslant 3$, there exists constants c_1, c_2 such that for all $x \in \mathbb{Z}^d$, the two point function of model 0.3 obeys :

$$c_2\, \bar{m}^{-2}(\beta)[\,\beta\, |x|^{d-2}]^{-1} \leqslant \langle \sin\phi_0 \sin\phi_x\rangle \leqslant c_1[\beta\, |x|^{d-2}]^{-1}$$

This theorem shows that the low temperatures two point function $S(x) \equiv \langle \sin\phi_0 \sin\phi_x\rangle$ of the rotator model behaves exactly like $\beta^{-1}|x|^{-(d-2)}$ up to the critical point ($d \geqslant 3$) ! The proof of the theorem starts with the corresponding of inequalities (7) :

$$c_2'\, \bar{m}^{-2}(\beta)[\beta \sum_e (2 - 2\cos p_e)]^{-1} \leqslant \tilde{S}(p) \leqslant c_1'[\beta \sum_e (2-2\cos p_e)]^{-1}$$

and uses monotonicity properties of $\langle \sin\phi_0 \sin\phi_x \rangle$ that we can get from correlation inequalities [36,40,37].

3.2.5 Using the criterium of [2], it is easy to check that model 0.1 satisfies F.K.G. inequalities. By a suitable modification of the methods of[30,42],we use them in [7] to obtain a bound on the decay of all truncated correlation functions in terms of the two point function ($d \geqslant 3$). However these bounds are not optimal. It appears more convenient to study general correlation functions in momentum space.

3.3 <u>Momentum space bounds</u>.

3.3.1 <u>Notations</u> : If A is a "subset of $\bar{\Lambda}$ and B is a "subset" of $\Lambda \times \{e_1,\dots,e_d\}$ then $A + x \equiv \{i + x | i\in A\}$ and $B + x \equiv \{(i+x,\xi)|(i,\xi)\in B\}$. δ will denote an integral operator on $L^2(\mathbb{Z}^d)(d \geqslant 3)$ with kernel :

$$(2\pi)^{-3}\int \exp(ip.x)(-\Delta)^{-1}(p) * (\Delta)^{-1}(p)\, d^dp,$$

$*$ means convolution product and $(\Delta)^{-1}(p)=[\sum_e(2-2\cos p_e)]^{-1}$. We shall use the notation $\langle A;B \rangle = \langle AB \rangle - \langle A\rangle\langle B\rangle$.

3.3.2 <u>Theorem</u> [13] : For any $f\in L^2(\mathbb{Z}^d)$,

a. $$0 \leqslant \sum_{x,y} f(x)\, f(y) \langle \prod_{i\in A+x} \phi_i \,;\, \prod_{j\in A+y} \phi_j \rangle \leqslant c(f,Df)$$

where $D = (-\Delta)^{-1}$ if $|A|$ (the cardinality of A) is odd, $D = \delta$ if $|A|$ is even, and $d \geqslant 3$.

b. $$0 \leqslant \sum_{x,y} f(x)f(y) \langle \prod_{(i,\xi)\in B+x} \nabla_i^\xi\phi \,;\, \prod_{(i,\xi)\in B+y} \nabla_i^\xi \phi \rangle \leqslant c\, \|f\|_{L^2}.$$

3.3.3 Theorem [13] : in any dimension d,

$$< \nabla_0^e \phi \, \nabla_x^e \phi >^{\sim} (p) = 2(1 - \cos p_e)[2 \sum_e (1 - \cos p_e)]^{-1}$$
$$[1 - 3\lambda w + \lambda^2 \, c(p,\lambda)]$$

where $w = <(\nabla_0^e \phi)^2>_{\lambda=0}$ and $c(p,\lambda) \leqslant c$ uniformly in λ and p.

3.3.4 The proof of th. 3.3.2 is obtained using B-L inequalities in an inductive way. The method of duplicated variables allows us to do this elegantly. Instead of going through the whole construction of the induction, we want to explain the idea of combining the method of duplicated variables with B-L inequalities which originates in [5]. The most convenient way of doing this is to establish th. 3.3.2 for $|A| = 2$, namely to prove

3.3.5 $\sum_{x,y} f(x) \, f(y) < \phi_x^2 ; \phi_y^2 > \leqslant c(f,\delta f)$

Proof : Consider the unnormalized density

$$(9) \quad \exp \sum_{i,\xi} [-1/2 \, (\nabla_i^\xi \phi)^2 - \lambda/4 \, (\nabla_i^\xi \phi)^4 - 1/2(\nabla_i^\xi \phi')^2 - \lambda/4(\nabla_i^\xi \phi')^4]$$

where ϕ' is a duplication of ϕ. Using the variables $\psi_i^+ = 1/2(\phi_i + \phi_i')$, $\psi^- = 1/2(\phi_i - \phi_i')$, (9) becomes :

$$(10) \quad G(\psi^+,\psi^-) \equiv \exp 2 \sum_{i,\xi} [- \tfrac{1}{2} \, (\nabla_i^\xi \, \psi^+)^2 - \lambda/4(\nabla_i^\xi \, \psi^+)^4$$
$$-3/2 \, \lambda (\nabla_i^\xi \, \psi^+)^2 (\nabla_i^\xi \, \psi^-)^2 - 1/2(\nabla_i^\xi \, \psi^-)^2 - \lambda/4(\nabla_i^\xi \, \psi^-)^4]$$

< > will also denote the normalized measure associated to (10). We finally introduce

$$H(\psi^+,\psi^-) = \exp 2 \sum_{i,\xi} [-1/2(\nabla_i^\xi \, \psi^-)^2 - \lambda/4(\nabla_i^\xi \, \psi^-)^4$$
$$-3/2 \, \lambda \, (\nabla_i^\xi \, \psi^-)^2 (\nabla_i^\xi \, \psi^+)^2 \,]$$

and use the notation $\phi(f) = \sum_x \phi_x \, f(x)$.

This gives (with a slight abuse of notations because one should really consider $< >_{\Lambda,m}$ and, at the end, take the limits $\Lambda \uparrow \infty$ $m \downarrow o$):

$$(11) \quad \langle \phi^2(f);\phi^2(f) \rangle = \sum_{x,y} f(x)\, f(y) \langle \psi^+(x)\psi^+(y)\ \psi^-(x)\psi^-(y) \rangle$$

$$= \sum_{x,y} f(x)\, f(y) \int d\psi^+\ \psi^+(x)\psi^+(y).$$

$$\{\int d\psi^-\ \psi^-(x)\psi^-(y)\ H(\psi^+,\psi^-)[\int d\psi^-\ H(\psi^+,\psi^-)]^{-1}\}\int d\psi^-\ G(\psi^+\psi^-)\ Z_\Lambda^{-1}$$

We used the notations $d\psi^\pm \equiv \prod_{i\in\Lambda} d\psi_i^\pm$ and $Z_\Lambda = \int d\psi^-\ d\psi^+\ G(\psi^+,\psi^-)$. Now for any configuration of ψ^+,

$$2 \sum_{i,\xi} [\lambda/4\ (\nabla_i^\xi\ \psi^-)^4 + 3/2\ \lambda\ (\nabla_i^\xi\ \psi^+)^2(\nabla_i^\xi\ \psi^-)^2]$$

is convex. Therefore using B-L inequalities we have :

$$(11) \leq \sum_{x,y} f(x)\ f(y)\ C(x,y) \int d\psi^+\ \psi^+(x)\ \psi^+(y) \int d\psi^-\ G(\psi^+,\psi^-)\ Z_\Lambda^{-1}$$

$$= \sum_{x,y} f(x)\ f(y)\ C(x,y) \langle \phi(x)\ \phi(y) \rangle$$

A second application of B-L inequalities implies 3.3.5

3.3.6 Proof of th. 3.3.3 . It combines the I.P. formula (2) with the bounds of th. 3.3.2 . The I.P. formula gives :

$$\langle \nabla_0^e\ \phi\ \nabla_x^e\ \phi \rangle = \nabla_0^e\ \nabla_x^e C(o,x) - \lambda \sum_{i,\xi} \nabla_0^e\ \nabla_i^\xi\ C(o,i) \langle \nabla_x^e\ \phi(\nabla_i^\xi\phi)^3 \rangle$$

$$(11) \quad = \nabla_0^e\ \nabla_x^e\ C(o,x)$$

$$- 3\lambda \sum_{i,\xi} \nabla_0^e\ \nabla_i^\xi\ C(o,i)\ \nabla_x^e\ \nabla_i^\xi\ C(x,i) \langle (\nabla_i^\xi\ \phi)^2 \rangle$$

$$+ \lambda^2 \sum_{i,\xi} \sum_{j,\eta} \nabla_o^e\ \nabla_i^\xi\ C(o,i)\ \nabla_x^e\ \nabla_j^\eta\ C(x,j) \langle (\nabla_i^\xi\ \phi)^3(\nabla_j^\eta\ \phi)^3 \rangle$$

By the ordinary integration by parts formula,

$$\sum_{i,\xi} \nabla_0^e\ \nabla_i^\xi\ C(oi)\ \nabla_x^e\ \nabla_i^\xi\ C(x,i)$$

$$= \sum_i (-\ \Delta(\nabla_0^e\ C(o,i))\nabla_x^e\ C(x,i)$$

$$(12) = \nabla_0^e\ \nabla_0^e\ C(o,x)$$

Using (12) and taking the Fourier transform of both sides of (11), we get :

$$(13)\ S^{ee}(p) = [1-3\lambda < (\nabla_0^e \phi^2)>][2-2\cos p_e][\sum_e (2-2\cos p_e)]^{-1}$$

$$+ \lambda^2 \sum_{\xi,\eta} (\exp i\, p_\xi - 1)(\exp i\, p_\eta - 1)[\sum_e (2-2\cos p_e)]^{-1} B^{\xi,\eta}(p,\lambda)$$

ere $B^{\xi,\eta}(p,\lambda)$ is the Fourier transform of $<(\nabla_0^\xi \phi)^3 (\nabla_x^\eta \phi)^3>$ which well defined by th.3.3.2 . Now,

$$(14)\ (\exp ip_\xi - 1)(\exp i\, p_\eta - 1)\ [\sum_e (2-2\cos p_e)]^{-1} \leqslant c$$

d by th.3.3.2 and Schwarz inequality,

$$(15)\ B(p_\xi, p_\eta, \lambda) \leqslant c\ ,$$

iformly in λ and p.(13),(14) and (15) imply the result; note that $(\nabla_0^e \phi)^2 >$ in (13) can be replaced by $< (\nabla_0^e \phi)^2 >_{\lambda=0}$ because rturbation theory for local functions is asymptotic (th.2.1.1).

3.3.7 Remarks : Theorem 3.3.2 implies the bound $\leqslant < \phi_0^{2n+1}\ \phi_x^{2n+1}>^\sim (p) \leqslant c_1 [2 \sum_e (1-\cos p_e)]^{-1}$. On the other hand Mermin-Wagner argument [35,18] gives $\phi_0^{2n+1} ; \phi_x^{2n+1}>^\sim (p) \geqslant c_2\ [2 \sum_e (1 - \cos p_e)]^{-1}$. These two bounds show at the low momentum behaviour of these correlations is (up to a onstant) the same as in the purely Gaussian case (λ=o). The same emark applies to even truncated correlations $< \phi_0^{2n} ; \phi_x^{2n}>^\sim (p)$. This uggest that th.3.3.2 should be useful to study the scaling limit f this model. As predicted by renormalization group arguments this imit should be Gaussian.

The dielectric constant ε is a non local quantity; in other words, cannot be expressed as the expectation value of a function of $\{\phi_x\}_{x \in B}$ with B being a fixed finite region of $\mathbb{Z}^d$. Therefore the symptotic expansion of ε (as $\lambda \to 0$) does not follow from section 2. his is the reason why we had to obtain strong bounds on the decay f correlation to control its expansion (up to second order). The ielectric constant is a quantity very similar to the mass of a massive heory.

The question arising from the above remarks can be solved

completely in a simplified version of model 0.1 . This is the object of the next section.

4. The Scaling Limit.

In this last section we shall solve problem 0.8 for a simplified version of model 0.1 described by the Hamiltonian :

$$(16)\quad \beta H_\Lambda = 1/2 \sum_{\substack{i\in\Lambda \\ \xi}} (\nabla_i^\xi \phi)^2 - \lambda/4 \sum_{\substack{i\in\Lambda \\ \xi}} (\partial_\xi^2 \phi_i)^4$$

where $\partial_\xi^2 \phi_i = 2\phi_i - \phi_{i+e_\xi} - \phi_{i-e_\xi}$ and $\Lambda \subset \mathbb{Z}^d$ $(d \geqslant 3)$. $\langle\ \rangle$ will denote any state associated to (16) via the construction given for model 0.1 (see section 1). The model described by (16) is considerably simpler than model 0.1 because the two point function $w(ox) = \langle \partial_\xi^2 \phi_0 \partial_\xi^2 \phi_x \rangle$ has a rather fast decay ($\sim |x|^{-d-2}$) (at least in the Gaussian approximation) ; as will be seen below w(xo) is the relevant quantity to consider for this model. For weak coupling this model has been considered by Malyshev and Tirozzi [34], by Federbush [11] and more recently by Magnen and Sénéor [32]. We now want to show that *for any coupling* λ, the scaling limit (to be defined below) of this model is Gaussian.

4.1 Definition of the scaling limit [1,25,38].

For any integer a > 1 one defines "*block variables*".

$$A_x^a = \phi(B_x)/\langle \phi^2(B_0)\rangle^{1/2}$$

where B_x is the characteristic function of a cube of side length a centered at the lattice point ax. The weak limit of the random fields A_x^a when $a \to \infty$ is called the *scaling limit*.

4.2 Theorem [14]

$$\text{(i)}\quad \lim_{a\to\infty} \langle A_x^a A_0^a \rangle = \int \psi(E_0)\, \psi(E_x)\, d\mu(\psi)$$

where $d\mu(\psi)$ is the Gaussian measure associated to the massless free field in the continuum ($\int\psi(x)\,\psi(y)\,d\mu = \int d^d p\, \exp[ip(x-y)]\, p^{-2}$), and E_x is the characteristic function of a cube of side length 1 centered at the point x.

(ii) $\lim_{a\to\infty} <(A_o^a)^{2n}> = 2n\,!/(n!2^n)$

4.3 Idea of the proof of th. 4.2

4.3.1 Part (i) is based on the bounds

$$(17)\quad <\phi_0\phi_x>^{\sim}(p) \equiv S(p)$$
$$\geq [\sum_e (2-2\cos p_e)]^{-1}$$
$$- 3\lambda \sum_e (2-2\cos p_e)^2[\sum_e(2-2\cos p_e)]^{-2}<(\partial_e^2 \phi_0)^2>$$

$$(18)\quad S(p) \leq [\sum_e(2-2\cos p_e)]^{-1}$$

(17) can be obtained using the I.P. formula (see [14]) and (18) is just B-L inequality. Up to the λ-term, the upper and lower bounds on S(p) coincide. However this λ-term is less singular (as $p \to o$) than the Gaussian term $[\sum_e (2-2\cos p_e)]^{-1}$. Therefore in the scaling limit it will be negligible compared to the Gaussian two point function. We refer to [14] for more details.

4.3.2 For part (ii) let us restrict ourselves to the case n = 2. The result will follow from upper and lower bounds on the fully truncated four point function u_4 that both converge to zero as $a \to \infty$. u_4 is defined by :

$$u_4(i,j,k,l) \equiv <\phi(i)\,\phi(j)\,\phi(k)\,\phi(l)>$$
$$- <\phi(i)\,\phi(j)><\phi(k)\phi(l)>$$
$$- <\phi(i)\,\phi(k)><\phi(j)\phi(l)>$$
$$- <\phi(i)\,\phi(l)><\phi(j)\phi(k)>$$

$$u_4(f,f,g,g) \equiv \int f(i)\,f(j)\,g(k)\,g(l)\,u_4(i,j,k,l)\,di\,dj\,dl\,dk.$$

4.3.3 Lemma :

$$u_4\,(f,f,g,g) \leq 3\lambda \int[|f^{\sim}(p)|^2\,H(p)\,d^d p]\,<(\partial_e^2\phi)^2>\,<\phi^2(g)>$$

where $H(p) = \sum_e (1-\cos p_e)^2[\sum_e (1-\cos p_e]^{-2}$

Proof : It combines B-L inequalities and the method of duplicated variables. Using the variables ψ^+, ψ^- of section 3.3, (see also [29])

$$u_4(i,j,k,l) = 8(< \psi_i^- \psi_j^- \psi_k^+ \psi_l^+ > - < \psi_i^- \psi_j^- > < \psi_k^+ \psi_l^+ >)$$

As in (3.3.5), B-L inequalities imply :

$$<\psi^+(g)^2 \psi^-(f)^2 > \leq \sum_{i,j} f(i)\, f(j)\, C(i,j) < \psi^+(g)^2>$$

Therefore

$$8(< \psi^+(g)^2 \psi^-(f)^2> - < \psi^+ (g)^2> < \psi^-(f)^2>)$$
$$\leq 8 < \psi^+(g)^2 > \sum_{i,j} f(i)\, f(j)\, [\, C(i,j) - < \psi_i^- \psi_j^- > \,]$$
$$= < \phi^2(g)> \sum_{i,j} f(i)\, f(j)\, [C(i,j) - <\phi_i \phi_j >]$$

If we now use the lower bound (17) that we had on the full two point function $\tilde{S}(p)$ we get

$$\leq < \phi^2(g) > 3\lambda \int |\tilde{f}(p)|^2 H(p)\, d^d p <(\partial_e^2 \phi)^2>,$$

which proves the lemma.

4.3.4 The upper bound. Using the lemma,

$$\lim_{a\to\infty} u_4(B,B,B,B)/< \phi^2(B)>^2$$
$$\leq \lim_{a\to\infty} 3\lambda <(\partial_e^2 \phi)^2 > \int|\tilde{B}(p)|^2 H(p)\, d^d p < \phi^2(B)>^{-1}$$

As in 4.3.1 H(p) is less singular about p = o than $[\sum_e (2-2\cos p_e)]^{-1}$ which imply that

$$(19) \quad \lim_{a\to\infty} \int|\tilde{B}(p)|^2 H(p)\, d^d p <\phi^2(B)>^{-1} = 0$$

For details we refer to [14] but (19) essentially follows from a power counting argument.

4.3.5 The lower bound. Using the I.P. formula,

$$\text{0)}\quad u_4(B,B,B,B) \equiv \langle \phi^4(B) \rangle^c = 3\langle \phi^2(B)\rangle \sum_{i,j} B(i)\, B(j)[C(i,j)-\langle\phi_i\phi_j\rangle]$$
$$- 3\lambda \sum_e \langle \phi^3(B)[\sum_i (\partial_e^2\, \phi_i)^3\, h_i^e] \rangle$$

ere $h_i^e = \sum_x B(x)\, \partial_e^2\, C(x,i); \tilde{h^e}(p) = B(p)(2-2\cos p_e)[\sum_e(2-2\cos p_e)]^{-1}$

B-L inequalities, the first term of the r.h.s. of (20) is positive,
we have :

$$\langle \phi^4(B)^c\rangle \geqslant -3\lambda \langle \phi^3(B) \sum_{i,e} (\partial_e^2\, \phi_i)^3\, h_i^e \rangle$$

ing Schwarz inequality we get the bounds :

$$\langle \phi^4(B) \rangle^c \geqslant -3\lambda \langle \phi^6(B)\rangle^{1/2} [\sum_{i,jee'} \langle(\partial_e^2\, \phi_i)^3(\partial_{e'}^2\phi_j)^3\rangle\, h_i^e\, h_i^{e'}]^{1/2}$$

$$\text{1)}\quad \langle \phi^4(B)\rangle^c \geqslant -3\lambda \langle \phi^6(B) \rangle^{1/2}[\int|\tilde{B}(p)|^2 \sum_{ee'} (2-2\cos p_e)(2-2\cos p_{e'})$$
$$[\sum_e (2-2\cos p_e]^{-2} K^{ee'}(p)]^{1/2}$$

ere $K^{ee'}(p) = \langle(\partial_e^2\, \phi_0)^3(\partial_{e'}^2\phi_x)^3\rangle^{\sim}(p)$. If we now divide both sides
f (21) by $\langle \phi^2(B) \rangle^2$ and use the following estimates :

i) $K^{ee'}(p) \leqslant c$ which follows from th. 3.3.2

ii) $\langle \phi^6(B)\rangle^{1/2} \langle \phi^2(B)\rangle^{-3/2} \leqslant c$ which can be obtained as a
orollary to an extension of lemma 4.3.3 see [14], we get the
stimate :

$$\text{22)}\quad \langle(A_0^a)^4\rangle^c \geqslant -3\lambda\, c[\int|\tilde{B}(p)|^2|H'(p)|\, d^d\, p] \langle \phi^2(B) \rangle^{-1}]^{1/2}$$

here $H'(p) = \sum_{ee'} (2-2\cos p_e)(2-2\cos p_{e'})[\sum_e(2-2\cos p_e)]^{-2}$. As before
simple scaling argument shows that the r.h.s of (22) goes to zero
hen $a \to \infty$ because $H'(p)$ is less singular about $p = o$ than
$\sum_e(2-2\cos p_e)]^{-1}$ (for details see [14]). This finishes the proof
f th.4.2 (ii) for the case n = 4. The ideas presented above can
asily be extended to get the result for all n(see [14]).

5. Acknowledgements :

Many results presented in this article have been obtained in collaboration with J. Bricmont, J.L. Lebowitz, E.H. Lieb and T. Spencer. I am deeply indebted to them all. I would like also to thank J. Fröhlich for his constant interest and encouragements.

6. References.

[1] Baker, G.A., Krinsky, S. : Renormalization group structure for translationally invariant ferromagnets. J. Math. Phys. 18, 590-607 (1977)

[2] Battle, G.A., Rosen, L. : The F.K.G. inequality for Yukawa Quantum field theory, J. Stat. Phys. 22, 128 (1980).

[3] Brascamp, H.J., Lieb, E.H., Lebowitz, J.L. : The statistical mechanics of anharmonic lattices. Bull. Int. Statist. Inst. 46, Invited Paper n° 62 (1975).

[4] Brascamp, H.J., Lieb, E.H. : On extensions of the Brunn-Minkowski and Prekopa-Linder theorems, including inequalities for log concave functions, and with an application to the diffusion equation. J. Funct. Anal. 22, 366 (1976)

[5] Brascamp, H.J., Lieb, E.H. : Some inequalities for Gaussian measures. In : Functional Integration and its Applications (ed. A.M. Arthurs), Clarendon, Oxford (1975)

[6] Bricmont, J., Fontaine, J.R., Lebowitz, J.L., Spencer, T. : Lattice systems with a continuous symmetry. I Perturbation theory for unbounded spins. Commun. Math. Phys. 78, 281 (1980)

[7] Bricmont, J., Fontaine, J.R., Lebowitz, J.L., Spencer, T. : Lattice systems with a continuous symmetry. II Decay of correlations. Commun. Math. Phys. 78, 363 (1981)

[8] Bricmont, J., Fontaine, J.R., Lebowitz, J.L., Lieb, E.H., Spencer, T.: Lattice systems with a continuous symmetry. III Low temperature asymptotic expansion for the plane rotator model. Commun. Math. Phys. 78, 545 (1981).

[9] Mc Bryan, O., Spencer, T. : On the decay of correlations in SO(n)-symmetric ferromagnets. Commun. Math. Phys. 53, 299 (1977)

[10] Federbush, P. : A mass zero cluster expansion. Part 1. The expansion. Commun. Math. Phys. 81, 327 (1981)

[11] Federbush, P. : A mass zero cluster expansion. Part 2. Convergence. Commun. Math. Phys. 81, 341 (1981)

[12] Fontaine , J.R.: Low fugacity asymptotic expansion for classical lattice dipole gases. J. Stat. Phys. 26, 767 (1981)

[13] Fontaine , J.R. : Bounds on the decay of correlations for $\lambda(\nabla\phi)^4$ models. To appear in Commun. Math. Phys. (1982)

[14] Fontaine , J.R. : Scaling limit of some critical models. Preprint UCL-IPT-82-21.

15] Fröhlich, J. : Continuum (scaling) limits of lattice field theories (triviality of $\lambda\phi^4$ in $d \underset{(=)}{\geq} 4$ dimensions). Nucl. Phys. B200, [F54], 281 (1982).

16] Fröhlich, J., Israel, R., Lieb, E.H., Simon, B. : Phase transitions and reflection positivity. I General Theory and long range lattice models. Commun. Math. Phys. 62, 1 (1978).

17] Fröhlich, J., Simon, B., Spencer, T. : Infrared bounds, phase transitions and continuous symmetry breaking.

18] Fröhlich, J., Spencer, T. : On the statistical mechanics of classical Coulomb and dipole gases. J. Stat. Phys. 24, 617 (1981).

19] Fröhlich, J., Spencer, T. : The Kosterlitz-Thouless transition in two dimensional abelian spin systems and the Coulomb gas. Commun. Math. Phys. 81, 527 (1981)

20] Gawedzki, K., Kupiainen, A. : A rigorous block spin approach to massless lattice theories. Commun. Math. Phys. 77, 31(1980)

21] Gawedzki, K., Kupiainen, A. : Renormalization group study of critical lattice model. I Convergence to the line of fixed points. Commun. Math. Phys. 82, 407, 1981.

22] Gawedzki, K., Kupiainen, A. : Renormalization group study of critical lattice model.II The Correlation functions. Commun. Math. Phys. 83, 469 (1982).

23] Gawedzki, K., Kupiainen, A. : Renormalization group for a critical lattice model. Effective interactions beyond the perturbation expansion or bounded spins approximation. IHES/P/82/15.

24] Glimm, J., Jaffe, A., Spencer, T. : The particle structure of the weakly coupled $P(\phi)_2$ model and other applications of high temperature expansions. In : Constructive Quantum Field theory (eds. G. Velo and A. Whightman). Lecture notes in physics, vol. 25. Berlin, Heidelberg, New York : Springer 1973.

25] Glimm, J., Jaffe, A. : Particles and scaling for lattice fields and Ising models. Commun. Math. Phys. 51, 1 (1976)

26] Goldstone, J. : Field theories with superconductor" solutions. Nuovo Cimento 19, 154 (1961)

27] Hegerfelt, G. : Correlation inequalities for Ising ferromagnets with symmetries. Commun. Math. Phys. 57, 259 (1977)

28] Kac, M. : On the partition function of a one-dimensional gas. Phys. Fluids 2, 8 (1959).

29] Lebowitz, J.L. : G.H.S. and other inequalities. Comm. Math. Phys. 35, 87 (1974).

30] Lebowitz, J.L. : Bounds on the correlations and analyticity properties of ferromagnetic Ising spin systems. Commun. Math. Phys. 28, 313 (1972).

31] Ma, S.K. : Modern Theory of Critical Phenomena. London-Amsterdam-Don Mills, Ontario - Benjamin (1976).

32] Magnen, J., Sénéor, R. : A note on cluster expansions. Preprint, Ecole Polytechnique - Palaiseau.

33] Magnen, J. Sénéor, R. : The infra-red behaviour of $(\nabla\phi)^4_3$.Preprint Ecole Polytechnique - Palaiseau.

[34] Malyshev, V.A., Tirozzi, B. : Renormalization group convergence for small perturbations of Gaussian randon fields with slowly decaying correlations. Preprint Univ. of Rome.

[35] Mermin, N.D., Wagner, H. : Absence of ferromagnetism or antiferromagnetism in 1 or 2 dimensional isotropic Heisenberg models. P.R. Lett. 17, 1183 (1966).

[36] Messager, A., Miracle-Sole, S. : Correlation functions and boundary conditions in the Ising ferromagnet. J. Stat. Phys. 17, 245 (1977)

[37] Messager, A., Miracle-Sole, S., Pfister, C.E. : Correlation inequalities and uniqueness of the equilibrium state for the plane rotator ferromagnetic model. Commun. Math. Phys. 58, 19 (1978).

[38] Newman, C.M. : Normal fluctuations and the F.K.G. inequalities. Commun. Math. Phys. 74, 119 (1980)

[39] Park, Y.M. : Lack of screening in the continuous dipole systems. Commun. Math. Phys. 70, 161 (1979)

[40] Schrader, R. : New correlation inequalities for the Ising model and $P(\phi)$ theories. Phys. Rev. B15, 2798 (1977)

[41] Sokal, A.D. : An alternative constructive approach to the ϕ_3^4 quantum field theory, and a possible destructive approach to ϕ_4^4. Princeton University thesis to appear in Ann. Inst. Henri Poincaré.

[42] Simon, B. : The $P(\phi)_2$ Euclidean (quantum) field theory. Princeton N.J. : Princeton University Press (1974).

RIGOROUS RENORMALIZATION GROUP AND ASYMPTOTIC FREEDOM

by

K. GAWĘDZKI*) and A. KUPIAINEN**)

*) Department of Mathematical Methods of Physics, Warsaw University, 00682 Warsaw, Poland and Institut des Hautes Etudes Scientifiques, 91440 Bures-sur-Yvette, France.

**) Research Institute for Theoretical Physics, University of Helsinki, 00170 Helsinki 17, Finland.

1. Introduction.

In this lecture, we will present a rigorous non-perturbative analysis of the Wilson-Kadanoff [5],[18] block spin (BS) renormalization group (RG) for the case of certain critical lattice systems of unbounded spins such as the dipole gas and the like. Our aim is to establish the asymptotic freedom (AF) of these models in the infrared, i.e. to prove that the correlations become canonical Gaussian at long distances. This problem is the first of a series to which we propose to apply the rigorous RG. Some of these are :

(1) The infrared behavior of the abelian lattice gauge models in $d = 4$ in the Coulomb phase and of the $d = 2$ low temperature plane rotator [6], [7] .

(2) The solid-on-solid model of surface roughening [8].

(3) The infrared AF of the ϕ^4 lattice model in $d \geq 4$ at the critical temperature and of the many components ϕ^4 or Heisenberg models in $d \geq 4$ in the broken symmetry phase (for $2 < d < 4$ these models do not exhibit the infrared AF although in the latter ones the critical exponent $\eta = 0$ by the Goldstone picture).

(4) The $d = 2$ Heisenberg model : mass generation and ultraviolet AF , existence of the continuum limit.

(5) The $d = 4$ nonabelian lattice gauge theory : the problems as under (4).

The behavior in which we are interested of all the above models is believed, mostly on the bases of perturbative arguments, to exibit the common feature : the AF . In the RG language, this means that it is governed by the flow of the RG transformation in the vicinity of a massless Gaussian fixed point. In some of the models ((3) in $d = 4$, (4), (5)), the approach of the effective couplings towards triviality is slow (logarithmic) in the others it is fast (by power law). Up to now, we have been able to treat some cases in the second class. We are quite confident, however, that the methods of these lectures apply also to ϕ^4 in $d = 4$ which is in the first class; in fact the hierarchical version of this model has already been treated [1], [9].

The present lectures can be viewed as a realization of Wilson's idea to consider the RG as a transformation in a space of Hamiltonians of physical systems. We establish a quite general setting for this transformation and control its iterations. The Hamiltonians under consideration are of the form $H(\phi) = 1/2 \sum_x (\nabla\phi_x)^2 + V(\nabla\phi)$ and we show that every "small" V is driven by the RG to a Gaussian fixed point. This will be used to study the long distance behavior of the correlations.

We have tried to keep the lectures pedagogical and thus more technical details are skipped. The interested reader will find them in the original publications. The organization of these lectures is as follows.

In Section 2, we introduce the BS transformation and work out its realization in terms of the integral over fluctuations.

Section 3 is devoted to the study of a hierarchical model mimicking the behavior of more realistic statistical mechanical systems. The characteristic feature of the hierarchical model is its simplicity under the BS transformation which becomes a simple recursion for one-spin potentials. The recursion is introduced and studied in Sections 3.1 and 3.2. It is shown that under its iterations the one-spin potential is driven to the line of Gaussian fixed points. In Section 3.3, we extend the RG method to the hierarchical model correlation functions. The two-point function is shown to be free at long distances.

Section 4 extends the analysis of Section 3 to more realistic lattice models whose Hamiltonians depend on the gradient of the spin-field. The properties of the (non-local) effective interactions generalizing those of the hierarchical model single-spin potentials are stated (Section 4.2). In Section 4.3, it is briefly explained why they are stable under the BS transformation. Finally, in Section 4.4 we sketch how the RG method works for the correlation functions yielding the existence of the thermodynamical limit, the decay of the correlations and other information.

2. THE BLOCK-SPIN RENORMALIZATION GROUP.

2.1. Block Spin Transformation.

Consider a model of classical statistical mechanics of unbounded spins ϕ_x on a d-dimensional periodic lattice Λ (we take Λ to be the periodic cube of side L^N in $\mathbb{Z}^d$ centered at the origin and will suppress Λ in the notation). Let $H(\phi)$ be a translationally invariant Hamiltonian on Λ. The Gibbs state with periodic boundary conditions in finite volume Λ is determined by

$$< - >_H = 1/Z \int - \exp[-H(\phi)]\, D\phi \tag{1}$$

with $D\phi \equiv \prod_{x\in\Lambda} d\phi_x$ and the partition function $Z = \int \exp[-H(\phi)] D\phi$ assumed to be non-zero. The use of the periodic boundary conditions is not essential in what follows but results in technical simplifications due to the preservation of the translational invariance. We shall comment below on the existence of the thermodynamical limit.

The RG transformation consists of finding the effective Hamiltonian RH on the unit sub-lattice of $L^{-1}\Lambda$ describing the same physics as H except for fluctuations with wave length less than L, being disregarded by RH. For this purpose, divide Λ into cubes b_{Lx} (blocks) of L^d points centered at $Lx \in \Lambda$ and consider the average spins ϕ^{av} in the blocks :

$$\phi^{av}_{Lx} = L^{-d} \sum_{z\in b_{Lx}} \phi_z \ . \tag{2}$$

ϕ^{av} describes the "low momentum" part of ϕ, i.e. it forgets about the fluctuations within the blocks. To get the spins on the unit lattice, we define the rescaled block-spin field ϕ^1 by

$$\phi^1_x = L^{\alpha}\phi^{av}_{Lx} \equiv (C\phi)_x \ . \tag{3}$$

The scale factor L^{α} will be crutial in what follows. $RH(\phi^1)$ is now defined in finite periodic volume by integrating over ϕ with ϕ^1 held fixed :

$$\exp[-RH(\phi^1)] \; := \; \text{const} \int \exp[-H(\phi)]\delta(\phi^1 - C\phi)\, D\phi \quad . \tag{4}$$

Here $\delta(\phi^1 - C\phi) \equiv \prod_{x:Lx\in\Lambda} \delta(\phi^1_x - (C\phi)_x)$ and the constant in (4) will be chosen later (it is obviously immaterial as far as the distribution of ϕ^1 is concerned). In the cases which we shall study, the thermodynamical limit of (4) may be taken. The meaning of the exponent α in (3) is heuristicly as follows. For the infinite volume two-point function we expect for large x ($>> L$)

$$\begin{aligned} G(Lx;H) &\equiv \langle \phi_o \phi_{Lx} \rangle_H \approx \langle \phi_o^{av} \phi_{Lx}^{av} \rangle_H = \\ &= L^{-2\alpha} \langle \phi_o^1 \phi_x^1 \rangle_{RH} = L^{-2\alpha} G(x;RH) \ . \end{aligned} \tag{5}$$

Hence

$$G(L^n x;H) \approx L^{-2n\alpha} G(x;R^n H) \ . \tag{6}$$

Assume now that $R^n H$ stays "bounded" as $n \to \infty$ so that

$$0 < C_o \leq |G(x,R^n H)| < C_1 < \infty \tag{7}$$

for $|x| \sim O(1)$. Then (6) implies

$$G(x;H) = O\left(\frac{1}{|x|^{2\alpha}}\right) \quad \text{as} \quad |x| \to \infty \quad . \tag{8}$$

H for which (7) holds is called critical. Of course H may be critical only for one value of α . α determines the <u>critical exponent</u> η for the correlation function of the critical H : $\alpha = 1/2(d-2+\eta)$. In very vague terms at this point, we may consider some "space of Hamiltonians" M where R acts. Let R have some fixed point H^* in M , i.e. $RH^* = H^*$. Let $\mathcal{D}(H^*)$ be the stable manifold of H^* i.e. $R^n H \xrightarrow[n\to\infty]{} H^*$ for $H \in \mathcal{D}(H^*)$ (e.g. in the sense that $\langle - \rangle_{R^n H}$ converge). Then $\mathcal{D}(H)$ forms a <u>critical surface</u> in M and all $H \in \mathcal{D}(H)$ have the same long distance behavior : the principle of <u>universality</u>. The above discussion is of course extremely qualitative. Our purpose in these lectures is to show that a rigorous analysis along these lines is indeed possible, yielding rigorous information on a wide class of critical models. In particular, we wish to discuss the flow of the RG transformation in the vicinity of the massless Gaussian fixed point.

2.2. Gaussian Fixed Point.

We shall discuss in this paper only the scaling where the field has its canonical dimension, i.e.

$$\alpha = \frac{d-2}{2} \quad \text{or} \quad \eta = 0 \quad . \tag{9}$$

The corresponding Gaussian fixed point H^* is most easily found by picking any massless Gaussian H and computing $\lim_{n\to\infty} R^n H$. Let us work directly in the infinite volume limit (we disregard here the problem of the zero mode which will be addressed later).

Let $< - >_H$ be given by the Gaussian measure $d\mu_G(\phi)$ with covariance G . Let $\hat{G}(p)$, the Fourier transform of G , satisfy : $\hat{G}(p)^{-1}$ is analytic in $([-\pi,\pi]+i]-\varepsilon , \varepsilon[)^d \subset \mathbb{C}^d$ $(\varepsilon > 0)$ with $\widetilde{G}(p)^{-1}/p^2 \xrightarrow[p^2\to 0]{} 1$ and $|\widetilde{G}(p)^{-1}/p^2|$ is bounded from above and below. Thus $\widetilde{G}$ is "anything" having the massless pole $1/p^2$. RH is now obtained by computing its generating functional. Formally

$$\frac{1}{Z^1}\int \exp[-RH(\phi^1) + i(f|\phi^1)]D\phi^1 = \int \exp[i(f|C\phi)]d\mu_G(\phi)$$

$$= \exp[-1/2(f|CGC^+f)] \quad . \tag{10}$$

Thus, RH is again Gaussian with the covariance CGC^+ . Hence R^nH has the covariance $G_n = C^nG(C^+)^n$. From (2) and (3)

$$(G_n)_{xy} = L^{-n(d+2)} \sum_{\substack{u\in b^n_{L^n x} \\ v\in b^n_{L^n y}}} G_{uv} \; , \tag{11}$$

where $b^n_{L^n x}$ is the block of size L^n centered at $L^n x$. The reader may now easily check that as $n \to \infty$

$$G_n \to G^*$$

(e.g. pointwise for the kernels in $d \geq 3$), where

$$G^*_{xy} = \int_{\square(x)} du \int_{\square(y)} dv \int \frac{d^dp}{(2\pi)^d} e^{ip(x-y)} (\lim_{n\to\infty} L^{-2n}\widetilde{G}(L^{-n}p)) =$$

$$= \int_{\square(x)} du \int_{\square(y)} dv \, (-\Delta)^{-1}(x-y) \tag{12}$$

with Δ being the continual Laplacian on $\mathbb{R}^d$ and $\square(x)$ the unit square centered at x . $H^*(\phi) = 1/2(\phi|G^{*-1}\phi)$ is the Gaussian fixed point of the BS RG . Explicitely, in the momentum space,

$$\widetilde{G}^*(p) = \sum_{M\in\mathbb{Z}^d} (p + 2\pi M)^{-2} \prod_{\mu=1}^{d} \frac{4\sin^2(p_\mu/2)}{(p_\mu + 2\pi M_\mu)^2} \quad . \tag{13}$$

2.3. Fluctuation Fields.

Let us now consider a non-Gaussian H . We take H of the form

$$H(\phi) = 1/2(\phi|G^{-1}\phi) + V(\phi) \tag{14}$$

i.e. a perturbation of a Gaussian. G could be here any critical covariance of the form described in the previous section. For definiteness and to stay conventional we take

$$(\phi|G^{-1}\phi) = \sum_{x\in\Lambda} (\nabla\phi_x)^2 , \tag{15}$$

∇ being the lattice gradient.

The RG transformation (4) clearly involves integration over the degrees of freedom corresponding to fluctuations within the blocks. Let us realize this assertion explicitely. Given the BS configuration ϕ^1 , choose the most probable ϕ, provided that the latter is distributed according to $d\mu_G(\phi)$. This is obviously done by minimizing $1/2(\phi|G^{-1}\phi)$ with the constraint $C\phi = \phi^1$. The minimum ϕ_c is easily computed :

$$\phi_c = GC^+(CGC^+)^{-1}\phi^1 \equiv A\phi^1 \quad . \tag{16}$$

Now, write general ϕ as $\phi_c + \tilde{Z}$. Note that $C\tilde{Z} = C(\phi-\phi_c) = 0$ so that $\tilde{Z}$ describes the fluctuations within the blocks. The point of this decomposition is that it makes the Gaussian part of (14) split

$$1/2(\phi|G^{-1}\phi) = 1/2(\phi_c|G^{-1}\phi_c) + 1/2(\tilde{Z}|G^{-1}\tilde{Z}) \quad . \tag{17}$$

The first term on the RHS of (17) is by (16)

$$1/2(\phi^1|(CGC^+)^{-1}\phi^1) = 1/2(\phi^1|G^{-1}\phi^1) \quad . \tag{18}$$

(17) leads to the factorization of the Gaussian measure $d\mu_G(\phi)$:

$$d\mu_G(\phi) = d\mu_{G_1}(\phi^1) \cdot 1/N\ \delta(C\tilde{Z})\ \exp[-1/2(\tilde{Z}|G^{-1}\tilde{Z})]D\tilde{Z} \tag{19}$$

with

$$\phi = A\phi^1 + \tilde{Z} \quad . \tag{20}$$

It is convenient to choose a parametrization of the hyperplane $C\tilde{Z} = 0$ where the fluctuation measure of (19) lives. A general solution of $C\tilde{Z} = 0$ can be obtain e.g. by taking an arbitrary field $\overline{Z}$ on $\mathbb{Z}^d \setminus L\mathbb{Z}^d$ and setting

$$\tilde{Z}_x = (Q\overline{Z})_x \equiv \begin{cases} \overline{Z}_x & \text{for } x \in \mathbb{Z}^d \setminus L\mathbb{Z}^d \\ -\sum_{y\in b_x \setminus \{x\}} Z_y & \text{for } x \in L\mathbb{Z}^d \end{cases} \quad . \tag{21}$$

Then

$$1/N\ \delta(C\tilde{Z})\exp[-1/2(\tilde{Z}|G^{-1}\tilde{Z})]D\tilde{Z} = 1/N'\exp[-1/2(\overline{Z}|Q^{+}G^{-1}Q\overline{Z})]D\overline{Z}$$

$$\equiv d\mu_{\Gamma}(\overline{Z}) \tag{22}$$

where

$$\Gamma^{-1} = Q^{+}G^{-1}Q\ . \tag{23}$$

The BS transformation of (4) may be written now as

$$\exp[-RH(\phi^{1})]D\phi^{1} = \text{const}\ d\mu_{G_1}(\phi^{1}) \int \exp[-V(A\phi^{1}+Q\overline{Z})]d\mu_{\Gamma}(\overline{Z}) \tag{24}$$

which is the desired realization of the RG as an integration over fluctuations.

Let us pause for a moment to discuss various kernels introduced so far. The reader might have been worried by certain manipulations (e.g. (16)) with G^{-1} which has a zero mode. We resolve these problems by introducing an infrared regulator to G. We define

$$G^{-1} = -\Delta + \xi P_o \tag{25}$$

where P_o is the projection onto the zero mode of Δ, i.e. onto the constant functions on Λ.

With $\xi > 0$ all the operators introduced so far are well defined. Moreover, the BS transformation is independent of ξ. This is quite obvious since we are integrating over fluctuations orthogonal to constants. Explicitely, first $P_oQ = 0$ and so Γ does not depend on ξ. Similarly for A we only need to show that $P_o\ A\ P_o$ is ξ independent. But

$$C1 = L^{\frac{d-2}{2}}\ ,\ C^{+}1 = L^{-\frac{d+2}{2}} \tag{26}$$

and so

$$CP_oC^{+} = L^{-2}P_o\ . \tag{27}$$

Thus,

$$P_oG_1P_o = L^{-2}\xi^{-1}P_o \tag{28}$$

and

$$A1 = \xi^{-1}\ L^{-\frac{d+2}{2}}\ L^{2}\xi = L^{-\frac{d-2}{2}} \tag{29}$$

verifying the ξ independence. (29) shows in fact that the kernel A is integrable uniformly in the volume Λ. Actually, by Fourier analysis, one can show that A has an exponential fall-off uniform in Λ (see [10]). Because of C^+ in the definition (16) of A, y in A_{xy} is localized near $L^{-1}x$ (recall (3)). Hence if we define for $x \in L^{-1}\Lambda$, $y \in \Lambda$

$$A_{xy} = L^{\frac{d-2}{2}} A_{Lxy} \tag{30}$$

then

$$A1 = 1 \tag{31}$$

and

$$|A_{xy}| \leq C \exp[-\varepsilon d(x,y)] . \tag{32}$$

It turns out that ε and C may be taken L-independent.

The covariance Γ has an exponential fall-off too (see [10]). Intuitively this is obvious since Γ involves only high momentum fluctuations.

Now (20) may be written as

$$\phi_x = L^{-\frac{d-2}{2}} (A\phi^1)_{L^{-1}x} + (Q\overline{Z})_x \equiv L^{-\frac{d-2}{2}} \psi^1_{L^{-1}x} + Z_x . \tag{33}$$

Due to (31), ψ^1 is a correctly normalized field in the infrared, realizing the scaling involved in the BS transformation.

Our principal objective, according to the discussion in Section 2.1, is to study the BS transformation of V as induced from (24):

$$\exp[-R\mathcal{H}(\phi^1)]D\phi^1 = \text{const } d\mu_{G_1}(\phi^1) \exp[-V^1(\psi^1)] , \tag{34}$$

$$\exp[-V^1(\psi^1)] = \int \exp[-V(L^{-\frac{d-2}{2}} \psi^1_{L^{-1}\cdot} + Z)]d\mu_\Gamma(\overline{Z}) . \tag{35}$$

(35) involves an integration over the field $\overline{Z}$ with massive covariance Γ. This is the point of the RG: to reduce the study of a critical theory to that of a succession of non-critical problems. In analyzing (35), we shall partly use techniques for non-critical theories such as the high temperature (cluster) expansions. One has to extract carefully from (35) contractive properties of the transformation $V \to V^1$. This is the crutial part of the problem. To gain some insight into it, we first discuss (35) in a hierarchical approximation which essentially consists of replacing the exponentially decaying kernels by finite range ones in such a way that the map $V \to V^1$ becomes a transformation of finite number of variables instead of involving infinite (as $\Lambda \to \mathbb{Z}^d$) number of them.

3. HIERARCHICAL MODEL

3.1. Recursion Relation.

In a certain sense the hierarchical model is an approximate solution to the problem of finding the "local part" of V^1. Let us explain this in more detail. To start with, assume that V is a local function of ϕ

$$V(\phi) = \sum_{x\in\Lambda} v(\phi_x) . \tag{1}$$

Then, due to non-locality of Γ, TV is not local any more, in general. Computing V^1 perturbatively from (2.35), we obtain formally, e.g. for v a polynomial,

$$V^1 = \sum_{n=1}^{\infty} \frac{(-1)^{n+1}}{n!} < \prod_1^n V(L^{-\frac{d-2}{2}} \psi^1_{L^{-1}\cdot} + Z) >^T_\Gamma =$$

$$= \sum_{m=0}^{\infty} \sum_{x_1,\ldots,x_{2m}} K_m(x_1,\ldots,x_{2m}) \prod_{i=1}^{2m} \phi^1_{x_i} , \tag{2}$$

where the K_m's are given in terms of connected graphs with Γ lines and $2m$ A legs. Define the zero momentum part of K_m as k_m

$$k_m = \sum_{x_2,\ldots,x_{2m}} K_m(x_1,\ldots,x_{2m}) . \tag{3}$$

The local part of V^1 is then given by

$$V^1_{loc}(\psi^1) = \sum_x v^1_{loc}(\phi^1_x) \tag{4}$$

with v^1_{loc} expressed by the formal power series

$$v^1_{loc}(\varphi) = \sum_{m=0}^{\infty} k_m \varphi^{2m} . \tag{5}$$

Non-perturbatively,

$$v^1_{loc}(\varphi) = L^d \, |\Lambda|^{-1} \, V^1 |_{\phi^1_x \equiv \varphi} \tag{6}$$

or

$$\exp[-v^{1}_{loc}(\varphi)] = \{\int \exp[-\sum_{x\in\Lambda} v(L^{-\frac{d-2}{2}}\varphi + Z_x)]d\mu_{\Gamma}(\bar{Z})\}^{L^d/|\Lambda|} \quad . \tag{7}$$

v^1_{loc} gives in some sense the leading infrared part of V_1. One may take in (7) $|\Lambda| = L^d$. This should be a qualitatively reasonable approximation of the big Λ case since the Z field has covariance $Q\Gamma Q^+$ with exponential decay on a scale of the order L. For $|\Lambda| = L^d$, the right hand side of (7) becomes an integral involving only $L^d - 1$ degrees of freedom given by the $\bar{Z}$ field within one block. To simplify the problem as far as possible, we shall still replace the one-block-fluctuation field $Z = Q\bar{Z}$ by a specific fluctuation pattern setting $Z_x = B_x z$, L even, $B_x = \pm 1$ with $\sum_x B_x = 0$ and taking z distributed according to $d\nu(z) \equiv (2\pi)^{-1/2}\exp[-1/2\, z^2]dz$. All this produces the following approximate recursion for the local potential :

$$\exp[-tv(\varphi)] = \text{const} \int \exp[-1/2\, L^d(v(L^{-\frac{d-2}{2}}\varphi+z)$$
$$+ v(L^{-\frac{d-2}{2}}\varphi-z))]\, d\nu(z) \; . \tag{8}$$

For $L = 2$ this is similar to Wilson's approximate recursion [18],[19] which is

$$\exp[-t_W v(\varphi)] = \text{const}\,\{\int \exp[-v(2^{-\frac{d-2}{2}}\varphi+z)-v(2^{-\frac{d-2}{2}}\varphi+z)]$$
$$\cdot\, d\nu(z)\}^{2^{d-1}} \; . \tag{9}$$

(8) is also a d-dimensional generalization of the RG transformation for the potential of Dyson's hierarchical model [2], [4], [5] with canonical scaling. Alternatively, one obtains the RG transformation for the Dyson model directly from (8) by making a suitable choice of L and d $(L^d = 2\, ,\; L^{-\frac{d-2}{2}} = \frac{1}{\sqrt{c}})$.

In these lectures we are actually interested in models where V depends on ϕ only through $\nabla\phi$. In this case

$$V(\nabla\phi) = \sum_x v(\nabla\phi_x) \tag{10}$$

would be considered local. An example is provided by the dipole gas in the sine-Gordon picture with

$$v(\chi) = \lambda\sum_{\mu}(1-\cos(\gamma\chi_{\mu}))$$

or the anharmonic crystal with

$$v(\chi) = \lambda \sum_\mu \chi_\mu^4 .$$

For such V the BS transformation reads

$$\exp[-V^1(\nabla\psi')] = \int \exp[-V(L^{-\frac{d}{2}} \nabla\psi^1_{L^{-1}\cdot} + \nabla Z)] d\mu_\Gamma(\bar{Z}) . \tag{11}$$

Now an approximate recursion may be obtained similarly as in (8) (with additional inessential replacement of the vector field χ by a scalar φ) yielding

$$\exp[-tv(\varphi)] = \text{const} \int \exp[-1/2\, L^d(v(L^{-\frac{d}{2}}\varphi+z)+v(L^{-\frac{d}{2}}\varphi-z))]d\nu(z) . \tag{12}$$

(12) mimics again the scaling properties of the BS transformation (11). The following two subsection are devoted to the analysis of (12). Later, we shall see that the recursion (12) (as well as (8)) is a RG transformation for a statistical mechanical (hierarchical) model which mimics the massless models with $H(\phi) = 1/2(\phi|-\Delta\phi) + V(\nabla\phi)$. We shall show how the knowledge of the properties of (12) yields the information about the long distance behavior of the correlation functions for this hierarchical model.

3.2. Stability of the (Line of) Gaussian Fixed Points.

In what follows, we shall restrict the transformation t of (12) to even potentials v , $v(0) = 0$, and shall choose the constant in (12) so that also $tv(0) = 0$. First, note that t possesses a line of fixed points $v(\varphi) = 1/2s\varphi^2$ for arbitrary $s > L^{-d}$. The differential dt of t , the linearized RG transformation, at $v = 0$ is given by

$$dt(\delta v)(\varphi) = L^d \int \delta v(L^{-\frac{d}{2}}\varphi+z)d\nu(z) - L^d \int \delta v(z)d\nu(z) . \tag{13}$$

Its spectrum in the space of even polynomials is easily found to be $\{L^{-nd}\}_{n=0}^\infty$ with $n = 0$ corresponding to the marginal φ^2 direction along the line of fixed points and L^{-nd} to a polynomial of degree 2n+2. Apart from the marginal direction, we thus expect the fixed point to be stable. In other words a potential sufficiently close to the fixed-point line should go under iterations of t to a point on the line. This will be proven in the present subsection.

The linearized transformation (13) corresponds to the first term in the expansion of (12) in powers of v . It is clear that such an expansion in general cannot be convergent. For instance, if we take $v(\phi) = \lambda\varphi^4$, we encounter the familiar divergences of the perturbation theory : $tv(\varphi)$ will have an expansion $\sum_{n=1}^{\infty} c_{2n}\varphi^{2n}$ with $c_{2n} \sim (2n)!$, i.e. the expansion is divergent even for φ small. The reason for this is also clear. It does not make any sense to expand the exponent of (12) once the argument of v, $L^{-d/2}\varphi \pm z$ is large in absolute value. This may happen even for $|\varphi|$ small since z is unbounded. Thus, three regions in (φ,z) space enter naturally. For $|\varphi|$, $|z|$ small, the perturbative approach should be correct. For $|\varphi|$ small (12) should receive contribution from large $|z|$ region which are small due to small probability of big $|z|$ in the Gaussian measure $d\nu(z)$, provided that $\exp[-v(\psi)]$ is bounded appropriately for large $|\psi|$. But such bounds are essential for the stability of the model. This brings us to the third region : $|\varphi|$ large. In a system of unbounded spins the Gibbs factor $\exp[-v(\varphi)]$ can diverge at most like a Gaussian for $|\varphi|$ large. We shall show that this property iterates under t.

Let us consider the region of small $|\varphi|$, $|\varphi| < B$, first. B will be specified later. Divide the integral in (12) to small and large $|z|$ using the characteristic function $\chi(z)$ of $\{z : |z| < \varepsilon B\}$ (ε will also be specified later on). Then

$$\exp[-tv(\varphi)] = N^{-1} \int \exp[-w(\varphi,z)]\chi(z)d\nu(z) + $$
$$+ N^{-1} \int \exp[-w(\varphi,z)]\ (1-\chi(z))\ d\nu(z) \tag{14}$$

with

$$w(\varphi,z) = 1/2\ L^d(v(L^{-\frac{d}{2}}\varphi+z)+v(L^{-\frac{d}{2}}\varphi-z)) \tag{15}$$

and

$$N = \int \exp[-w(0,z)]\ d\nu(z)\quad . \tag{16}$$

As argued above, for the first term on the RHS of (14), we expect the perturbation theory to be applicable. Assume that v has a convergent Taylor series at $\varphi = 0$. The linearized transformation suggests that the Taylor coefficients of the logarithm of the first term go down except for that of φ^2. It just seems natural to consider potentials $v(\varphi)$ analytic in $|\varphi| < B$. The second integral in (14) is of the order of $\exp[-C\varepsilon^2 B^2]$ provided that the stability bounds for v guarantee that $\exp[-w] \leq C\exp[1/4\ z^2]$, say. For B large, this gives a small correction to the first term (which e.g. for φ^4 becomes non-perturbative if

$B \sim \lambda^{-\delta}$). However, to assume that this contribution is not only small for $|\varphi| < B$ but also analytic, we need to assume that $\exp[-v(\psi)]$ is analytic in a strip around the reals : $|\mathrm{Im}\,\psi| < B$. In this set-up the convergence to the fixed point will come from the expansion of the analyticity region and the contraction of the Taylor coefficients, both due to the scaling $L^{-d/2}$ involved in the transformation.

Let us state the assumptions for v under which the convergence of $t^n v$ to the fixed-point line is to be proven. We assume for the Gibbs factor g ($=\exp[-v]$) the following

(A) $g(\varphi)$ is even , $g(0) = 1$, analytic in the strip $|\mathrm{Im}\varphi| < B$, satisfying there $|g(\varphi)| \leq \exp[\kappa|\varphi|^2]$.

(B) For $|\varphi| < B$, $g = \exp[-v]$ for analytic v ,

$$v(\varphi) = 1/2\ L^{-d}(c-1)\varphi^2 + \tilde{v}(\varphi)$$

where $\tilde{v}(0) = \frac{d^2}{d\varphi^2}\tilde{v}(0) = 0$, $L^{-d}|c-1| < \kappa$

and

$$|\tilde{v}| < \eta .$$

About the parameters, we shall assume that $L > L_o$, $\kappa < \kappa_o(L) < 1/2\ L^{-d}$, $B > B_o$, $\eta < \eta_o(B)$. $L > L_o$ is not very essential since we could effectively increase it by composing several transformations. Let us note that the dipole gas $(1-\cos(\gamma\varphi))$ and the anharmonic crystal $\lambda\varphi^4$ clearly satisfy our assumptions for v if λ is small enough. We shall prove the following :

<u>Theorem 1.</u> (a) Under the above assumptions on v ,
$\exp[-t^n v] \xrightarrow[n\to\infty]{} \exp[-1/2\ L^{-d}(c_\infty - 1)\varphi^2]$ uniformly on compacts in $\mathbb{R}$ (c_∞ depends on v).

(b) If v_λ is analytic in λ so is c_∞ .

To prove this result, take $B = n_o + n$, $\eta = \delta^{n_o+n}$, $0 < \delta < 1$. For n_o large enough, we shall show that tv satisfies (A) and (B) with $B = n_o+n+1$, $\eta = \delta^{n_o+n+1}$, $c \to c + \Delta c$, $|\Delta c| < \delta^{n_o+n}$ and $\kappa \to \kappa + O((n_o+n)^{-2})$. This will, upon iteration, prove (a). (b) will be a simple consequence of the Vitali Theorem.

Let us first show that $g^1(\varphi) \equiv \exp[-tv(\varphi)]$ is analytic for

$|\mathrm{Im}\varphi| < L^{d/2}B$, i.e. in an expanded region. Indeed, by the assumptions $\exp[-w(\varphi,z)]$ is analytic in this region and also bounded by

$$\exp[1/2\, L^d \kappa(|L^{-d/2}\varphi+z|^2 + |L^{-d/2}\varphi-z|^2)] = \exp[\kappa|\varphi|^2 + L^d\kappa|z|^2] \ .$$

Thus $\int \exp[-w]d\nu$ is analytic too if $\kappa < 1/2\, L^{-d}$ assuring stability. N of (16) is easily shown to be non-zero. The required analyticity of g' follows together with the bound

$$|g'(\varphi)| \le C \exp[\kappa|\phi|^2] \ . \tag{17}$$

Next, let $|\varphi| < (1-2\varepsilon)L^{d/2}(n_o+n+1)$. In the first integral of (14) $|L^{-d/2}\varphi\pm z| < B$, i.e. we are in the small field region of v . Write

$$g'(\varphi) = g_1'(\varphi)(1+g_2'(\varphi)) \tag{18}$$

where

$$g_1'(\varphi) = \int \exp[-w(\varphi,z)]\ \chi(z)\ d\nu(z) / \int \exp[-w(0,z)]\chi(z)d\nu(z) \tag{19}$$

and

$$g_2'(\varphi) = \int \exp[-w(\varphi,z)](1-\chi(z))d\nu(z)/(g_1'(\varphi) \int \exp[-w(0,z)]d\nu(z)) - (\phi = 0) \ . \tag{20}$$

This makes sense since $g_1'(\varphi) \neq 0$ as will be shown below. Notice that by (B)

$$\begin{aligned} g_1'(\varphi) &= \exp[-1/2\ L^{-d}(c-1)\varphi^2]\int \exp[1/2\ \ L^d(\tilde v(L^{-d/2}\varphi+z)+\tilde v(L^{-d/2}\varphi-z))](z) \\ &\cdot \exp[-1/2(c-1)z^2]d\nu(z) / \int \exp[-L^d\tilde v(z)]\chi(z)\exp[-1/2(c-1)z^2]d\nu(z) \\ &\equiv \exp[-1/2\ L^{-d}(c-1)\varphi^2]\tilde g_1'(\varphi) \ , \end{aligned} \tag{21}$$

where

$$|\tilde g_1'(\varphi)-1| \le L^d\ \delta^{n_o+n} \exp[C\delta^{n_o}] \ . \tag{22}$$

Hence,

$$\tilde v_1' \equiv -\log \tilde g_1' \tag{23}$$

is analytic and

$$|\tilde v_1'| \le L^d\ \delta^{n_o+n} \exp[C\delta^{n_o}] \ . \tag{24}$$

Now, since

$$\begin{aligned} \left|\int \exp[-w(\varphi,z)](1-\chi(z))d\nu(z)\right| &\le \exp[\kappa|\varphi|^2] \int (1-\chi(z)) \\ &\cdot \exp[L^d\kappa z^2]\ d\nu(z) \le \exp[-\varepsilon'(n_o+n)^2] \end{aligned} \tag{25}$$

for κ small enough and n_o big enough,

$$|g_2'(\varphi)| \leq C \exp[-\varepsilon'(n_o+n)^2] . \tag{26}$$

Thus,

$$\tilde{v}_2' \equiv -\log(1+g_2') \tag{27}$$

is analytic and

$$|\tilde{v}_2'| \leq C \exp[-\varepsilon'(n_o+n)^2] . \tag{28}$$

Evidently, we have

$$v' = 1/2\, L^{-d}(c-1)\varphi^2 + \tilde{v}_1' + \tilde{v}_2' \tag{29}$$

which is then analytic for $|\varphi| < (1-2\varepsilon)L^{d/2}(n_o+n+1)$ and bounded there by $L^d\, \delta^{n_o+n} \exp[C\, \delta^{n_o+n}]$. Finally, writing

$$\tilde{v}_1' + \tilde{v}_2' = 1/2\, L^{-d}\, \Delta c \varphi^2 + \tilde{v}' \tag{30}$$

with $\frac{d^2}{d\varphi^2}\tilde{v}'(0) = 0$, we obtain from the Cauchy integral formula :

$$|\Delta c| < \delta^{n_o+n} . \tag{31}$$

Moreover, for $|\varphi| < n_o+n+1$

$$|\tilde{v}'(\varphi)| \leq \sum_{m=4}^{\infty} \frac{1}{m!} \left| \frac{d^m}{d\varphi^m} (\tilde{v}_1'+\tilde{v}_2')(0) \right| |\varphi|^m$$

$$\leq L^d \delta^{n_o+n} \exp[C\delta^{n_o+n}] \sum_{m=4}^{\infty} [(1-2\varepsilon)L^{d/2}]^{-m} \leq \delta^{n_o+n+1} \tag{32}$$

where we have used the Cauchy bounds again.

We have thus shown that for $|\varphi| < n_o+n+1$ $g' = \exp[-v']$ where v' satisfies (B) with $B = n_o+n+1$, $\eta = \delta^{n_o+n+1}$ and $c \to c+\Delta c$, $|\Delta c| < \delta^{n_o+n}$ as desired. This implies that for $|\varphi| < n_o+n+1$

$$|g'(\varphi)| \leq \exp[\kappa |\varphi|^2]. \tag{33}$$

On the other hand for $|\varphi| \geq n_o+n+1$ and $|\text{Im}\varphi| < n_o+n+1$ by (17)

$$|g'(\varphi)| \leq \exp[(\kappa + C(n_o+n)^{-2})|\varphi|^2] \tag{34}$$

so that g' satisfies also (A) for $B = \delta^{n_o+n+1}$ and $\kappa \to \kappa + O((n_o+n)^{-2})$ This completes the inductive step.

3.3. <u>The Correlation Functions.</u>

As mentioned at the end of Sec. 3.1, (12) is a RG transformation for the local potential of a hierarchical model. Here, we shall show how the methods and the results of Section 3.2 can be used to carry out the long-distance analysis of the correlation functions of this model. As a by-product, we shall also prove the existence of the thermodynamical limit.

The hierarchical model in question is obtained by noticing the relation underlying (12) between the original spin-field ϕ, the blocked one ϕ^1 and the fluctuation field Z :

$$\phi_x = L^{-d/2}\,\phi^1_{[L^{-1}x]} + B_x\, Z_{[L^{-1}x]} \;. \tag{35}$$

If the BS field ϕ^1 at y is φ and the fluctuation field $Z_y = z$ then for x in the block of side L centered at Ly, i.e. for such x that the integral part of $L^{-1}x$, $[L^{-1}x] = y$, the original field $\phi_x = \varphi + B_x z$, where the standard fluctuation pattern $B_x = \pm 1$ so that $\sum_x B_x = 0$ (L even). Notice that indeed $\phi^1 = C\phi$ (see (2.2), (2.3)) for $\alpha = d/2$, justifying the name of "BS field" given to ϕ^1. Iterating (35), we may express ϕ in terms of a hierarchy of fluctuation fields Z^k :

$$\phi_x = \sum_{k=0}^{N-1} L^{-dk/2}\, B_{[L^{-k}x]}\, Z^k_{[L^{-k-1}x]} \;. \tag{36}$$

Take first the "free" state $\langle - \rangle$ (in periodic volume Λ) with all Z^k independent and distributed according to the Gaussian measure $d\nu$. Then ϕ is also Gaussian and may be obtained from a Hamiltonian $H_0(\phi)$ (taking also $+\infty$ value) according to (2.1). The perturbed state $\langle - \rangle_v$ is obtained by replacing $H_0(\phi)$ by $H(\phi) = H_0(\phi) + \sum_x v(\phi_x)$ in (2.1). The BS Hamiltonian $RH(\phi^1)$ is given by (2.4) which becomes now the Z° integration factoring over the blocks :

$$\exp[-RH(\phi^1)] = \text{const.} \prod_{k=1}^{N-1} \prod_{x_{k+1}} d\nu(Z^k_{x_{k+1}}) \cdot$$
$$\cdot \prod_{x_1} \int \exp[- \sum_{x:[L^{-1}x]=x_1} v(L^{-d/2}\phi^1_{x_1} + B_x Z^\circ_{x_1})]\, d\nu(Z^\circ_{x_1}) \;. \tag{37}$$

It is thus straightforward that

$$RH(\phi^1) = H_0(\phi^1) + \sum_x tv(\phi^1) + \text{const} \tag{38}$$

as has been claimed.

Return to the free case which is, aside from the volume change, a fixed point of the RG transformation. The free two-point function is

$$<\phi_x\phi_y> = \sum_{k=k_o}^{\infty} L^{-dk} B_{[L^{-k}x]} B_{[L^{-k}y]} = L^{-dk_o}[\pm 1$$

$$+ (1-L^{-d(N-1-k_o)})/(L^d-1)] \xrightarrow[N\to\infty]{} L^{-dk_o}[\pm 1+(L^d-1)^{-1}] \tag{39}$$

where k_o is the smallest integer such that $[L^{-k_o-1}x] = [L^{-k_o-1}y]$. (39) establishes the existence of the thermodynamical limit for the free model. Notice that L^{-dk_o} is typically of order $\frac{1}{|x-y|^d}$ so that (39) mimics the decay of the two-point function of $\nabla\phi$ in the massless Gaussian case. Our aim is to show that this long-distance decay persists in the hierarchical model perturbed by a (small) local potential v except for the overall factor c_∞^{-1} of Theorem 1 (field-strength renormalization, inverse dielectric constant).

Consider $<\phi_x\phi_y>_v$. Suppose that $[L^{-1}x]\neq[L^{-1}y]$. The Z° integral factors and since by parity $\int z\exp[-w]d\nu = 0$, we get

$$<\phi_x\phi_y>_v = L^{-d} <\phi_{[L^{-1}x]}\phi_{[L^{-1}y]}>_{tv} \tag{40}$$

so that the relation (2.5) is exact now. We may iterate (40) until k_o, obtaining

$$<\phi_x\phi_y>_v = L^{-dk_o} <\phi_{[L^{-k_o}x]}\phi_{[L^{-k_o}y]}>_{t^{k_o}v} = <G_{k_o+1}(\phi_u)>_{t^{k_o+1}v}. \tag{41}$$

In the last step, we performed the Z^{k_o} integration ($u = [L^{-k_o-1}x] = [L^{-k_o-1}y]$) and have defined

$$G_{k_o+1}(\varphi) = L^{-d(k_o+1)}\varphi^2 + L^{-dk_o} B_{[L^{-k_o}x]} B_{[L^{-k_o}y]} <z^2>_{k_o} \tag{42}$$

using the notation

$$<->_k = \int - \exp[-w_k(\varphi,z)]d\nu(z)/\int\exp[-w_k(\varphi,z)]d\nu(z) \tag{43}$$

(w_k is (15) with v replaced by $t^k v$). Subsequent Z^k-integrations allow us to iterate the second equality of (42) :

$$<\phi_x\phi_y>_v = <G_{k_o+\ell}(\phi_{[L^{-\ell+1}u]})>_{t^{k_o+\ell}v} \tag{44}$$

where

$$G_{k_o+\ell+1}(\varphi) = <G_{k_o+\ell}(L^{-d/2}\varphi+\cdot)>_{k_o+\ell} \tag{45}$$

in the notation of (43). The claim is that G_n, $n > k_o$, have the following properties, to be shown inductively :

(A_n) $G_n(\varphi)\exp[-t^n v(\varphi)]$ is analytic for $|\mathrm{Im}\,\varphi| < n_o+n$, even and bounded there by $C\exp[2\kappa_n|\varphi|^2]$.

(B_n) For $|\varphi| < n_o+n$

$$G_n(\varphi) = a_n+b_n\varphi^2 + \tilde{G}_n(\varphi) \tag{46}$$

with $\tilde{G}_n(0) = \frac{d^2}{d\varphi^2}\tilde{G}_n(0) = 0$ and

$$|\tilde{G}_n(\varphi)| < DL^{-dn}\delta^{n_o+n}, \tag{47}$$

$$|a_{n+1}-a_n-b_n c_n^{-1}| < 2DL^{-dn}\delta^{n_o+n}, \tag{48}$$

$$|b_{n+1}-L^{-d}b_n| < DL^{-dn}\delta^{n_o+n}, \tag{49}$$

where $L^{-d}(c_n-1) = \frac{d^2}{d\varphi^2}t^n v(0)$ and D is an n-independent constant.

The proof of these claims is similar to the one of Section 3.2, where we estimated $t^n v$. Let us thus be brief. Assume that (A_k) and (B_k) hold for $k \leq n$. The upper bound of (A_{n+1}) will follow as before with $\kappa_{n+1} = \kappa_n+O((n_o+n)^{-2})$. For (B_{n+1}), we split again to small and large z in (45) whenever $|\varphi| < (1-2\varepsilon)\,L^{d/2}(n_o+n+1)$:

$$G_{n+1}(\varphi) = \langle G_n(L^{-d/2}\varphi+\cdot)\chi(\cdot)\rangle_n + \langle G_n(L^{-d/2}\varphi+\cdot)(1-\chi(\cdot))\rangle_n . \tag{50}$$

The second term is $O(\exp[-\varepsilon'(n_o+n)^2])$ whereas the first one equals

$$\begin{aligned}\langle \chi(z)\,G_n(L^{-d/2}\varphi+z)\rangle_n &= (a_n+L^{-d}b_n\varphi^2)(1+O(\exp[-\varepsilon'(n_o+n)^2]))\\ &+ b_n\langle\chi(z)z^2\rangle_n + \langle\chi(z)\tilde{G}_n(L^{-d/2}\varphi+z)\rangle .\end{aligned} \tag{51}$$

Now

$$\langle\chi(z)z^2\rangle_n = c_n^{-1}(1+O(\delta^{n_o+n})) . \tag{52}$$

Similarly,

$$|\langle\chi\tilde{G}_n\rangle| < DL^{-dn}\delta^{n_o+n}(1+C\delta^{n_o+n}) . \tag{53}$$

We may now extract the constant a_{n+1} and the quadratic term $b_{n+1}\varphi^2$ from G_{n+1}. At the start of the induction (see (42)), we obtain

$$a_{k_o+1} = \pm L^{-dk_o}c_{k_o}^{-1}(1+O(\delta^{n_o+k_o})) , \tag{54}$$

$$b_{k_o+1} = L^{-d(k_o+1)}(1+O(\delta^{n_o+k_o})) . \tag{55}$$

(55) and (49) for $k \leq n$ give

$$b_n = L^{-dn}(1+O(\delta^{n_o+k_o})) \ . \tag{56}$$

(50) - (53) and (56) with the use of the Cauchy estimates yield easily (B_{n+1}) .

For the two-point function, we get from (48), (54) and (55)

$$< \phi_x \phi_y >_v = a_N = a_{k_o+1} + \sum_{n=k_o+1}^{N-1} (a_{n+1}-a_n) =$$

$$= \pm L^{-dk_o} c_{k_o}^{-1}(1+O(\delta^{n_o+k_o})) + \sum_{n=k_o+1}^{N-1} L^{-dn} c_n^{-1}(1+O(\delta^{n_o+k_o}))$$

$$+ \sum_{n=k_o+1}^{N-1} L^{-dn} O(\delta^{n_o+n}) \ . \tag{57}$$

Since $|c_n - c_\infty| \leq C\,\delta^{n_o+n}$, see (31), (57) produces finally

$$< \phi_x \phi_y >_v = L^{-dk_o} c_\infty^{-1}[\pm 1+(L^d-1)^{-1}]\ (1+O(\delta^{n_o+k_o})) + O(L^{-dN}) \ . \tag{58}$$

The existence of the thermodynamic limit of $< \phi_x \phi_y >_v$ follows immediately from (58) together with the infinite-volume relation (see (39))

$$\lim_{|x-y|\to\infty} \frac{<\phi_x\phi_y>_v}{<\phi_x\phi_y>_0} = \lim_{k_o\to\infty} c_\infty^{-1}(1+O(\delta^{n_o+k_o})) = c_\infty^{-1} \ , \tag{59}$$

showing that the long distance behavior of the perturbed two-point function is the same as of the free one, except for the field-strength renormalization.

The analysis of the infinite volume limit of general correlations and of their infrared behavior follows along the same lines as that of the two-point function and will not be pursued here.

4. RENORMALIZATION GROUP FOR $\nabla\phi$ MODELS.

4.1. Iteration of the BS Transformation.

The specific goal of these lectures is to study the iterations of the BS transformation (2.35) in the case when V depends on ϕ through its gradient. Then let us suppose that V is a functional of a general vector field $\chi \equiv (\chi_{\mu x})$ on the unit lattice Λ. We assume V to be even and invariant under lattice translations, rotations and reflections. For a vector field $\chi^1 \equiv (\chi^1_{\mu x})$ on the L^{-1}-lattice $L^{-1}\Lambda$, define

$$\exp[-V^1(\chi^1)] = \int \exp[-V(L^{-d/2}\chi^1_{L^{-1}\cdot}+\nabla Z)]d\mu_\Gamma(\overline{Z}) \ . \tag{1}$$

The Gibbs factor for the BS field ϕ^1 is now given by (see (2.34), (2.35))

$$\exp[-RH(\phi^1)]D\phi^1 = \text{const}\ d\mu_{G_1}(\phi^1)\ \exp[-V^1(\nabla\psi^1)]. \tag{2}$$

(1) stems from the fact that we find it convenient to treat $\nabla\psi^1$ as a field at least as important as $\nabla\phi^1$, in particular to study the properties of the effective interaction as a functional of $\nabla\psi^1$ (or more generally χ^1) rather than of $\nabla\phi^1$. The reason for this is the local nature of the relation between $\nabla\phi$ and $\nabla\psi^1$

$$\nabla\phi_x = L^{-d/2}\ \nabla\psi^1_{L^{-1}x} + \nabla Z_x \ , \tag{3}$$

as opposed to the one between $\nabla\phi$ and $\nabla\phi^1$, which are related by a kernel with an exponentially decaying tail. This locality is very convenient in the study of the behavior of the effective interaction for large fields.

Let us first examine the linearization of the RG transformation (1).

$$\delta V^1(\chi^1) = \int \delta V(L^{-d/2}\chi^1_{L^{-1}\cdot} + \nabla Z)\ d\mu_\Gamma(\overline{Z}) \ . \tag{4}$$

If we take

$$\delta V(\chi) = \int dx \prod_{i=1}^{2m} D_i^{k_i} \chi_{\mu_i x} , \tag{5}$$

where $D_i^{k_i}$ is a product of $k_i \geq 0$ (lattice) partial derivatives and the integral sign stands for the lattice Riemann sum then

$$\delta V^1(\chi^1) = L^{d(1-m)-\sum_i k_i} \int dx \prod_{i=1}^{2m} D_i^{k_i} \chi^1_{\mu_i x}$$

$$+ \text{ terms of order less than } 2m . \tag{6}$$

From (6), it follows that the RG transformation preserves (the shape of) the interaction if $m = 1$ and $k_i = 0$ (the marginal case) whereas, it has a contractive effect on the one of order 2 but with more derivatives or of order higher than two (irrelevant interactions). The only marginal interaction which respects the lattice Euclidean symmetries is proportional to $\sum_{\mu,x} (\chi_{\mu,x})^2 \equiv (\chi|\chi)$.

It is easy to check that if $V(\chi) = 1/2 \; s(\chi|\chi)$ then $V^1(\nabla\psi^1) = 1/2 \; s(\nabla\psi^1|\nabla\psi^1) + \text{const}$, not only in the first order approximation. Indeed, this follows by (1) since, with $G_1 \equiv CGC^+$,

$$(L^{-d/2}\nabla\psi^1|\nabla Z) = (\nabla A\phi^1|\nabla Q\overline{Z}) = (-\Delta A\phi^1|Q\overline{Z})$$

$$= (C^+G_1^{-1}\phi^1|Q\overline{Z}) = (G_1^{-1}\phi^1|CQ\overline{Z}) = 0 . \tag{7}$$

Addition of $V(\nabla\phi) = 1/2 \; s(\nabla\phi|\nabla\phi)$ to the Hamiltonian $1/2 \; (\phi|G^{-1}\phi)$ changes only its overall coefficient from 1 to $(1+s)$, (we disregard here for simplicity the infrared regulator ξ) . Hence the BS Hamiltonian becomes $1/2 \; (1+s)(\phi^1|G_1^{-1}\phi^1)$, in agreement with the computation of V^1 since $1/2 \; s(\nabla\psi^1|\nabla\psi^1) = 1/2 \; s(\phi^1|A^+(-\Delta)A\phi^1) = 1/2 \; s(\phi^1|G_1^{-1}\phi^1)$.

For the general case, we shall decompose V^1 into a marginal part proportional to $(\chi^1|\chi^1)$ and the irrelevant rest transverse to the marginal massless Gaussian direction. The first part will be absorbed into the change of the coefficient at the "free" part $1/2 \; (\phi^1|G^{-1}\phi^1)$ of the BS Hamiltonian. In the subspace transverse to the massless Gaussian direction the BS transformation will be proven to possess contractive properties, in agreement with the prediction of the linearized RG . To achieve the decomposition write

$$V^1(\nabla\psi^1) = V^1(0) + 1/2 \; (\nabla\phi^1|\widetilde{K}^1\nabla\phi^1) + \widetilde{V}^1(\nabla\psi^1) , \tag{8}$$

where $\widetilde{V}^1$ is the remainder of the Taylor expansion to the second order of $V^1(\chi^1)$ around $\chi^1 = 0$. Given kernel $\widetilde{K}^1$ (invariant under lattice

Euclidean symmetries), we may write uniquely by the momentum space decomposition

$$1/2\ (\nabla\phi^1|\tilde{K}^1\nabla\phi^1) = 1/2\ \Delta c(\phi^1|G_1^{-1}\phi^1) + 1/2\ (\nabla\nabla\phi^1|\tilde{K}_1^1\nabla\nabla\phi^1)\ . \tag{9}$$

The first expression on the RHS is just the desired marginal term. The second one may be cast into a clearly irrelevant form involving the $\nabla\nabla\psi^1$ field. To this end, notice that, since the average of ψ_x^1 over x with the integral part equal to y is $(CA\phi^1)_y = \phi_y^1$, it is easy to find local kernels k such that

$$\nabla\nabla\phi^1 = k\nabla\nabla\psi^1\ . \tag{10}$$

Using (10) the second term on the RHS of (9) may be rewritten as

$$1/2\ (\nabla\nabla\psi^1|k^+\tilde{K}_1^1 k\nabla\nabla\psi^1) \equiv 1/2\ (\nabla\nabla\psi^1|K^1\nabla\nabla\psi^1)\ . \tag{11}$$

Let us define

$$TV(\chi^1) = 1/2\ (\nabla\chi^1\ |\ K^1\nabla\chi^1) + \tilde{V}^1(\chi^1)\ . \tag{12}$$

With these transformations, (2) becomes

$$\exp[-RH(\phi^1)]D\phi^1 = \text{const}\ d\mu_{c_1^{-1}G_1}(\phi^1)\ \exp[-TV(\nabla\psi^1)] \tag{13}$$

where

$$c_1 = 1+\Delta c\ . \tag{14}$$

Upon iteration of the BS transformation, the above scheme stays essentially unchanged except for the fact that the fields ψ^n, which we use together with the block spin fields $\phi^n = C^n\phi$, live on finer and finer lattices $L^{-n}\Lambda$:

$$\psi_x^n = (A_n\phi^n)_x \tag{15}$$

where

$$(A_n)_{xy}\ ,\ x \in L^{-n}\Lambda\ ,\ y\ \ \text{integral}\ ,$$

are kernels with exponential decay in $d(x,y)$ (uniform in the volume and n), see [10]. The relation between the block-spin fields on the neighboring scales stays local for the fields ψ^n:

$$\nabla\psi_x^n = L^{-d/2}\nabla\psi_{L^{-1}x} + \nabla A_n\ Q\bar{Z}^n{}_x \equiv L^{-d/2}\nabla\psi^n_{L^{-1}x} + \nabla Z_x^n\ . \tag{16}$$

For a vector field χ^{n+1} on the L^{-n-1}-lattice $L^{-n-1}\Lambda$ we define

$$\exp[-V^{n+1}(\chi^{n+1})] = \int \exp[-T^nV(L^{-d/2}\chi^n_{L^{-1}\cdot} + \nabla Z^n)]\ d\mu_{c^{-1}\Gamma_n}(\bar{Z}^n)\ , \tag{17}$$

where the fluctuation covariance $\Gamma_n = Q^+ G_n Q$ is again exponentially decaying (uniformly in n) (fluctuations have effective mass gap !). The marginal contribution to the Gaussian piece of V^{n+1} receives again a special treatment leading to

$$V^{n+1}(\nabla\psi^{n+1}) = \text{const} + 1/2\ \Delta c_n(\phi^{n+1}|G_{n+1}^{-1}\phi^{n+1}) + 1/2(\nabla\nabla\psi^{n+1}|K^{n+1}\nabla\nabla\psi^{n+1}) + \tilde{V}^{n+1}(\nabla\psi^{n+1}) \ . \quad (18)$$

Finally,

$$\exp[-R^{n+1}H(\phi^{n+1})]D\phi^{n+1} = \text{const}\ d\mu_{c_{n+1}^{-1}G_{n+1}}(\phi^{n+1}) \cdot \exp[-T^{n+1}V(\chi^{n+1})] \quad (19)$$

where

$$T^{n+1}V(\chi^{n+1}) = 1/2\ (\nabla\chi^{n+1}|K^{n+1}\nabla\chi^{n+1}) + \tilde{V}^{n+1}(\chi^{n+1}) \quad (20)$$

and

$$c_{n+1} = c_n + \Delta c_n \ . \quad (21)$$

The task is to show that for sufficiently general V's , $T^n V$'s are indeed driven down and c_n's go to a limit.

4.2. Effective Interactions.

The first step is to find a class of effective interactions preserved by T . Let us rewrite (17) changing the variables $\bar{Z}^n$ to $Z^n = \Gamma_n^{-1/2}\bar{Z}^n$:

$$\exp[-V^{n+1}(\chi^{n+1})] = \int \exp[-T^n V(L^{-d/2}\chi^n_{L^{-1}\cdot} + \nabla Z^n)]d\mu_{c_n^{-1}}(Z^n) \quad (22)$$

where now

$$\nabla Z^n = \nabla A_n Q\Gamma_n^{1/2} Z^n \equiv E_n Z^n \ . \quad (23)$$

E_n is again a kernel with exponential decay uniform in n . From (22), (23), it is clear that even if we start with a local interaction $V(\chi) = \sum_{\mu,x} v(\chi_{\mu x})$, V^1 will contain general many-body interactions. If no disaster happens the many-body forces should, however, fall off exponentially on the scale of order one (see [14] for a discussion of an interesting case of such a disaster caused by low-lying Z excitations). For small χ^1 , one may analyze (22) with local V by means of a cluster expansion introducing step-wise decoupling of the kernel E_o . With v satisfying e.g. the assumptions of Section 3.2, this is standard and allows to express $\exp[-V^1(\chi^1)]$ in terms of a partition function of a polymer gas [16]

$$\exp[-V^1(\chi^1)] = \{ \sum_{\{X_i \text{ disjoint}\}} \prod_i \rho^1_{X_i}(\chi^1)\} \cdot \text{local expression} \tag{24}$$

with polymer activities $\rho^1_{X_i}$ being small and having an exponential tree decay in the size of the cluster $X_i \subset L^{-1}\Lambda$. Standard algebraic trick, see e.g. [16] , allows exponentiation of (24) and leads to the expression of V^1 in terms of a sum of many-body interactions

$$V^1(\chi^1) = \sum_Y V^1_Y(\chi^1) \tag{25}$$

where V^1_Y depends only on $\chi^1 \lceil Y$ and again has a tree decay in the size of Y . This way, given local V , the computation of V^1 for small χ^1 is a standard high temperature problem.

Complications appear if we allow for big χ^1 . If we attempt the same cluster expansion, there will be problems with its convergence in the region D where χ^1 is large. We may resolve them by decoupling E_0 only outside D . This leads to (24) again, with $\rho^1_{X_i} \to \rho^{1D}_{X_i}$. Now however, the activities $\rho^{1D}_{X_i}$ of the polymers with clusters X_i containing points (and consequently connected components) of the big χ^1 region are not necessarily small. This creates an obstacle for the exponentiation of the RHS of (24). What we can do, however, is to fix the clusters overlapping D and exponentiate the rest of the sum. This would give

$$\exp[-V^1(\chi^1)] = \sum_{\substack{\{X_i\} \\ X_i \cap D \neq \emptyset \\ X \supset D}} \prod_i \rho^{1D}_{X_i}(\chi^1) \exp[-\sum_{Y \cap X = \emptyset} V^1_Y(\chi^1)] \tag{26}$$

where $X = \bigcup_i X_i$.

What we finally search for, are the properties of TV^1 . Its non-Gaussian piece is given by the remainder $\widetilde{V}^1$ of the Taylor expansion $\delta^2 V^1$ to the second order of V^1 about $\chi^1 = 0$. Since $\delta^2 V^1$ is determined by small χ^1 behavior of V^1 we have

$$\delta^2 V^1 = \sum_Y \delta^2 V^1_Y \quad . \tag{27}$$

Hence,

$$\exp[-\widetilde{V}^1(\chi^1)] = \sum_{\{X_i\}} \prod_i \rho^{1D}_{X_i}(\chi^1) \exp[-\sum_{Y \cap X = \emptyset} V^1_Y(\chi^1)]$$

$$\cdot \exp[\sum_Y \delta^2 V_Y^1(\chi^1)]$$

$$= \sum_{\{X_i\}} \prod_i \rho_{X_i}^{1D}(\chi^1) \sum_{\substack{\{Y_\alpha\} \\ Y_\alpha \cap X = \emptyset}} \prod_\alpha (\exp[-V_Y^1(\chi^1)]-1) \sum_{\{Y_\beta\}} \prod_\beta (\exp[\delta^2 V_Y^1(\chi^1)]-1) . \tag{28}$$

Fixing the clusters of the Mayer expansion $\sum_{\{Y_\alpha\},\{Y_\beta\}}$ in (28) connected to X and resumming the rest of the expansion, we arrive at the analog of (26)

$$\exp[-\tilde{V}^1(\chi^1)] = \sum_{\substack{\{X_j\} \\ X_j \cap D \neq \emptyset \\ X \supset D}} \prod_j g_{X_j}^{1D}(\chi^1) \exp[-\sum_{Y \cap X = \emptyset} \tilde{V}_Y^1(\chi^1)] \tag{29}$$

where $X = \bigcup_j X_j$ now.

(29) expresses the locality properties of $\exp[-\tilde{V}^1]$. The point is that (29) iterates under right assumptions for g_X^D's and $\tilde{V}_Y$'s . It should be clear after the discussion of the hierarchical model that these assumptions specify both analyticity properties and bounds on the large field g_X^D and small field $\tilde{V}_Y$ contributions.

To discuss analyticity, let us introduce various open sets of complex vector fields χ^n . For $X,D \subset L^{-n}\Lambda$ put

$$K_n(X) = \{\chi^n \equiv (\chi_{\mu x}^n) , x \in X : |\chi_{\mu x}^n| < (n+n_o)^\nu \text{ and if } x+L^{-n}e_\nu \in X$$
$$\text{then } |\nabla_\nu \chi_{\mu x}| < C_1(n_o+n)^{\nu+d}\} , \tag{30}$$

$$B_n(D,X,a) = \bigcup_{\phi^n \text{real}}{}' (\nabla\psi^n\restriction_X + aK_n(X)) , \tag{31}$$

where $\bigcup'$ is restricted by taking ϕ^n s.t.

$$|\nabla_\mu \psi_{x'}^n| < (n_o+n)^\nu \exp[\alpha d(x,x')] \text{ for } x \notin D \text{ and every } x' , \tag{32}$$

$\alpha,\nu > 0$ are suitably fixed constants. $K^n(X)$ is the set of small fields χ^n (on X) . Fields in $B_n(D,X,1)$ may contain real $\nabla\psi^n$-type parts getting large inside D but at most exponentially in the distance from the boundary.

We are ready now to formulate the inductive assumptions for the effective interactions.

The Gibbs factor $\exp[-\widetilde{V}^n(\chi^n)]$ is assumed to be analytic in $\mathcal{B}_n(L^{-n}\Lambda, L^{-n}\Lambda, 1)$ (which generalizes the strip around the real axis of the hierarchical model case to the nonlocal one). For given D (a union of lattice blocks of size L^{N_0}), (29) with the superscript 1 replaced by n holds on $\mathcal{B}_n(D, L^{-n}\Lambda, 1)$ (X_j, Y are also built from L^{N_0}-blocks and each $X_j \cap D$ is a union of connected components of D; L^{N_0} is the scale of the high temperature expansion used to analyse (22), see below).

(A) g^{nD}_X is an even analytic functional on $\mathcal{B}_n(D,X,1)$. If $\chi^n = \nabla\psi^n + \widetilde{\chi}^n$ with real $\nabla\psi^n$ satisfying (32) and $\widetilde{\chi}^n \in K_n(X)$ then

$$|g^{nD}_X(\chi^n)| \leq \exp[\kappa_n(\int_{D\cap X} dx + \int_{\partial(D\cap X)} d\sigma(x)) \sum_\mu (\nabla\psi^n_{\mu x})^2$$
$$-2\alpha L(X) + E|D\cap X|] \tag{33}$$

where the integrals are to be understood as Riemann sums and $L(X)$ is the length of the shortest tree on the centers of the L^{N_0}-blocks building X.

(B) $\widetilde{V}^n_Y$ is an even analytic functional on $2K_n(Y)$ satisfying there

$$|\widetilde{V}^n_Y| \leq \delta^{n_0+n} \exp[-2\alpha L(Y)] . \tag{34}$$

Moreover $\widetilde{V}^n_Y(0) = \dfrac{d^2}{\delta\chi^n\delta\chi^n} \widetilde{V}^n_Y(0) = 0$.

Finally, we need assumptions on the Gaussian kernel K^n and the shift in the field-strength renormalization Δc_{n-1}, see (18) and (20).

(C) $$|\Delta c_{n-1}| \leq \delta^{n_0+n} \tag{35}$$

and

$$\| 1_\square K^n 1_{\square'} \| \leq \delta^{n_0+n} \exp[-2\alpha d(\square,\square')] \tag{36}$$

where $\|\cdot\|$ is the operator norm and $1_\square$ denotes the characteristic function of the unit cube $\square$. We have also to assume a compatibility condition between g^{nD}_X and $g^{nD_1}_X$ for $D_1 \supset D$ which will guarantee that, on $\mathcal{B}_n(D,L^{-n}\Lambda,1) \subset \mathcal{B}_n(D_1,L^{-n}\Lambda,1)$, (29) for D_1 may be obtained from (29) for D by Mayer-expanding $\exp[-\sum_Y \widetilde{V}_Y]$, fixing the clusters of this expansion connected to X_j's and components of D_1 and resumming the rest.

It is easy to observe that if $V(\chi) = \sum_{\mu,x} \widetilde{v}(\chi_{\mu,x})$ and v satisfies the assumptions of Section 3.2 for the initial $(n=0)$ potential

then

$$g_X^D(\chi) = \begin{cases} \prod_{\mu,x\in X\cap D} \exp[-\tilde{v}(\chi_{\mu,x})] & \text{if } X \text{ is a connected component of } D\,, \\ 0 & \text{otherwise} \end{cases}$$

and

$$\tilde{V}_Y(\chi) = \begin{cases} \sum_{\mu,x\in Y} \tilde{v}(\chi_{\mu,x}) & \text{if } Y \text{ is an } L^{N_0}\text{-block,} \\ 0 & \text{otherwise} \end{cases}$$

satisfy our assumptions for $n = 0$. In particular, the anharmonic cristal with $v(\varphi) = \lambda\varphi^4$ and the dipole gas with $v(\varphi) = \lambda(1-\cos(\gamma\varphi))$ do.

Remark. In fact, for the inductive step described below, a slightly refined version of (33) is needed. We shall neglect here this minor point of technical nature related to the presence of the boundary integral in (33).

4.3. The Inductive Step.

In this subsection, we shall sketch the major ideas of the proof that the inductive assumptions for T^nV imply those for $T^{n+1}V$. This will be done by expanding (22). The argument involves a play with large and small field regions. Its origin may be traced back to [3]. Some of the details may be found in [11] which treats a nonlocal hierarchical model. A complete exposition is contained in [12].

Suppose that we are given $\chi^{n+1} \in \mathcal{B}_{n+1}(D', L^{-n-1}\Lambda, (1-2\varepsilon)L^{d/2})$. Hence, the region D' where χ^{n+1} may get large is specified. The first step of the expansion is to insert into the RHS of (22) a partition of unity specifying the size of the fluctuation field Z :

$$1 = \sum_{P} 1_P(Z) \tag{37}$$

where $P = (p_x)$, $p_x = 0,1,\ldots$ and

$$1_P(Z) = \prod_x 1(\varepsilon^2(n_0+n)^\nu p_x \le |Z_x| < \varepsilon^2(n_0+n)^\nu(p_x+1))\,. \tag{38}$$

Define the large Z region

$$R = \bigcup_x \{x' \in L^{-n}\Lambda : d(x',x) < 2\alpha^{-1}\log(1+p_x)\}\,. \tag{39}$$

Inside R, Z may get big but again at most exponentially in the distance from the boundary. The crutial fact is that if χ^{n+1} is in $\mathcal{B}_n(D', L^{-n}\Lambda, (1-2\varepsilon)L^{d/2})$ then

$$\chi^n = L^{-d/2}\chi^{n+1}_{L^{-1}\cdot} + \nabla Z^n \tag{40}$$

is in $\mathcal{B}_n(D,L^{-n}\Lambda,1)$ if

$$D = LD' \cup \overline{R} \tag{41}$$

where $\overline{R}$ is the union of the L^{N_0}-blocks intersecting R. This follows roughly speaking since if both χ^{n+1} and Z^n are small, then so is χ^n. Now one might insert (29) (with $1 \to n$) under the integral on the RHS of (22) obtaining

$$\exp[-V^{n+1}(\chi^{n+1})] = \sum_{P} \sum_{\{X_j\}} \int \prod_j g^{nD}_{X_j}(\chi^n) \exp[-\sum_{Y\cap X=\emptyset} \tilde{V}^n_Y(\chi^n)$$
$$\cdot \exp[1/2\,(\nabla\chi^n|K^n\,\nabla\chi^n)]\,1_P(Z^n)\,d\mu_{c_n^{-1}}(Z^n)\,. \tag{42}$$

To decouple maximally the nonlocality of the integrand of (42), one Mayer-expands $\exp[-\sum_Y \tilde{V}^n_Y]$. The remaining nonlocality is due to the kernel E_n relating $\mathcal{Z}^n$ to Z^n (see (23)) and the kernel K^n. These are decoupled the standard way by a cluster expansion introducing interpolating parameters turning off the off-diagonal pieces of the kernels. What we end up with, fixing the clusters of the expansion first, doing all the rest of summations (which factor over the clusters) and then summing over the sets of clusters, is the polymer gas expression of the type (26) (with $1 \to n+1$).

The polymer activities $\rho^{n+1\,D'}_X$ with X disjoint from D' are at least $O(\delta^{n_0+n})$ since they contain $(\exp[-\tilde{V}^n_Y]-1)$ factor or one differentiated $\tilde{V}^n_Y$ factor or one K^n (for which we use (34) or (36)) or finally $1(|Z^n_x| \geq \varepsilon^2(n_0+n)^\nu)$ giving rise to the $\exp[-O((n_0+n)^{2\nu})] \leq \exp[-(n_0+n)^2]$ factor. By construction, these activities have also exponential tree decay in the size of X.

For the polymers intersecting D' the activities may get contributions from $g^{nD}_{X_j}$ for which the bound (33) should be used. Here the important fact is the essential marginality of $\kappa_n \int_{D\cap X_j} dx \sum_\mu (\nabla_\mu \psi^n_x)^2$. The latter is

$$\kappa_n \int_{D\cap X_j} dx \sum_\mu (L^{-d/2}\nabla_\mu \psi^{n+1}_{L^{-1}x} + \nabla_\mu Z^n_x)^2 = \kappa_n \int_{L^{-1}(D\cap X_j)} \sum_\mu (\nabla_\mu \psi^{n+1}_x)^2$$
$$+ \kappa_n \int_{D\cap X_j} dx \sum_\mu (\nabla Z^n_x)^2 + 2L^{-d/2}\kappa_n \int_{D\cap X_j} dx \sum_\mu \nabla_\mu \psi^{n+1}_{L^{-1}x} \nabla_\mu Z^n_x\,. \tag{43}$$

The first term on the RHS is $\kappa_n \int_{D'\cap L^{-1}X_j} dx \sum_\mu (\nabla_\mu \psi_x^{n+1})^2$ which we want to carry over to the next step plus

$$\kappa_n \int_{(D'\setminus L^{-1}\bar{R})\cap L^{-1}X_j} dx \sum_\mu (\nabla_\mu \psi_x^{n+1})^2 \leq \kappa_n (n_o+n+1)^{2\nu} L^{-d} |\bar{R} \cap X_j|$$

which, for small κ_n, may be absorbed into $\exp[-O(n_o+n)^{2\nu}|\bar{R}|]$ factor arising from small Gaussian probability of large values of Z^n. The second term on the RHS of (43) is a small perturbation of the Gaussian measure for the fluctuations Z^n. For $D \cap X_j = L^{-n}\Lambda$ the third term vanishes if we carry the derivative ∇_μ from Z^n onto $\nabla\psi^{n+1}$ by integrating by parts (we have seen this in (7) for $n = 0$). For $D \cap X_j \subsetneq L^{-n}\Lambda$ we shall be left with a boundary term. This may be taken care of, together with $\kappa_n \int_{\partial(D\cap X_j)} d\sigma(x) \sum_\mu (\nabla_\mu \psi_x^n)^2$ which serves this purpose (and which is irrelevant under RG). This is precisely the point where a slight generalization of (33), mentioned at the end of Section 4.2, is needed).

Once we have obtained the polymer gas expression of the type of (24) for $\exp[-V^{n+1}]$, the next steps are as the ones described for $n=0$ in Section 4.2 :

The exponentiation of the sum over the polymers not connected to D' which leads to V_Y^{n+1}'s and the step $V^{n+1} \to T^{n+1} V$ producing $\tilde{V}_Y^{n+1}, K^{n+1}$ and Δc_n.

The bound (34) for (n+1) follows now from the estimates for small χ^{n+1} activities plus the Cauchy type argument which allows to shrink $O(\delta^{n_o+n})$ to δ^{n_o+n+1} when going from V_Y^{n+1} to $\tilde{V}_Y^{n+1}$ and from $\chi^{n+1} \in (1-2\varepsilon)L^{d/2} K_{n+1}(Y)$ to $\chi^{n+1} \in 2K_{n+1}(Y)$.

The bound (33) for $n+1$ with $\kappa_{n+1} = \kappa_n$ and with E increased on combinatoric grounds follows from the bounds for large χ^{n+1} activities discussed above (see (43)). To bring down E, we use the relations between $g_X^{n+1\,D'}$'s for different D'. For D' being the minimal set such that (32) holds for $\nabla\psi^{n+1}$, i.e. for D' being the true large χ^{n+1} set, we may absorb the increase in E by increasing κ_{n+1} by $O((n_o+n)^{-2})$ in virtue of

$$\text{const} \int_{D'\cap X} dx \sum_\mu (\nabla_\mu \psi_x^{n+1})^2 \geq (n_o+n)^2 \, |D'\cap X| \,. \tag{44}$$

For bigger D', we express $g^{n+1\,D'}_{X}$ by $g^{n+1\,D'_{\min}}_{X_i}$ and show that (33) carries over. (44) holds since, although $\nabla\psi^{n+1}$ lives on a fine lattice, it cannot get too rough so that when $\nabla\psi^{n+1}$ gets large at some point, the integral of (44) has to obtain a big contribution from around this point.

The bound (36) for K^{n+1} expressing the irrelevant character of the $1/2\,(\nabla\nabla\psi|K\nabla\nabla\psi)$ term is easy to prove. Roughly speaking

$$1/2\,(\nabla\nabla\psi^{n+1}|K^{n+1}\nabla\nabla\psi^{n+1}) = 1/2\,(L^{\frac{-d-2}{2}}\nabla\nabla\psi^{n+1}_{L^{-1}\cdot}|K^{n}L^{\frac{-d-2}{2}}\nabla\nabla\psi^{n+1}_{L^{-1}\cdot}) + O(\delta^{2(n_0+n)}) \,, \tag{45}$$

the first term being the leading order, so that

$$K^{n+1}_{xy} = L^{d-2}\,K^{n}_{LxLy} + O(\delta^{2(n_0+n)}) \tag{46}$$

and

$$\|K^{n+1}\| \le L^{-2}\,\|K^{n}\| + O(\delta^{2(n_0+n)}) \,. \tag{47}$$

The exponential decay of $\|1_{\square}K^{n+1}1_{\square'}\|$ is easy to extract from the connectedness properties of the expansion terms. Finally, (35) is straightforward.

If we start with a local potential $V(\chi) = \sum_{\mu,x} v(\chi_{\mu x})$ then all our estimates are uniform in the volume. It is not difficult to see that the infinite volume limit of g^{nD}_{X}, $\tilde{V}^{n}_{Y}$ (uniformly on compacts) and of K^{n}, c_n exists producing infinite volume limit effective interactions defined for χ^{n} or ϕ^{n} with compact supports. Of course, our iterative bounds hold in infinite volume limit showing that the effective interactions go to zero when $n \to \infty$ so that the effective Hamiltonians tend to $1/2\,c_\infty(\phi|G_\infty^{-1}\phi)$ where $c_\infty = \lim_{n\to\infty} c_n$ and G_∞ is the Gaussian fixed point covariance described in Section 2.2.

The above analysis extends in a straightforward way to the free energy f since

$$f = -L^{-Nd}\log\int \exp[-V(\nabla\phi)]\,d\mu_G(\phi) = -L^{-Nd}\log\int d\mu_{G_1}(\phi^1)$$
$$\cdot\int \exp[-V(L^{-d/2}\nabla\psi^{1}_{L^{-1}\cdot}+\nabla Z)]\,d\mu_1(Z) = -L^{-Nd}\log\int \exp[-V^1(\nabla\psi^1)]$$
$$\cdot\, d\mu_{G_1}(\phi^1) = L^{-Nd}\,V^1(0) + 1/2(L^{-d}-L^{-Nd})\log c_1 - L^{-Nd}\log\int \exp[-TV^1(\nabla\psi^1)]$$

$$\cdot\ d\mu_{c_1^{-1}G_1}(\phi^1) = \ldots = L^{-Nd}(V^1(0)+V^2(0)+\ldots+V^N(0))$$
$$+1/2\ (L^{-d}-L^{-Nd})\log c_1 + 1/2\ (L^{-2d}-L^{-Nd})\log\frac{c_2}{c_1}+\ldots$$
$$\ldots +1/2\ (L^{-(N-1)d}-L^{-Nd})\log\frac{c_{N-1}}{c_{N-2}} \qquad (48)$$

in the limit where the infrared regulator ξ is taken to infinity. It is easy to see that $V^n(0) = O(\delta^{n_o+n})\ L^{(N-n)d}$. Similarly,

$$\frac{c_{n+1}}{c_n} = 1+c_n^{-1}\Delta c_n = 1+O(\delta^{n_o+n}) .$$

Hence the infinite volume limit f_∞ for the free energy exists.

4.4. The Two-Point Function.

As in the case of the hierarchical model, we shall concentrate on the two-point correlation function of the $\nabla\phi$ field. The method clearly extends to any correlations involving $\nabla\phi$ allowing to prove the existence of their thermodynamical limit and to control their decay. The correlations involving the ϕ fields (in $d \geq 3$) may be reduced by integration by parts to those with $\nabla\phi$ only. Our discussion will be even more sketchy than in the previous subsections, since no essentially new ideas enter in the study of the correlations.

Integrating the $Z°$ fluctuations in the two-point function first, we obtain

$$<\nabla\phi_x\nabla\phi_y>_V \equiv \int \nabla\phi_x\nabla\phi_y \exp[-V(\nabla\phi)]\ d\mu_G(\phi)\ /\int \exp[-V(\nabla\phi)]\ d\mu_G(\phi)$$
$$= \int d\mu_{G_1}(\phi^1)\int (L^{-d/2}\nabla\psi^1_{L^{-1}x}+\nabla Z_x)(L^{-d/2}\nabla\psi^1_{L^{-1}y}+\nabla Z_y)$$
$$\cdot \exp[-V(L^{-d/2}\nabla\psi^1_{L^{-1}\cdot}+\nabla Z)]d\mu_1(Z)/\int d\mu_{G_1}(\phi^1)\exp[-V(L^{-d/2}\nabla\psi^1_{L^{-1}\cdot}+\nabla Z)]d\mu_1(Z)$$
$$= \int G^1_{xy}(\nabla\psi^1)\exp[-TV(\nabla\psi^1)]d\mu_{c_1^{-1}G_1}(\phi^1)/\int\exp[-TV(\nabla\psi^1)]d\mu_{c_1^{-1}G_1}(\phi^1)$$
$$\equiv\ <G^1_{xy}>_{TV}\ , \qquad (49)$$

where by definition

$$G^1_{xy}(\chi^1) = \int (L^{-d/2}\chi^1_{L^{-1}x}+\nabla Z_x)(L^{-d/2}\chi^1_{L^{-1}y}+\nabla Z_y)\ \exp[-V(L^{-d/2}\chi^1_{L^{-1}\cdot}+\nabla Z)]$$

$$\cdot\ d\mu_1(Z)/\int \exp[-V(L^{-d/2}\chi^1_{L^{-1}\cdot}+\nabla Z)]\ d\mu_1(Z) \quad . \tag{50}$$

Further fluctuation integrals give

$$<\nabla\phi_x\ \nabla\phi_y>_V = <G^n_{xy}>_{T^nV} \quad , \tag{51}$$

where, inductively,

$$G^{n+1}_{xy}(\chi^{n+1}) = \int G^n(L^{-d/2}\chi^{n+1}_{L^{-1}\cdot}+\nabla Z^n)\ \exp[-T^nV(L^{-d/2}\chi^{n+1}_{L^{-1}\cdot}+\nabla Z^n)]$$

$$\cdot\ d\mu_{c_n^{-1}}(Z^n)/\int \exp[-T^nV(L^{-d/2}\chi^{n+1}_{L^{-1}\cdot}+\nabla Z^n)]d\mu_{c_n^{-1}}(Z^n) \quad . \tag{52}$$

The notation of (50) and (52) might be slightly abusive since we shall not be able to assure that the denominators never vanish. Instead, we shall analyze the products $G^n_{xy}\ \exp[-\tilde{V}^n]$ (we use the notation of (18) and (20)) jointly. Again the inductive formula will be analyzed by means of the expansion described in the previous section. To carry this analysis out, we need to know properties of G^n's exibiting their locality together with their behavior both for small and for big fields. The careful reader will not be surprised to learn that these are the following :

$G^n_{xy}\ \exp[-\tilde{V}^n]$ is analytic on $\mathcal{B}_n(L^{-n}\Lambda, L^{-n}\Lambda, 1)$. Given (the large field region) D , on $\mathcal{B}_n(D, L^{-n}\Lambda, 1)$,

$$\tilde{G}^n_{xy}\ \exp[-\tilde{V}^n] = \sum_{\substack{\{X_j\}\\ X_j\cap D\neq\emptyset\\ X\supset D}} \prod_j \tilde{g}^{nD}_{X_j}\ F^n_{\sim X}\ \exp[-\sum_{Y\cap X=\emptyset} \tilde{V}^n_Y] \tag{53}$$

$(X = \cup_j X_j)$, $\tilde{G}^n_{xy} = G^n_{xy} - G^n_{xy}(0)$.

$\tilde{g}^{nD}_{X_j}$ are analytic on $\mathcal{B}_n(D, X_j, 1)$ and fulfil (33) with $\kappa_n \to 2\kappa_n$, $E \to 2E$. $F^n_{\sim X}$ is analytic on $2K_n(\sim X)$. If $L^{-n}x, L^{-n}y \in X$ then $F^n_{\sim X} = 1$ (all the information about the entries of the correlation function is absorbed into $\tilde{g}^n$'s) . If $L^{-n}x \in X$, $L^{-n}y \notin X$ then

$$F^n_{\sim X} = \sum_{L^{-n}x\in Y\subset \sim X} F^n_{1Y} \quad . \tag{54}$$

If $L^{-n}x \notin X$, $L^{-n}y \in X$ then analogically

$$F^n_{\sim X} = \sum_{L^{-n}y\in Y\subset \sim X} F^n_{2Y} \quad . \tag{55}$$

If $L^{-n}x, L^{-n}y \notin X$ then

$$F^n_{\sim X} = \sum_{\substack{L^{-n}x\in Y_1\subset\sim X \\ Y_1\cap Y_2=\emptyset}} \sum_{L^{-n}y\in Y_2\subset\sim X} F^n_{1Y_1} F^n_{2Y_2} + \sum_{L^{-n}x,L^{-n}y\in Y\subset\sim X} F^n_{12Y} . \tag{56}$$

F^n_{iY} are odd functionals of X^n and satisfy

$$|F^n_{iY}| \le C\, L^{-(d/2\,+\varepsilon)n} \exp[-2\alpha\, L(Y)] . \tag{57}$$

$F^n_{12Y}(0) = 0$, and

$$|F^n_{12Y}| \le C\, L^{-(d+2\varepsilon)n} \exp[-2\alpha L(Y)] . \tag{58}$$

Finally,

$$|G^n_{xy}(0) - G^{n-1}_{xy}(0)| \le CL^{-(d+2\varepsilon)n} \exp[-\varepsilon d(L^{-n}x, L^{-n}y)] . \tag{59}$$

These properties of $G^n_{xy}\exp[-\widetilde{V}^n]$ iterate (modulo minor technical modifications, see Remark at the end of Section 4.2) under the expansion (convergent uniformly in the volume) of the same type as for the effective potentials. For the details, we refer the reader to [13]. The value of the two-point function (in the limit when the infrared regulator is taken to infinity) is $G^N_{xy}(0)$. Similar arguments as for the effective potentials give now the existence of the thermodynamical limit (if the starting V is local). Moreover,

$$|G^N_{xy}(0)| \le C \sum_{n=1}^{\infty} L^{-(d+2\varepsilon)n} \exp[-\varepsilon d(L^{-n}x, L^{-n}y)] \le C\, |x-y|^{-d+2\varepsilon} . \tag{60}$$

Extracting from G^n_{xy} also a term $\propto X^n_{L^{-n}x} X^n_{L^{-n}y}$ allows to show that the two-point function at long distances becomes c_∞^{-1} times the free one $\nabla\nabla G_{xy} = \sum_{k=0}^{\infty} L^{-dk} < \nabla Z^k_{L^{-k}x}\, \nabla Z^k_{L^{-k}y} >_0$ plus the corrections decaying faster than $d(x,y)^{-d}$. These corrections may also be studied by the present method.

Let us notice that the effective interactions on the next scale as well as the effective functionals for the correlations (G^n_{xy} for the two-point function) are given by a convergent expansion involving the entries of the previous step, both in finite and in infinite volumes. Superposition of these expansions gives a convergent expansion in terms of the original potential v for the infinite volume limit effective interactions on any scale, for the field strength renormalization c_∞ , free energy f_∞ as well as for the correlation functions. Although we work with periodic volumes in order to be able to use translation invariance, the infinite-volume objects should be independent of (sufficiently regular) boundary conditions as usually when they are given by convergent expansions.

Similarly as in the hierarchical model case, if the initial potential v_λ is analytic in some parameter λ , then so are the effective interactions and effective functionals for the correlations (in finite and infinite volumes) in the region of λ where v_λ fulfils our assumptions. Hence, also the field-strength renormalization c_∞ , the free energy f_∞ and the correlation functions themselves are analytic in λ . For example for $\lambda\varphi^4$ the analyticity holds for $\mathrm{Re}\lambda > 0$ and small λ and for the dipole gas $\lambda(1-\cos(\gamma\varphi))$ for small activity λ . As a result, we obtain the convergence of the Mayer expansion for a small activity dipole gas, the result not known previously (see [17]).

This ends the discussion of the $\nabla\phi$-type models. We believe that the main ideas involved in the non-perturbative analysis of the RG presented here : the separation of small and big fields combined with analyticity techniques provide the right tools to study other asymptoticly free models enumerated in Introduction. The big and totally open at the moment remains the question whether there exists an extension of these methods which would allow to treat rigorously non-Gaussian fixed points of the RG .

REFERENCES

[1] Bleher, P.M. : Usp. Mat. Nauk. 32 (1977), 243.

[2] Bleher, P.M., Sinai, Ja. G. : Commun. Math. Phys. 33 (1973), 23.

[3] Benfatto, G. et al. : Comm. Math. Phys. 71 (1980), 95.

[4] Collet, P., Eckmann, J.-P. : A renormalization group analysis of the hierarchical model in statistical physics, Lecture Notes in Phys. 74, Springer 1978.

[5] Dyson, F.J.: Commun. Math. Phys. 12 (1969), 91.

[6] Fröhlich, J., Spencer, T. : Commun. Math. Phys. 81 (1981), 527.

[7] Fröhlich, J., Spencer, T. : Commun. Math. Phys. 83 (1982), 411.

[8] Fröhlich, J., Pfister, C.E., Spencer, T. : On the statistical mechanics of surfaces, IHES/P/82/20 preprint.

[9] Gawędzki, K., Kupiainen, A. : Triviality of ϕ_4^4 and all that in a hierarchical model approximation, IHES/P/82/30 preprint, submitted to J. Stat. Phys.

[10] Gawędzki, K., Kupiainen, A. : Commun. Math. Phys. 77 (1980), 31.

[11] Gawędzki, K., Kupiainen, A. : Renormalization group for a critical lattice model ..., IHES/P/82/15 preprint, to appear in Commun. Math. Phys.

[12] Gawędzki, K., Kupiainen, A. : Block spin renormalization group for dipole gas and $(\nabla\phi)^4$, University of Helsinki preprint.

[13] Gawędzki, K., Kupiainen, A. : in preparation

[14] Göpfert, M., Mack, G. : Commun. Math. Phys. 84 (1982), 545.

[15] Kadanoff, L.P. : Physics 2 (1965), 263 .

[16] Kunz, H. : Commun. Math. Phys. 59 (1978), 53.

[17] Malyshev, V.A. : Teor. Mat. Fiz. 45 (1980), 235.

[18] Wilson, K.G., Kogut J.B. : Phys. Rep. 12 C (1974), 75.

[19] Wilson, K : Phys. Rev. B 4 (1971), 3184.

ON INFRARED SUPERRENORMALIZATION

J. MAGNEN

R. SENEOR

Centre de Physique Théorique
Ecole Polytechnique
91128 Palaiseau Cédex - France

A518.0982

Septembre 1982

CONTENTS

5) THE CONVERGENCE OF THE EXPANSION.

a) The convergence factors due to the renormalization.

b) The power decrease of the strongly connected functions.

c) The convergence of the expansion.

1) INTRODUCTION

a)

This article presents a general approach to the thermodynamic limit of infrared superrenormalizable massless theories (in particular they have no anomalous dimension).

Even if nowhere in the text the renormalization group [R-G] is explicitely mentionned, it can be remarked how close are our methods to the ones of the [R-G] analysis [B-C-G-N-O-P-S], [G-K]. To the dimensional analysis corresponds the power counting of perturbation graphs, and the scaled [R-G] transformations are replaced by a phase space analysis.

On the example of $(\nabla\phi)^4_3$ we give the main arguments leading to the convergence of a massless cluster expansion. We emphasize on the main ideas, specially on the finite renormalizations and on the way they are performed. A detailed version of this work is given in [$M-S_1$].

The model $(\nabla\phi)^4_3$ is an idealization of an anharmonic crystal and has been already studied [B-F-L-S], [F_1],[F_2], [G-K]. The dipole gaz is an other model of physical interest where our expansion should work as well ; it applies also to more divergent models like φ^4_d, d>4 (with an ultraviolet cut off).

Finally we use the strong analogy between infrared and ultraviolet problems - for example φ^4_3 - to remark that in the frame of the phase space expansion of [G-J], one can compute inductively (for each ultraviolet cut off) a mass counterterm following our procedure for getting - here - the finite two-point function renormalization.

b) <u>The power counting</u> .

The infrared power counting of the model φ^n in d dimensions and with propagators $C(x,y) \xrightarrow[|x-y|\to\infty]{} \frac{1}{|x-y|^\alpha}$ (or $\tilde{C}(p) \xrightarrow[p\to 0]{} 1/p^{d-\alpha}$) is : $\gamma = \frac{n\alpha}{2} - d$.

If $\gamma>0$ the interaction is superrenormalizable and has no anomalous dimension (for a scale invariant theory this corresponds to be ultraviolet non renormalizable).

The fact of having $\gamma>0$ means that there is more decreasing powers of the propagators than what is necessary to localize each vertex.

c) <u>The model</u> : $(\nabla\phi)^4_3$

Its Schwinger functions are :

$$\langle F(\varphi)\rangle = \lim_{\substack{\chi\to 1 \\ M_\rho\to o}} \int F(\varphi)\, e^{-\lambda\int(\nabla\varphi)^4(x)\,\chi(x)\,dx}\, d\mu(\rho)/Z(\chi,\rho) \qquad (1.1)$$

where $d\mu$ is the Gaussian measure of covariance $\widetilde{\mathbb{C}}(p) = \frac{1}{p^2(p^2+1)^2}$; the factor $1/(p^2+1)^2$ is an ultraviolet cut off. The ρ means just that we have put moreover an infrared cut off M_ρ. Finally $\chi(x)$ is a space cut off, and $Z(\chi,\rho)$ is the normalization.

In all the following we pose $A = \nabla\varphi$, and we study a vector field theory with propagator :

$$C_{\alpha,\beta}(x,y) = \int A_\alpha(x)\, A_\beta(y)\,d\mu = \int d^3p\, \frac{p_\alpha p_\beta}{p^2(p^2+1)^2}\, e^{ip(x-y)}.$$

d) <u>The necessity of a phase space analysis.</u>

To prove the existence of the infinite volume limit we do a cluster expansion [G-J-S]. In its simplest form it consists of considering the space as an union of unit cubes and then perturbing the coupling between distinct cubes.

For each cube the result is :

- either there is no coupling of this cube with the other ones
- or there is a propagator between two vertices, one in the cube, the other one in the complementary (thus coupling the cube with an other one)

The theory being infrared superrenormalizable, the Gaussian integration over the product of the perturbation vertices is superficially convergent up to a finite number of divergent subgraphs.

If the propagator were absolutely summable, it could be used to distinguish different unit cubes :

$$\sum_{\text{unit cubes}\,\Delta}\ \sup_{y\in\Delta} |C(x,y)| < o(1)$$

then there would be no divergent subgraphs and the cluster expansion must converge ; if C(x,y) decreases sufficiently this has been proved [M-S_2].

Our propagator is not absolutely summable. The low momentum part of the propagator doesn't make difference between different unit cubes, it is as if all vertices were at the same point. Because the perturbation theory of the A^4 theory is divergent we obtain a divergent cluster expansion.

We are then led to renormalize the divergent subdiagrams ; to do it we combine the cluster expansion with a phase space analysis à la Glimm and Jaffe [G-J] in which each range of momenta is analysed separately.

e) <u>A phase space expansion</u>.

We decompose the field over the values of the momenta : $C = \Sigma C_i$, $A = \oplus A_i$, A_i being roughly of momenta between M_i and M_{i-1}, where $\{M_i\}_{i=1,\dots}$ is a decreasing sequence with $\lim_{i\to\infty} M_i = 0$ and $|C_i(x,y)| \leqslant e^{-M_i|x-y|} M_{i-1}^3$,

The natural localizations for fields A_i are cubes of size M_i^{-3}.

For each C_i we make a cluster expansion on a cubic lattice $\mathcal{D}_i$ of mesh M_i^{-1}, it is a perturbation relatively to the coupling between different cubes.

The exponential of the interaction couples the frequencies ; for each i we perturb the coupling between :

$$\sum_{j<i} A_i \text{ and } \sum_{j\geqslant i} A_i \text{ in each cube } \Delta \in \mathcal{D}_i .$$

A perturbation term of index i must be small depending on i because we have to sum over the perturbation steps, that is on i.

The perturbation terms are characterized by the cubes in which the vertices are produced. A divergent subdiagram can only be a diagram whose vertices have fields with much larger index than the cubes of production (see the above section) ; each subdiagram formed with small cubes (compared to the inverse of the momenta of the low momentum fields) can be considered as a point vertex.

The divergent subgraphs are 2 and 4-point functions.

"Proof" : let us consider vertices $\int_\Delta A_i^n(x)dx$, $\Delta\in\mathcal{D}_j$ produced at step j, $i \geqslant j$. Because in each cube the expansion is finite, let us suppose that there is one vertex by cube of $\mathcal{D}_j$. To estimate such a product of vertices, we have two possibilities :

1) We do the Gaussian integration. For the low momenta M_i, it is almost as if all the vertices were at the same point.

The perturbation theory for the interaction A^n, $n \geqslant 4$ is divergent. The Gaussian integration is thus not a convenient way to estimate the low momentum fields (we recall that because the propagator is not absolutely summable the vertices A^2 give rise to divergences).

2) We dominate the fields by the exponential of the interaction. A field in $\Delta\in\mathcal{D}_j$ and of momentum smaller than* $\Delta^{-1/3}$ is s.t.

$A(x)\chi_\Delta(x) \simeq \frac{\chi_\Delta(x)}{\Delta}\int_\Delta A(x)dx$ and

$(1/\Delta\int_\Delta A(x)dx)^n \exp -\int A^4(x)dx \leqslant O(1)\ (\Delta^{-1/4})^n$.

A vertex $\int_\Delta A^n(x)dx$ is then estimated by $\Delta^{1-n/4}$

Conclusion : if n = 2,4 the vertices give rise to divergences : they have to be renormalized

if $n>4$ the vertices are small depending on Δ.

f) The finite renormalization.

To obtain a convergent expansion we renormalize the 2 and 4-point subdiagrams (that is subdiagrams with 2 or 4 low momentum fields) ; we do this by substracting the zero momentum value of the (amputated) subdiagrams.

The presentation is the following : we give us the value of the zero momentum wave function and 4-point function and we compute (inductively) what is the corresponding theory i.e. the bare wave function and the bare coupling constant.

Let F(x,y) be a 2-point subdiagram. Summing over all its translated gives a translation invariant 2-point function F(x-y).

---

* In all the following Δ stands also for the volume of the cube Δ.

This allows to compute $\widetilde{F}(k)$ and to perform the renormalization : $\widetilde{F}(k)-\widetilde{F}(0)$. To sum over all translations means that we have subdiagrams with independent localizations. We achieve this by decomposing each product of graphs in sum of products of truncated functions. By definition each truncated function is independent of the other ones.

The interaction of the model (in contrast to (1.1)) is :

$$-(\lambda+\delta(\lambda))\int A^4(x)\,dx - \frac{\alpha(\lambda)}{2} A^2(x)\,dx \tag{1.2}$$

In chapter 2 and 3 we ignore δ and α. They are computed in chapter 4, where their insertion in the expansion is also discussed. Their momentum cutoff dependence is such that they are small enough so that they change nothing in the bounds of chapter 3.

2) THE EXPANSION STEPS.

a) Cutoffs and notations.

We define first the sequence of momentum cutoffs.

For $M>1$ and $\xi>0$ to be chosen later we define :

$$M_0=\infty,\ M_1=M^{-1},\ \dots,\ M_i=M^{-(1+\varepsilon)^{i-1}},\ \dots$$

Then :

$$C_{\alpha,\beta;i}(x,y)=\int\frac{(e^{-M_{i-1}^{-2}p^2}-e^{-M_i^{-2}p^2})}{p^2\,(p^2+1)^2}\,e^{ip(x-y)}\,p^\alpha p^\beta d^3p$$

We introduce an infrared cutoff ρ by β defining the cutoff covariance to be :

$$C=\sum_{i=1}^{\rho} C_i \text{ and } A=\bigoplus_{i=1}^{\rho} A_i$$

The basic bound on C_i is :

$$|C_i(x,y)| \leqslant O(1)\, e^{-M_i(x-y)}\, M_{i-1}^3 \text{ for } i>1 \text{ and} \tag{2.1}$$

$$|C_1(x,y)| \leqslant O(1)\, e^{-M_1(x-y)}$$

We define now the functions that localize in space.

Let $\mathfrak{D}_1, \dots, \mathfrak{D}_i, \dots$ be a sequence of lattices, $\mathfrak{D}_{i+1}$ is a sublattice of $\mathfrak{D}_i$, and the cubes of $\mathfrak{D}_i$ are of size :

$$K M_i^{-1} < \Delta^{1/3} \leqslant 8\ K M_i^{-1} \text{ where } K>1 \text{ is to be chosen later.} \tag{2.2}$$

The localization functions are in principle the characteristic functions of the cubes of the lattices. Since the renormalization produces gradients acting on fields which are by construction localized we are led to define smooth localization functions. Then the action of gradients on these functions is well defined.

For $\Delta \in \mathcal{D}_i$ let $c(\Delta)$ be the center of Δ.

Let $\eta(x) \in C^0_\infty$ s.t.

$$\begin{cases} \eta(x) = 1 & 0 \leqslant |x| \leqslant 1/4 \\ 0 \leqslant \eta(x) \leqslant 1 & 1/4 \leqslant |x| \leqslant 3/4 \\ \eta(x) = 0 & 3/4 \leqslant |x| \end{cases}$$

and $\int \eta(x)dx = 1$

and $\{\eta(x+k)\}$, $k \in Z$ is a partition of the unity.

then $\chi^0_\Delta(x) = \chi^0([x-c(\Delta)]\Delta^{-1/3})$, $\chi^0(x) = \prod_{\alpha=0}^{2} \eta(x^\alpha)$

so that

$$\nabla^p \chi^0_\Delta(x) = \nabla^p \chi^0([x-c(\Delta)]\Delta^{-1/3}) = \Delta^{-\frac{p}{3}} (\nabla^p \chi^0)([x-c(\Delta)]\Delta^{-1/3})$$

The bad feature of such localization functions is that they possibly overlap. We define then for each face b a smooth characteristic function χ^o_b of the overlap. Let b be the face common to Δ and Δ' (which are cubes of the same lattice) :

$$\chi^o_b(x) = \chi^0[(x - \frac{c(\Delta) + c(\Delta')}{2})\frac{\Delta^{-1/3}}{3}]$$

Finally we shall consider the theory in a volume Λ which is an union of cubes of $\mathcal{D}_\rho$. The existence of the limit $\Lambda \to \infty$ is a consequence of the expansion.

b) The scaled cluster expansion.

For each i we do a cluster expansion relatively to $C_{\alpha,\beta;i}$ in the lattice $\mathcal{D}_i$. The exponential decrease of the covariance enables to sum over the cubes of $\mathcal{D}_i$:

let $\Delta, \Delta' \in \mathcal{D}_i$, from (2.1) $|C_i(x,y)\chi_\Delta(x)\chi_{\Delta'}(y)| \leqslant M^3_{i-1} \exp{-M_i d(\Delta,\Delta')}$

(d is for distance) and $\sum_{\Delta' \in \mathcal{D}_i} \exp{-M_i d(\Delta,\Delta')} \leqslant O(1)$

The phase space factors M^3_{i-1} will be used to control the volume integration in the cubes, and will leave eventually a convergent factor $M_i^{\varepsilon_0}$

Let $\mathcal{D}_i{}^*$ be the set of faces of $\mathcal{D}_i$. To each $b \in \mathcal{D}_i{}^*$ we associate a variable $0 \leqslant s_b \leqslant 1$ and we define :

$$C_{\alpha,\beta;i}(s_b) = s_b\, C_{\alpha,\beta;i} + (1-s_b)\, C^b_{\alpha,\beta;i}$$

where C^b_i is C_i whith almost decoupling boundary conditions on b, essentially :

$$C^b_{\alpha,\beta;i}(x,y) = \sum_{\Delta',\Delta''} \frac{\mathbb{C}^b_i(\Delta',\Delta'')}{\mathbb{C}_i(\Delta',\Delta'')}\, \chi^o_{\Delta'}(x)\, \chi^o_{\Delta''}(y)\, C_{\alpha,\beta;i}(x,y) \qquad (2.3)$$

where for

$$\Gamma\subset\mathcal{B}^*_i, \mathbb{C}^\Gamma_i(\Delta',\Delta'') = \int_{\Delta'\times\Delta''} \mathbb{C}^\Gamma_i(u,v)\,du\,dv, \quad \mathbb{C}_i(x,y) = \int \frac{e^{-M^{-2}_{i-1}p^2} - e^{-M^{-2}_i p^2}}{p^2} e^{ip(x-y)} d^3p$$

and $\mathbb{C}^\Gamma_i(u,v)$ is $\mathbb{C}_i$ with Dirichlet boundary conditions on Γ see [G-J-S] and [M-S$_1$].

The above formula is not quite satisfactory because $\mathbb{C}^\Gamma_i/\mathbb{C}_i$ is not (obviously) of positive type ; in fact we symetrize using an intermediate point see [M-S$_1$].

Finally :

$$C_{\alpha,\beta;i;\{s\}} = \sum_{\Gamma\subset\mathcal{B}^*_i} \prod_{b\in\Gamma} s_b \prod_{b\notin\Gamma} (1-s_b)\, \tilde{C}^\Gamma_{\alpha,\beta;i}$$

and we have the bound

$$\left| \prod_{b\in\gamma} \frac{d}{ds_b} C_{\alpha,\beta;i}\{s\}(x,y) \right| \leqslant e^{-\frac{M_i}{2} d(x,y,\gamma)}\, M^{(3-\varepsilon_2)}_{i-1} \qquad (2.4)$$

for $\varepsilon_2 > 0$ as small as we want if M is big enough, and where $d(x,y,\gamma)$ is the length of the smallest path going from x to y and hitting all the elements of γ.

However since our localization functions χ^o_Δ in (2.3) are smooth it is clear that zero boundary condition on a closed surface [i.e. $s_b = 0$ for all b forming the surface] is not sufficient to decouple the inside from the outside.

To achieve this we are led to eliminate the field A_i on the faces with zero boundary conditions. We put for each is $b\in\mathcal{B}^*_i$ an s_b dependance in the field :

$$A_i(s_b)(x) = \prod_{b\in\mathcal{B}^*_i} [s\,\chi^o_b(x) + 1 - \chi^o_b(x)]\, A_i(x)$$

so that for the frequence i the decoupling is achieved with $s = 0$ conditions.

The cluster expansion is applied on a Schwinger function considered as a function of $\{s_b\}$, $b\in\mathcal{B}_i{}^*$:

$$F(1) = \sum_{\Gamma \subset \mathcal{B}_i^*} \int \prod_{b\in\Gamma} ds_b \, F(\$)\Big|_{\substack{s_b = 0 \\ b\notin\Gamma}} , \quad \partial^\Gamma = \prod_{b\in\Gamma} \frac{d}{ds_b} \tag{2.5}$$

and see [G-J-S] for $b \in \mathcal{B}_i^*$:

$$\int (.) \frac{d}{ds_b} d\mu(s) = \Sigma_{\alpha,\beta} \int \frac{d}{ds_b} C_{\alpha,\beta;i}(x,y) \frac{\delta}{\delta A_{\alpha,i}(x)} \frac{\delta}{\delta A_{\beta,i}(y)} (.) \, d\mu(s)\,dx\,dy$$

$\frac{\delta}{\delta A(x)}$ can derive either already produced vertices, or the exponential. In this last case by formula (2.3) these new vertices are naturally localized in cubes of $\mathcal{B}_i$ by the smooth localization functions $\overset{\circ}{\chi}_\Delta$, $\Delta \in \mathcal{B}_i$.

For a given $\Gamma \subset \mathcal{B}_i^*$ let $X_1^i \ldots, X_p^i, \ldots$ be the connected components of $R^3 \setminus (\mathcal{B}_i^* \setminus \Gamma)$; the coupling - if any - between them is done through the other frequencies.

c) The expansion on the coupling between low and high frequencies.

The different frequencies are coupled by the exponential of the interaction. In $\Delta \in \mathcal{B}_i$ the low frequency field is $\sum_{k \geq i} A_k$

the high frequency field is $\sum_{j<i} A_j$

decoupling them is to suppress the mixed terms in the interaction.

If we do that, we say that we decouple (in Δ) the smaller cubes (attached to the high frequencies) from the bigger cubes (attached to the low frequencies).

To each $\Delta \in \mathcal{B}_i$ we associate $0 \leq t_\Delta \leq 1$ and put a t_Δ dependence in the exponential of the interaction in Δ [i.e. in $\overset{\circ}{\chi}_\Delta$]. From this, follows that a t_Δ perturbation step can only derive the exponential of the interaction. The choice of not putting a t_Δ dependence on the already produced vertices is done for simplicity.

The dependence is given by :

$$\exp \{- \lambda \int \overset{\circ}{\chi}_\Delta(x) [(\sum_{j<i} A_j(x) + t_\Delta \sum_{k\geq i} A_k(x))^4 + (1-t_\Delta^4)(\sum_{k\geq i} A_k(x))^4] dx$$

$$+ \text{ terms with no } t_\Delta \text{ dependence } \} \tag{2.6}$$

At step* i-1 the new t dependence is only on

$$\chi^{o}_{\Delta}(x)\,(\sum_{j<i} A_j(x) + t_{\Delta} \sum_{k\geq i} A_k(x))^4$$

giving for $\Delta' \in \mathcal{D}_{i-1}$:

$$\chi^{o}_{\Delta}(x)\chi^{o}_{\Delta'}(x)\Big[[\sum_{j<i-1} A_j(x) + t_{\Delta'}(A_{i-1}(x) + t_{\Delta} \sum_{k\geq i} A_k(x))]^4$$

$$+ (1-t^4_{\Delta'})[A_{i-1}(x) + t_{\Delta} \sum_{k\geq i} A_k(x)]^4 \Big]$$

In formula (2.6) if $t_{\Delta} = 0$ then there is no coupling term between high and low momenta.

An expansion step is :

$$\int(\cdot)\exp\,(\text{Interaction})\ d\mu = F(1) = F(0) + \int_0^1 (\cdot)\,dt_{\Delta} \frac{d}{dt_{\Delta}} \exp\,(\text{inter.})\,d\mu$$

and $\frac{d}{dt_{\Delta}}$ Interaction =

$$-\lambda \int \chi^{o}_{\Delta}(x) \sum_{\alpha=1}^{3} O(1)\,(\sum_{j<i} A_j(x))^{\alpha} (\sum_{k\geq i} A_k(x))^{4-\alpha} dx$$ so that a perturbation vertex has a smooth support and contains both high and low momenta fields.

d) <u>A finite scaled perturbation expansion.</u>

We have two reasons for doing such an expansion :

- our bounds are not very good. In fact we shall obtain by vertex localized in $\Delta \in \mathcal{D}_i$ a factor $M_i^{3\varepsilon_o} = \Delta^{-\varepsilon_o}$ for some finite $\varepsilon_o > 0$. If we produce N vertices in Δ we obtain a factor $\Delta^{-N\varepsilon_o}$. When N is small this factor can be not small enough. We then do a perturbation expansion relatively to $\chi^{o}_{\Delta}(x)A_i(x)$ up to a sufficiently big order ζ. Then we are sure that the last term of the expansion is small like $\Delta^{-\zeta\varepsilon_o}$. In the other terms $\chi^{o}_{\Delta}(x)A_i(x)$ has been put to zero in the interaction ; what remains are only explicit perturbation graphs [for what concerns $\chi^{o}_{\Delta}(x)A_i(x)$], for which we can do exact computations.

---

* We do the expansion beginning with the biggest i. After the i step the next one is i-1. The term $(1-t^4_{\Delta})(\sum_{k\geq i} A_k(x))^4$ contains no coupling for i-1 between high and low momenta ; it contains only low momenta.

- The principle of the cluster expansion is that for each $i, \Delta\in\mathcal{D}_i$, the measure on $A_i(x)$ in Δ is : either decoupled

or we have a small factor.

However in our framework if $t_\Delta = 0$ and $s_b = 0$, $b\in\mathcal{D}_i^*\cap\Delta$ we have no perturbation attached to Δ, though $A_i(x)$ in Δ is not decoupled (it can be coupled to the lower momenta). The perturbation expansion relatively to A_i fills this gap.

To each $\Delta\in\mathcal{D}_i$ we associate $0\leqslant u_\Delta\leqslant 1$ and we introduce for each field A a u_Δ dependence :

$$A(u_\Delta)(x) = \sum_{j\neq i} A_j(x)+\sum_{\substack{\Delta'\neq\Delta\\ \Delta'\in\mathcal{D}_i}} A_i(x)\overset{o}{\chi}_{\Delta'}(x)+u_\Delta A_i(x)\overset{o}{\chi}_\Delta(x)$$

The expansion step in Δ is (for ζ big enough to be chosen later) :

$$F(1) = F(0)+F'(0)+\ldots+\frac{1}{(\zeta-1)!}F^{(\zeta-1)}(0)+\int_o^1\frac{(1-u_\Delta)^{\zeta-1}}{(\zeta-1)!}F^{(\zeta)}(u_\Delta)du_\Delta \qquad (2.7)$$

e) The expansion.

We consider the theory in Λ with $s_b = 0$ for all $b\in(\underset{i}{\cup}\mathcal{D}_i^*)\cap\partial\Lambda$

We do first the expansion in the cubes of $\mathcal{D}_\rho$, then in the cubes of $\mathcal{D}_{\rho-1}, \ldots \mathcal{D}_1$.

For each i we do :

I the s_b, $b\in\mathcal{D}_i^*$ expansion : the scaled cluster expansion

II the $t_\Delta, \Delta\in\mathcal{D}_i$ expansion : the frequencies coupling expansion

III the $u_\Delta, \Delta\in\mathcal{D}_i$ expansion : the scaled perturbation expansion

We obtain a sum of terms. For simplicity in all the following we consider the vertices produced by d/ds_b derivations acting on the exponential as d/du_Δ, $(b\in\Delta)$ derivations.

f) The strong connectedness.

Definition : For one term of the expansion two cubes $\Delta\in\mathcal{D}_i$, $\Delta'\in\mathcal{D}_{i'}$, $i\leqslant i'$ are said to be strongly connected if :

A) They are in the same lattice, adjacent (let b be the common face), and $s_b \neq 0$

B) $i' = i+1$, $\Delta\subset\Delta'$, $t_{\Delta'} \neq 0$ and $t_\Delta \neq 0$ or $u_\Delta \neq 0$ or there is a vertex of the cluster expansion in $\mathcal{D}_i$ localized in Δ.

This defines strongly connected sets (s.c.s). If G is a s.c.s. and if $X_1^i, \dots X_j^i, \dots$ are the connected components of $G \cap \mathcal{D}_i$, then for each X_ℓ^i either $G = X_\ell^i$

or X_ℓ^i has a cube strongly connected to a cube of $\mathcal{D}_{i+1}$ or $\mathcal{D}_{i-1}$

Definition : a field $A_i(x), x \in \Delta \in \mathcal{D}_i$ is said to belong to the strongly connected set containing Δ.

We precise the interest of s.c.s. in two lemmas :

lemma : the integration over the fields belonging to s.c.s. factorizes.

A term of the expansion is a product of Schwinger functions attached to the s.c.s.

The coupling between the s.c.s., if any, comes from the perturbation vertices having fields in several s.c.s.

lemma : for K_1 and M big enough :

$$\sum_{G \supset \Delta_o} \prod_{\text{cubes}} e^{-K_1} \prod_{\text{vertices}} \Delta^{-\varepsilon_1} \leqslant 1 \tag{2.8}$$

where the sum is over all s.c.s. containing a cube Δ_o, and the second product is over all perturbation vertices in G, Δ meaning for each vertex the cube of production, and the first product is over all cubes of G.

Remark :

1) an other way of stating (2.8) is

$$\sum_{G \supset \Delta_o} (.) \leqslant \sup_{G \supset \Delta_o} \prod_{\text{cubes}} e^{+K_1} \prod_{\text{vertices}} \Delta^{\varepsilon_1} \quad (.)$$

e^{K_1} by cube and Δ^{ε_1} by vertex (produced in Δ) are combinatoric factors - see [G.J] - accounting for the sum over the s.c.s. containing one cube. We shall use this language extensively in the following.

2) In all the following $\varepsilon_1 > 0$ is small enough s.t. the expansion converges. Moreover to simplify we write ε_1 for $0(1)\ \varepsilon_1$.

Proof of the lemma

Suppose that we know one cube $\Delta \in G$, $\Delta \in \mathcal{D}_i$, then we can sum on all the possible connected components of $\mathcal{D}_i$ containing G :

$$\sum_{X^i \supset \Delta} \prod_{\text{cubes}} e^{-K_1/2} \leqslant 1 \tag{2.9}$$

This is a consequence of the fact that a cube has at most (26) neighbours.

Knowing $\Delta \in G$, $\Delta \in \mathcal{D}_i$ we can also sum on all the possibilities :

- for $\Delta' \in \mathcal{D}_{i+1}$, $\Delta \subset \Delta'$ to be in G, a factor 2 is sufficient (by cube $\Delta' \in G$)
- for $\Delta'' \in \mathcal{D}_{i-1}$, $\Delta'' \sqsubset \Delta$ to be in G, a factor 2 for the cube Δ and a factor Δ''^{ε_1} is sufficient to sum on all the possible Δ'' :

$$\sum_{\text{subsets of } \Delta \cap \mathcal{D}_{i-1}} \prod_{\Delta'' \in \text{ subset}} \Delta''^{-\varepsilon_1} = (1+\Delta''^{-\varepsilon_1})^{\frac{\Delta}{\Delta''}} \leqslant (1+\Delta''^{-\varepsilon_1})^{\Delta''^{2\varepsilon}} \leqslant 2 \tag{2.10}$$

for M big enough and $\varepsilon_1 > 2\varepsilon$.

We have taken in account all the possible sums and putting together all the combinatoric factors we obtain the lemma.

The vertices produced by derivation relatively to cubes belonging to a s.c.s. set G are said internal to it. The internal vertices of G which have at least one field not belonging to G are of particular interest ; let $y_1, \ldots, y_q$ be their coordinates.

The vertices not internal to G but having fields belonging to G are said external ; let $x_1, \ldots, x_p$ be their coordinates. We consider G as a function $G\ (x_1, \ldots ;\ y_1 \ldots)$
A field $A_i(x)$ of an external vertex is localized in the cube $\Delta \ni x$, $\Delta \in \mathcal{D}_i$ and not in the cube of production of the vertex. The other fields are considered as localized in their cube of production.

g) The expansion in terms of diagrams.

Two s.c.s. are possibly connected by the fact that there is at least one vertex with fields in both sets.

A connected diagram is formed of s.c.s. which are connected, as said above.

Let us consider the sum over all the possible connected diagrams H containing one cube Δ_0 and formed of N cubes, our fundamental bound is :

$$\sum_{H \supset \Delta_0} |H| \leqslant \text{cst} \ e^{-NK_1} \tag{2.11}$$

where the cst accounts for a possible dependence on external legs ; it is equal to 1 for a vacuum diagram.

Then via a Kirkwood - Salsburg argument [G-J-S] one obtains the convergence of the infinite volume limite.

3) THE BOUNDS ON THE STRONGLY CONNECTED SETS.

We bound the integration on the fields of a strongly connected set G.

a) The domination by the exponential of the interaction.

The principle has been given in section (1.e). However the procedure is complicated by the t dependence which changes in the interaction but not in the already produced vertices.

we consider a vertex $\frac{d}{dt_\Delta}\lambda A^4 = \lambda \sum_\alpha O(1)\ (\sum_{k<i} A_k)^\alpha (\sum_{j\geqslant i} A_j)^{4-\alpha}\quad (\Delta\in\mathfrak{D}_i)$

$$\tag{3.1}$$

we write (this decomposition is to be found in [G-J])

$$A_{\geqslant i} = \sum_{j\geqslant i} A_j(x) = \bar{A}_{\geqslant i} + \delta A_{\geqslant i}\ ,\ \bar{A}_{\geqslant i} = \frac{1}{\Delta}\int \chi^o_\Delta(x)\ A_{\geqslant i}(x)\,dx \text{ and}$$

$$\delta A_{\geqslant i} = A_{\geqslant i} - \bar{A}_{\geqslant i} \tag{3.2}$$

We shall dominate the low momenta part $[\lambda^{1/4}\bar{A}_{\geqslant i}]^\beta$, $\beta\leqslant 3$ by the

term $\exp - \int \mathcal{P}_i(x)$, $\mathcal{P}_i(x) = \sum_{\Delta\in\mathfrak{D}_i} (1-t_\Delta^4)\chi^o_\Delta(x)\lambda(A_{\geqslant i}(x))^4$ (3.3)

using $\lambda^{1/4}\bar{A}_{\geqslant i} \leqslant \frac{\Delta^{-1/4}}{(1-t_\Delta^4)^{1/4}}\left[(1-t_\Delta^4)\int_\Delta A^4_{\geqslant i}(x)\,dx\right]^{1/4} \leqslant \frac{\Delta^{-1/4+\varepsilon_1}}{(1-t_\Delta^4)^{1/4}}\ \times$

$$\exp \Delta^{-\varepsilon_1}\int_\Delta \mathcal{P}_i(x)\,dx \tag{3.4}$$

Where $\int_\Delta$ means $\int\chi^o_\Delta$. We shall use the fact that $(1-t_\Delta^4)^{-3/4}$ is integrable.

The δA field can be considered as well localized (see next section).

However $(1-t_\Delta^4)^{-\eta/4}$ is non integrable for $\eta > 3$, the consequence is that we cannot dominate with $\mathcal{P}_i(x)$ other low momenta fields produced in Δ.

- let us consider a non derived field from a $d/_{du}$ or $d/_{ds}$ vertex produced in $\Delta\in\mathfrak{D}_i$:

$$A(x) = \sum_j A_j(x) = \bar{A} + \delta A$$

$$\bar{A}_\Delta = \frac{1}{\Delta}\int_\Delta A(x)\,dx \text{ and } \delta A = A - \bar{A}. \tag{3.5}$$

We write the interaction at the point x : $\sum_j \mathcal{P}_j(x)$, $x\in\mathcal{D}_j$; the field in $\mathcal{P}_j$, $j<i$ has not the same t dependence as the field A(x) in $\bar{A}_\Delta$; we can rewrite it :

$$A(x) = \sum_{j<i-1} A_j(x)+t_{\Delta_{i-1}} A_{\geqslant i-1}(x)+(1-t_{\Delta_{i-1}})A_{\geqslant i-1}(x) \tag{3.6}$$

where Δ_{i-1} is the cube of $\mathcal{D}_{i-1}$ containing x.

We dominate $\frac{1}{\Delta}\int_\Delta (1-t_{\Delta_{i-1}})A_{\geqslant i-1}(x)dx$ (Δ_{i-1} depends on x) by $\int_\Delta \mathcal{P}_{i-1}(x)$.

The field sum of the first term in the r.h.s. of (3.6) has now the correct $t_{\Delta_{i-1}}$ dependence we can then rewrite it as :

$$\sum_{j<i-2} A_j(x)+t_{\Delta_{i-2}} A_{\geqslant i-2}(x)+(1-t_{\Delta_{i-2}})A_{\geqslant i-2}(x) \tag{3.7}$$

with $A_{\geqslant i-2}(x) = A_{i-2}(x)+t_{\Delta_{i-1}} A_{\geqslant i-1}(x)$ ...

A field $\lambda^{1/4}\bar{A}_\Delta$ is decomposed as :

$$\sum_{j\leqslant i} \frac{1}{\Delta}\int_\Delta \sum_{\substack{\Delta'\subset\Delta \\ \Delta'\in\mathcal{D}_j}} (1-t_{\Delta'})\chi^{0}_{\Delta'}(x)\lambda^{1/4} A_{\geqslant j}(x)dx \tag{3.8}$$

using Hölder inequality it is bounded by (3.3)

$$\lambda^{1/4}\bar{A}_\Delta \leqslant \sum_{j\leqslant i} \Delta^{-1/4}\Big[\sup_{\Delta'} \frac{(1-t_{\Delta'})^4}{(1-t^4_{\Delta'})}\int_\Delta \mathcal{P}_j(x)dx\Big]^{1/4}$$

$$\leqslant \sup_{j\leqslant i} \Delta^{-1/4+\varepsilon_1}\Big(\int_\Delta \mathcal{P}_j(x)dx\Big)^{1/4} \tag{3.9}$$

where we have used Δ^{ε_1} as combinatoric factor to control the sum over j.

Let $\eta(\Delta)$ be the number of $\frac{dC}{ds}$ vertices produced in Δ, then the number of fields to be dominated in Δ is smaller than :

$$3\eta(\Delta)+3(7+\zeta) \quad (\zeta \text{ was defined in section 2.d}) \tag{3.10}$$

We obtain :$\left(\prod_i \prod_{\Delta\in\mathcal{D}_i} \bar{A}_\Delta \text{ fields}\right) \leqslant$

$$\prod_i \prod_{\Delta\in\mathcal{D}_i} \prod_{j\leqslant i} \Big(\Delta^{-\varepsilon_1}\int_\Delta \mathcal{P}_j(x)dx\Big)^{\frac{3}{4}\eta(\Delta)+0(1)} \prod_{\text{by } \bar{A}_\Delta \text{ field}} \Delta^{-1/4+\varepsilon_1} \tag{3.11}$$

$$\leqslant \Big(\prod_i \prod_{\Delta\in\mathcal{D}_i} 0(1)(\eta(\Delta)!)^{3/4}\Big)\Big(\exp \sum_k M_k^{\varepsilon_1}\int \mathcal{P}_k(x)dx\Big) \prod_{\text{by } \bar{A}_\Delta \text{ field}} \Delta^{-1/4+\varepsilon_1}$$

(where we have used that if $\Delta \in \mathcal{D}_i$ $\sum_{j \leqslant i} \Delta^{-\varepsilon_1} < M_i^{\varepsilon_1}$ if M is big enough).

The exponential factor in (3.4,11) is absorbed by the exponential of the interaction and we use the bound :

$$\exp - (1-2M_i^{\varepsilon_1}) \sum_i \mathcal{P}_i \leqslant 1 \tag{3.12}$$

We then apply a Schwarz inequality obtaining on a s.c.s. G the bound :

$$|G| \leqslant \prod_{\bar{A}\text{ legs}} \Delta^{-1/4(1-2\varepsilon_1)} \left(\int (\Pi\lambda^{1/4}\delta A \prod_{\substack{\text{high}\\ \text{index fields}}} \lambda^{1/4} A)^2 \, d\mu\right)^{1/2} \tag{3.13}$$

where Δ is for a $\bar{A}$ leg the cube of production.

We do explicitely the Gaussian integration in (3.13). It is equal to the sum over all possible diagrams obtained by contractions on the fields.

b) A bound on the number of diagrams generated by the Gaussian integration.

We contract first the fields localized in the smallest cubes. Consider a field in $\Delta \in \mathcal{D}_i$, it contracts with a field in $\Delta' \in \mathcal{D}_j$ and by convention $j \geqslant i$. We choose j with a factor :

$$M_j^{-\varepsilon_1} \tag{3.14}$$

Then we choose Δ' with a factor :

$$O(1)(1+M_i d(\Delta,\Delta'))^{3+1/2} \tag{3.15}$$

[in fact $\sum_{\Delta' \in \mathcal{D}_j} (1+M_j d(\Delta,\Delta'))^{-3-1/2} \leqslant O(1)$]

It remains to choose the field in Δ' :

- it can be a field produced in Δ' then a factor $3\eta(\Delta')+O(1)$ is sufficient.

- it can be also a field from an external vertex produced in $\Delta'' \in \mathcal{D}_\ell$. We choose ℓ with a factor $\Delta''^{\varepsilon_1/3} \simeq M_\ell^{-\varepsilon_1}$ and then the field with a factor $3\eta(\Delta'')+O(1)$. (3.16)

Finally the combinatoric factors for the contractions are (3.14,15,16) :

$\Delta^{\varepsilon_1}\eta(\Delta)$ by contracted field produced in Δ

and (3.17)

$(1+M_i d(\Delta,\Delta'))^{3+1/2}$ by propagator of index i between Δ and Δ'

c) Bounds on the propagators.

In formula (3.5) we have decomposed each field in $\Delta\in\mathcal{D}_i$:

$A(x) = \bar{A}_\Delta + \delta A(x)$. We decompose δA in two parts

$\sum_{j\leq i} \{A_j(x) - \frac{1}{\Delta}\int_\Delta A_j(x)dx\}$ that we call the high index part and what we call from now on the δA part :

$\sum_{j>i} \{A_j(x) - \frac{1}{\Delta}\int_\Delta A_j(x)dx\} = \delta A.$

Using a Schwarz inequality we dissociate the integration on high index fields from the one of δA fields.

A bound on the propagator between two high index fields ($A_j(x)$ produced in $\Delta\in\mathcal{D}_i$, $j\leq i$) is for $\Upsilon\subset\mathcal{D}^*_j$:

$$|\partial^\Upsilon C_j(x,y) \leq M_j^{3-4\varepsilon} \exp\{-M_j/8|x-y|-M_j/8\,d(x,y,\Upsilon)\},j>1 \tag{3.18}$$

$$|\partial^\Upsilon C_1(x,y) \leq 0(1) \quad \exp\{-M_1/8\ |x-y| - M_1/8\ d(x,y,\Upsilon)\}$$

where $d(x,y,\Upsilon)$ = the length of the smallest path going from x to y and hitting each element of Υ.

For a detailed proof see [M-S$_1$].

For $j = i$ note that $M_j^{-3/2(1-\varepsilon_1)} = \Delta^{-1/2(1-\varepsilon_1)}$

We bound now a propagator $C_j(x,y)$ between two δA fields. For δA localized in $\Delta\in\mathcal{D}_i$ the vertex function is (the propagator is $C_j(p)$) :

$$\{\tilde{\chi}^0_\Delta(k+p) - \frac{\tilde{\chi}^0_\Delta(k)\tilde{\chi}^0_\Delta(p)}{\Delta}\} \tag{3.19}$$

where k is the sum of all the other momenta of the vertex, and

$\sum_{j>i} C_j(p)$ is the propagator of the δA fields.

We write the difference in (3.19) as an integration on a suitable function, we obtain :

$$\int_0^1 d\alpha \frac{d}{d\alpha} \frac{\tilde{\chi}^o_\Delta (k+\alpha p)\tilde{\chi}^o_\Delta((1-\alpha)p)}{\Delta}$$

$$= \int_0^1 d\alpha \frac{\Delta^{1/3}}{\Delta} \vec{p} \, [(\vec{\nabla}\tilde{\chi}^o_\Delta)(k+\alpha p)\tilde{\chi}^o_\Delta((1-\alpha)p) \tag{3.20}$$

$$+ \tilde{\chi}^o_\Delta(k+\alpha p)(\vec{\nabla}\tilde{\chi}^o_\Delta)((1-\alpha)p)]$$

We consider a propagator between two vertex function of type (3.20). Suppose that the two δA fields are localized in $\Delta\in\mathcal{D}_i$ and $\Delta'\in\mathcal{D}_{i'}$, $j < i \leqslant i'$, we thus get an extra factor $|\Delta^{1/3} p \, \Delta'^{1/3} p|$. Instead of the bound $\{M^3_{j-1} \exp - M_j \text{ dist}\}$ for the propagator we obtain :

$$M_{j-1}^5 \Delta^{1/3} \Delta'^{1/3} \exp - M_j \text{ dist} \leqslant \frac{0(1) M_j^{\varepsilon_1} M_j^{3(1-\varepsilon_1)}}{(1+\Delta^{-1/3}\text{dist})^{3+1/2}} \tag{3.21}$$

A factor $M_j^{\varepsilon_1}$ is used to control the sum over j.

$$\text{or} \quad 0(1) M_j^{\varepsilon_1} \Delta^{-1/2(1-\varepsilon_1)} \Delta'^{-1/2(1-\varepsilon_1)} (1+\Delta^{-1/3}\text{dist})^{-3-1/2} \tag{3.22}$$

d) Two geometric bounds. (see [G-J-S])

lemma : Let G be a s.c.s. and χ a connected component of $G\cap\mathcal{D}_i$. Each derivation d/ds_b, $b\subset\chi$, derives either a field of a vertex or a covariance. Let $\partial^{\gamma_1} C_i, \ldots, \partial^{\gamma_p} C_i$ be the derived covariances, and $b_i, \ldots, b_q$ be the faces deriving some fields, then for any $K_1 > 0$, for M big enough we have :

$$6 \text{ (number of cubes of } \chi) \geqslant q + \sum_\ell \text{ number of faces of } \gamma_\ell \geqslant$$

$$\text{(number of cubes of X)} - 1 \tag{3.23}$$

and

$$\sum_{p,q} \sum_{\substack{\gamma_1\ldots\gamma_p \subset \mathcal{D}^\bullet_i \cap \chi \\ \gamma_i\cap\gamma_j=\emptyset, i\neq j}} [\prod_{\ell=1}^p M_i^{\varepsilon_1} \exp - \frac{M_i}{16} d(\gamma\ell)] M_1^{q\varepsilon_1} \leqslant \prod_{\Delta\in\chi} e^{-K_1} \tag{3.24}$$

where $d(\gamma)$ is the length of the smallest path hitting the elements of γ.

Proof : consider the subset formed of one element of each γ_i. A factor $\prod_{\Delta\in\chi} 2$ accounts for the sum over this subset.

So that knowing one element b of γ we have to sum over γ :

$\exp -\frac{M_i}{32} d(\gamma) \leqslant 0(1)$, this because the mesh of $\mathcal{D}_i$ is KM_i^{-1}, and
ɔb
ecause one face has 32 neighouring faces ; $d(\gamma)$ is bigger of KM_i^{-1} at
east each 33 faces. A consequence of (3.23) is that

$(1)\Sigma_\ell M_i K^{-1} d(\gamma_\ell)+p+q+1 \geqslant$ number of cubes of χ, which gives the r.h.s. of
3.24). The second bound shows how one controls the accumulation factors
$\eta(\Delta)!)^{0(1)}$:

emma : let us consider in $\mathcal{D}_i$ all the derived propagators $\partial^{\gamma_\ell} C_i(x_\ell,y_\ell)$
enerated by the cluster expansion (some x_ℓ or y_ℓ can be equal, if the
ropagators contract to the same vertex)

$$|\prod_\ell \exp -\frac{M_i}{16} d(x_\ell,y_\ell,\gamma_\ell)| \leqslant 0(1) \prod_{\Delta\in\mathcal{D}_i} 0(1)\eta(\Delta)^{-0(1)\eta(\Delta)} \tag{3.25}$$

here as said before $\eta(\Delta)$ = the number of distint vertices in Δ created
y the $\partial^\gamma C$.

roof : for each ℓ, let $b_\ell \in \gamma_\ell$.
ow for each $b\in\mathcal{D}_i^*$ there is only one derivation d/ds_b ; then if $H(\Delta)$ is
he subset of the b_ℓ such that $\partial^{\gamma_\ell} C_i$ contracts in Δ and if $\eta(\Delta) > N \gg 1$
hen at least $\frac{\eta(\Delta)}{2}$ elements of $H(\Delta)$ are such that $\frac{M_i}{K}$ dist $(\Delta,b) > \frac{\eta(\Delta)}{100}^{1/3}$
so that :

$$\prod_{b\in H(\Delta)} \exp -\frac{M_i}{16} d(x_\ell,y_\ell,\gamma_\ell) \leqslant \exp -\frac{K}{16.2.100} \eta(\Delta)^{1+1/3} \leqslant 0(1)\eta(\Delta)^{-0(1)\eta(\Delta)}$$

e) <u>The bound on a vertex $\Delta\in\mathcal{D}_i$</u>

Let us consider first the volume integration of an internal vertex
produced in $\Delta\in\mathcal{D}_i$. The vertex is :

$$[(\bar{A})^\alpha(\delta A)^\beta(A_j)^\gamma](x) \quad j\leqslant i \text{ , } A_j \text{ stands for a high index field}$$

$$\alpha+\beta+\gamma = 4. \ \alpha+\beta \geqslant 1, \ \gamma \geqslant 1$$

For each j let $V_j = G\cap\mathcal{D}_j$ (G is s.c.s. containing Δ)
then the volume of integration is $|\bigcap_{j,V_j\neq\emptyset} \Delta\cap V_j|$ (3.26)

The bound on a vertex is (i.e. the bound by vertex, without distance
factors):

$$|\bigcap_{V_j\neq\emptyset} (V_j\cap\Delta)| \prod_{j\leqslant i} M_j^{3/2(1-\varepsilon_1)} \prod_{\bar{A}} \Delta^{-1/4(1-\varepsilon_1)} \prod_{\delta A} \Delta^{-1/2(1-\varepsilon_1)} \tag{3.27}$$

moreover we have by external field of the vertex, localized in $\Delta'\in\mathcal{D}_j$ a factor

$$M_j^{3/2(1-\varepsilon_1)} = \Delta'^{-1/2(1-\varepsilon_1)} \quad \text{(see section 1.f)} \tag{3.28}$$

Using the inequality $x^p \leqslant (\frac{0(1)}{a})^p \, p! \, e^{ax}$

we transform the bound (3.28) on the fields with high index $j \leqslant i$, produced in Δ :

$$\Pi M_j^{3/2(1-\varepsilon_1)} \leqslant [\Pi(V_j \cap \Delta)^{-1/2(1-\varepsilon_1)} \Delta^{\varepsilon_1}] \eta(\Delta)! \exp\{\Delta^{-\varepsilon_1} M_j^3(V_j \cap \Delta)\} \tag{3.29}$$

and

$$\exp\{\Delta^{-\varepsilon_1} M_j^3(V_j \cap \Delta)\} \leqslant \exp \Delta^{-\varepsilon_1} \sum_{\substack{\Delta' \subset V_j \cap \Delta \\ \Delta' \in \mathcal{D}_j}} 0(1) \qquad (\Delta \in \mathcal{D}_i) \tag{3.30}$$

The product of the factor (3.30) over all index i, all cubes $\Delta \in \mathcal{D}_i$ gives :

$$\exp\{\sum_{\Delta \in \mathcal{D}_i} 0(1) M_i^{\varepsilon_1} \sum_{\substack{\Delta' \subset V_j \cap \Delta \\ \Delta' \in \mathcal{D}_j}} 0(1)\} \leqslant \prod_{\Delta' \in G \cap \mathcal{D}_j} 0(1) \tag{3.31}$$

This factor is associated with the one of formula (3.24) ; the bound on a vertex (3.27) becomes :

$$|\bigcap_{V_j \neq \emptyset} (V_j \cap \Delta)| \prod_{j \leqslant i} \Delta^{\varepsilon_1} (V_j \cap \Delta)^{-1/2(1-\varepsilon_1)} \prod_{\bar{A}} \Delta^{-1/4(1-\varepsilon_i)} \prod_{\delta A} \Delta^{-1/2(1-\varepsilon_1)} \tag{3.32}$$

In particular a vertex with all fields in the same s.c.s. is small like $\Delta^{-1/4+\varepsilon_1}$, Δ being the cube of production.

f) <u>The bound on a strongly connected set G</u> .

Putting to gether the result of the chapter 3, in particular formulas (3.18,22,24,25,32) we see that for $\varepsilon_0 = \frac{1}{16} - 0(1)\varepsilon_1$

$$|G| \leqslant \prod_{\Delta \in G} 0(1) e^{-K_1} (\eta(\Delta)!)^{-0(1)} \prod_{\text{internal Vertices}} \Delta^{-\varepsilon_0} \times$$

$$\prod_{\substack{\text{vertices with}\\ \text{external legs}}} \prod_{\bar{A}} \Delta^{-1/4(1-\varepsilon_1-4\varepsilon_o)} \prod_{\delta A} \Delta^{-1/2(1-\varepsilon_1-2\varepsilon_o)} \prod_{\substack{\text{high index}\\ \text{field of } G}} (V_j \cap \Delta)^{1/2(1-\varepsilon_1)} \tag{3.33}$$

$$\prod_{\substack{\text{field of external vertices}\\ \text{localized in } \Delta\in G \text{ and}\\ \text{produced in } \Delta'\notin G}} \eta(\Delta')^{1/2}\, \Delta^{-1/2(1-\varepsilon_1)}$$

where Δ is the cube of localization of the field and j its index. The first factor in the r.h.s. of (3.33) comes from (3.24) ; for i = 1, the factor $M_1^{\varepsilon_1}$ is given by the factor $\lambda^{1/2}$ that we can associate to each propagator (one $\lambda^{1/4}$ by field); thus we choose λ s.t.

$\lambda^{1/2} < M_1^{\varepsilon_1}$.

The bound on s.c.s. G is given by the lowest orders of our perturbation expansion because each new internal vertex gives a small factor $\Delta^{-\varepsilon_o}$ (see 3.33).

Thus the order of the best bound on G is given by the terms where $u_\Delta = 0$ for each Δ of G. Let us look more precisely to this point :

we define $N_G = \{i,\ \exists\Delta\in\mathcal{D}_i\ , \Delta\in G\}$ and $|N_G| = \#\{N_G\}$

- if $|N_G| \geqslant 100/3\varepsilon_o$, then because there is at least one internal vertex by index if $N_G > 1$ we have $|G| < \prod_{i\in N_G} M_i^{3\varepsilon_o}$

If $k = \sup\{i, i\in N_G\}$, then $M_{\sup\{1,k-100/3\varepsilon_o\}} < M_k^{(1-\varepsilon_2)}$ for any $\varepsilon_2 > 0$

and for ε small enough (depending on ε_o and ε_2) ; we have

$$|G| < M_k^{100(1-\varepsilon_2)}. \tag{3.34}$$

- if $|N_G| < 100/3\varepsilon_o$ and if there is at least one cube $\Delta\in G$ for which $u_\Delta \neq 0$ (i.e. in formila (2.7), G is made with the last term of the r.h.s.), then there is in Δ , ζ derivation relatively to u_Δ , that is at least $\zeta/4$ vertices produced so that

$$\Big|\prod_{\substack{\text{internal}\\ \text{vertices}}} \Delta^{-\varepsilon_o}\Big| \leqslant \Delta^{-\varepsilon_o\zeta/4} \leqslant M_k^{(1-\varepsilon_2)\frac{3\varepsilon_o\zeta}{4}}$$

because if $\Delta\in\mathcal{D}_i$ then $i > k - 100/3\varepsilon_o$

so that for ζ large enough $|G| < M_k^{100(1-\varepsilon_2)}$.

Definition : a strongly connected set G is said to be flat if :

$$|N_G| < 100/3\varepsilon_o \text{, and if for each cube } \Delta \in G,\ u_\Delta = 0 \tag{3.35}$$

$$\text{if} \quad j = \inf N_G,\ k = \sup N_G \quad \text{then : } M_j^{1+\varepsilon_o} \leqslant M_k \leqslant M_j \tag{3.36}$$

If g is flat, then by construction all fields in G are high index fields. G is then equal to a sum of perturbation diagrams explicitely computable.

Remark : The order of the bound on a sum of strongly connected sets is given by the lowest order terms of the flat contributions in this sum.

4) THE FINITE RENORMALIZATIONS :

a) The fixed point procedure.

To do the renormalization we proceed by induction. We suppose at step k, that we know the counterterms :

$\alpha^{(k-1)}(s,t,u)$ and $\delta^{(k-1)}(s,t,u)$ such that the two-point function and the four point-function of the theory with $A_j = 0$, $j \geqslant k$, have a vanishing zero momentum value. The two-point function is given by

$$< \frac{\delta}{\delta A(x)} \cdot \frac{\delta}{\delta A(y)} >$$

and the four-point function by :

$$< \prod_{i=1}^{4} \frac{\delta}{\delta A(x_i)} > - \sum_{\substack{\text{partition } \Pi=(\Pi_1,\Pi_2) \\ \text{of}(1,2,3,4)}} < \prod_{i\in \Pi_1} \frac{\delta}{\delta A(x_i)} > < \prod_{j\in \Pi_2} \frac{\delta}{\delta A(x_j)} > \tag{4.1}$$

We then compute the correction to α and δ due to the field A_k.

In the theory with $A_j = 0$, $j > k$ we compute the zero momentum value of the two and four-point functions :

$\alpha_1^{(k)}$ and $\delta_1^{(k)}$ (in the following we omit the subscript k). They are the value at $s \equiv t \equiv u \equiv 1$ of functions :

$\alpha_1(x,s,t,u)$ and $\delta_1(x,s,t,u)$

We then consider the theory with a new term in the measure :

$$\exp - \int \delta_1(x,s,t,u)\ A^4(x)dx - 1/2 \int \alpha_1(x,s,t,u)\ A^2(x)dx$$

We compute in the new theory the zero momentum value of the two and four-point functions: α_2 and δ_2 ...

We show that $|\alpha_1| < \lambda M_k^{5/2+2\varepsilon_1}$ and $|\delta_1| < \lambda^2 M_k$ for some $\varepsilon_2 > 0$.

Taking advantage of the renormalization due to α_1 and δ_1 one shows that for some $a < 1$

$$|\alpha_2| < a\lambda M_k^{5/2+2\varepsilon_2} \qquad \text{and } |\delta_2| < a\,\lambda^2 M_k \tag{4.2}$$

more generally

$$|\alpha_i| < a^{i-1}\lambda\, M_k^{5/2+2\varepsilon_2} \qquad \text{and } |\delta_2| < a^{i-1}\,\lambda^2 M_k$$

Then $\alpha^{(k)} = \alpha^{(k-1)} + \sum_i \alpha_i$, $\delta^{(k)} = \delta^{(k-1)} + \sum_i \delta_i$

and $\alpha = \lim_{k\to\infty} \alpha^{(k)}$ $\qquad \delta = \lim_{k\to\infty} \delta^{(k)}$

Remark : from the bounds (4.2) we see that an αA^2 vertex produced in $\Delta \in G$ is such that :

$$|\alpha| \leq 0(1)\Delta^{5/6+2/3\varepsilon_2} \qquad \text{if } \alpha \text{ is derived.}$$

$$|\alpha(x)| \leq 0(1)\ M_j^{5/2+2\varepsilon_2} \quad \text{if } j = \inf\{\ell, x \in \Delta' \in \mathcal{D}_\ell,\ \Delta' \in G\} \tag{4.3}$$

So that the bound on α is almost the bound that one obtains for two fields in G. Thus a αA^2 vertex for what concerns the convergence of the expansion behaves as a A^4 vertex.

This is obviously true for a δA^4 vertex. Note that the dominant contribution to α is negative so that the presence of α modifies the bound (3.12) because the polynomial of the interaction is not always positive.

If G is a s.c.s. then a consequence of (4.2) is that $|\alpha|_G(x)| \leq M_j^{5/2+2\varepsilon_2}$ where j(x) is defined in (4.3), and the restriction to G means that all s,t,u not corresponding to internal derivation of G are put to zero. (4.4)

$$\text{Thus } \exp - \sum_i \mathcal{P}_i \leq \prod_{\Delta\in G} \exp - 0(1)\int_\Delta \{A^4(x) - \Delta^{5/6} A^2(x)\}dx \leq \prod_{\Delta\in G} 0(1)$$

This factor is compensated by the one of (3.24) .
Finally the presence of αA^2 and δA^4 vertices change qualitatively nothing to our bounds, and just modify some finite constants.

b) An approximate renormalization scheme.

To simplify we consider mainly only the two point function

renormalization. In the remaining of this chapter we are concerned with the calculus of the function $\alpha(x,s,t,u)$ (resp δ). The necessity of putting s,t,u dependances in α (to have the bounds 4.3) obliges us to follow the s,t,u dependances in the process of computing α (resp δ).

In the expansion of: $\int \frac{\delta}{\delta A(x)} \frac{\delta}{\delta A(y)}$ exp (interaction) $d\mu$, consider a two-point connected diagram : $H(x,y)$ [H is formed of s.c.s. which are connected by vertices.] ; it is a connected term in the sum given by :

$$\prod_{b,\Delta\in H} \left(\frac{d}{ds_b}\right)^{\mathcal{S}_b} \left(\frac{d}{dt_\Delta}\right)^{\mathcal{G}_\Delta} \left(\frac{d}{du_\Delta}\right)^{\mathcal{U}_\Delta} \frac{\delta}{\delta A(x)} \frac{\delta}{\delta A(y)} \text{ exp interaction } d\mu \Big|_H$$

the restriction symbol is defined in (4.4), and $\mathcal{U}_\Delta$ is the number of derivation relatively to u_Δ in H , ...

Let $\mathcal{H}$ be the class of all the translated diagrams of H (relatively to the biggest lattice having a cube in H).

$$\text{we define} \quad \mathcal{H}(x,y) = \sum_{\substack{H'\in\mathcal{H} \\ x,y\in H'}} \frac{H'(x,y)}{Z_{H'}} , \quad Z_{H'} = Z_H = \int e^{\text{interaction}} \Big|_{H'}$$

then

$$\alpha_{1,\mathcal{H}} = \int \mathcal{H}(x,y)\,dy$$

$$\alpha_{1,H}(s,t,u) = \prod_{b\in H} (\Delta_b)^{\mathcal{S}_b} \prod_{\Delta\in H} \frac{(u_\Delta)^{\mathcal{U}_\Delta}}{u_\Delta!} (t_\Delta)^{\mathcal{G}_\Delta}\, \alpha_{1,\mathcal{H}}$$

an approximate choice for $\alpha_1(s,t,u)$ is :

$$\alpha_1(x,s,t) = \sum_{H\ni x} \alpha_{1,H}(s,t,u)$$

In the following we make work this scheme.

There are two difficulties with this approximation :

- <u>the first problem</u> : $\mathcal{H}(x,y)$ is not translation invariant, thus $\alpha_{1,\mathcal{H}}$ is in fact dependant on x. This because $\mathcal{H}$ contains only discrete translated of H.

We make $\mathcal{H}(x,y)$ translation invariant ; we define

$$\bar{\mathcal{H}}(x-y) = \int_{[-1/2,+1/2]^3} \mathcal{H}(x+a_k t,\, y+a_k t)\,dt$$

where if $k = \inf\{i, \exists\Delta\in\mathcal{D}_i ,\ \Delta\in H\}$ then $a_k = \Delta^{1/3}$, $\Delta\in\mathcal{D}_k$.

$$(\mathcal{H}-\bar{\mathcal{H}})(x,y) = \int_{[-1/2+1/2]^3} dt \int_0^1 ds_1 \int_0^1 ds_2\ s_1\ a_k^2 [\vec{t}(\vec{\nabla}_x+\vec{\nabla}_y)]^2\ H(x+s_1 s_2 a_k t,\ y+s_1 s_2 a_k t)$$

$\mathcal{H}-\bar{\mathcal{H}}$ contains two gradients and is thus considered as renormalized.

n the following we set $\mathcal{H} = \overline{\mathcal{H}}$.

- the second problem comes from the fact that to renormalize we compare $\alpha_{1,\mathcal{H}}$ to $\mathcal{H}$ and thus we have to be able to sum over all the translated of each diagram. In particular to compute $\alpha^{(k)} - \alpha^{(k-1)}$ we must be able to renormalize all the diagrams with $A_j = 0$, $j \geq k$, that is to take advantage of the renormalization counterterms $\alpha^{(k-1)}$ and $\delta^{(k-1)}$.

c) The renormalization.

Let $H(x,y)$ be a two point diagram ($A_j = 0$, $j \geq k$).

Let $n_k \in \Delta^{1/3}\,\mathbb{Z}^3$, $\Delta \in \mathcal{D}_k$, and H^{n_k} be the translation of H , obtained by the translation n_k of all the cubes of H , then :

$$\sum_{n_k} H^{n_k}(x,y) = \mathcal{H}(x-y) \qquad (4.5)$$

The Fourier transform is $\tilde{\mathcal{H}}(p)$ and $\tilde{\mathcal{H}}(0) = \int \mathcal{H}(x-y)\,dy$.

The renormalization is the substraction :

$$\tilde{\mathcal{H}}(p) - \tilde{\mathcal{H}}(0) = p^2 \int_0^1 (1-t)\, \tilde{\mathcal{H}}''(pt)\,dt \qquad (4.6)$$

In the expansion different diagrams don't overlap : the same cube cannot be in two diagrams ; thus the sum on the translated of each diagram are not independant. Moreover the expansion is done in the space cut-off Λ. Thus the sum (4.5) for a diagram doesn't appear in the expansion.

We consider successively the two restrictions :

i) for a diagram H we add and substract : $\sum_{\substack{n_k \\ H^{n_k} \not\subset \Lambda}} H^{n_k}(x,y) = \mathcal{H}_{\partial\Lambda}(x,y)$

$\mathcal{H}_{\partial\Lambda}(x,y)$ completes the sum over the translated of H.

The error term $-\mathcal{H}_{\partial\Lambda}(x,y)$ is a boundary term ; it binds x and y to $\partial\Lambda$; the integration on x and y instead of being over Λ (i.e. in three dimensions), is approximately only on $\partial\Lambda$ (i.e. in two dimensions).

In two dimensions, the Gaussian integration with a propagator decreasing as fast as $1/|x-y|^3$ is possible (because there is a finite number of vertices by cube of $\mathcal{D}_1$, ...). The propagator is summable, thus there is no divergent subdiagrams ; the renormalization is not necessar

ii) in order to suppress the limitation forbidding the overlaps of diagrams, we decompose each product in sum of products of independant truncated functions.

For example for two diagrams : (H , H_1 , H_2 are connected diagrams)

$$S(x_1 , x_2 , x_3 , x_4)= \sum_H H(x_1 ,x_2 , x_3 , x_4)+\sum_{H_1 \cap H_2=\emptyset}[H_1(x_1 , x_2)H_2(x_3 , x_4)$$

$$+ H_1(x_1 , x_3)H_2(x_2 , x_4)+H_1(x_1 , x_4)H_2(x_2 , x_3)]$$

$$=\Sigma\, S^T(x_1 , x_2 , x_3 , x_4)+ S^T(x_1 , x_2)\; S^T(x_3 , x_4))+ \ldots$$

where $S^T(x_1 , x_2) = \sum_H H(x_1 , x_2)$

$$S^T(x_1 , x_2 , x_3 , x_4) = S(x_1 , x_2 , x_3 , x_4)-S(x_1 , x_2)S(x_3 , x_4)- \ldots$$

We give now a more precise description of this decomposition in truncated functions.

d) The decomposition in independant truncated functions.

At a given step ℓ we suppose that all the diagrams $\underline{A_j = 0}$, $j \geqslant \ell-1$ are renormalized. In this section we decompose the diagrams of the theory $\underline{A_j = 0\ j>\ell-1}$ in truncated functions, and give the s,t,u dependance associated to them in the computation of the counterterms. The expansion can be written as :

$$\int \Omega e^{\text{interaction}} d\mu\Big|_{\substack{A_j = 0\\ j>\ell-1}} =S(\Omega)=\Sigma[S_{\leqslant \ell-1}(x_1 ,\ldots,x_n)S_{>\ell-1}(x_2 \ldots,x_n)](\Omega$$

where $S_{\leqslant \ell-1}$ is the product of the diagrams with $A_j = 0$, $j>\ell-1$

Ω is a set of variables,

$x_1 , \ldots, x_n$ are the coordinates of the vertices produced by derivation of index $\geqslant \ell-1$ but having fields of index $<\ell-1$.

$$S = \sum_\Gamma S_\Gamma\, Z_{\sim\Gamma} \tag{4.7}$$

where S_Γ is the product of the non vacuum diagrams.

$$S_\Gamma = (\Pi \text{ derivations in } \Gamma)S\Big|_{s,t,u\, =\, o \text{ outside } \Gamma}$$

$$Z_{\sim\Gamma} = Z\Big|_{s,t,u\, =\, o \text{ in } \Gamma}$$

We decompose S in truncated functions :

$$S(\Omega) = \Big(\sum_{\substack{\Pi\in\mathcal{P}(\Omega)\\ \Pi=\{\omega_1\ldots\omega_q\}}} \prod_{i=1}^{q} \mathcal{S}^T(\omega_i)\Big)Z, \text{ where the sum is over the partitions of } \Omega.$$

$$\mathcal{S}^T(\omega) = \sum_{\Gamma} \mathcal{S}^T_\Gamma(\omega) \tag{4.8}$$

and

$$\mathcal{S}^T_\Gamma(\omega) = \sum_{\substack{\gamma_1\ldots\gamma_n\\ \cup\gamma_i=\Gamma}} \sum_{\substack{\Pi\in\mathcal{P}(\omega)\\ \Pi=(\tau_1\ldots,)}} (\Pi S_{\gamma_i}(\tau_i))\hat{k}^T(\gamma_1,\ldots,\gamma_n)$$

where in the sum the γ can overlap and :

$$\hat{k}(\gamma_1,\ldots,\gamma_n) = \begin{cases} Z_{\sim(\cup\gamma_i)}/Z & \text{if the } \gamma_i \text{ are all disjoint}\\ 0 & \text{if there is some overlap in the } \gamma\text{'s.}\end{cases} \tag{4.9}$$

and $\hat{k}^T$ is the corresponding truncated function :

$$\hat{k}^T(\gamma_1\ldots\gamma_n) = \sum_q \frac{(-1)^{q-1}}{q} \sum_{\substack{\Pi\in\mathcal{P}(1,\ldots,n)\\ \Pi=(W_1,\ldots,W_q)}} \prod_{i=1}^{q} \hat{k}(\{\gamma_\ell\}_{\ell\in W_i}) \tag{4.10}$$

and $S_\Gamma = \Pi S_{\gamma_i}$ is the decomposition of S_Γ in connected diagrams.

We have then that :

$\mathcal{S}^T(x_1, x_2) = \mathcal{S}^T(x_1-x_2)$ (and similarly for a 4-point function).

Let us call $\mathcal{S}^T_{\leqslant \ell-1} = \mathcal{S}^T\big|_{A_j=0,\ j\geqslant \ell-1}$

then by induction hypothesis $\tilde{\mathcal{S}}^T(0) = 0$ thus (see also 4.6) :

$$\mathcal{S}^T_{\leqslant\ell-1}(x_1, x_2) = -\nabla^2_{x_1}\int_0^1 \mathcal{S}^T_{\leqslant\ell-1}\left(\frac{x_1-x_2}{t}\right)\frac{[x_1-x_2]^2}{t}(1-t)\,dt$$

$$= -\nabla^2_{x_1}\sum_\Gamma\int_0^1 \mathcal{S}^T_{\Gamma\,\leqslant\ell-1}(x_1, x_2)\frac{(x_1-x_2)^2}{t}(1-t)\,dt$$

$$\mathcal{S}^T_{\leqslant\ell-1}(x_1, x_2) = \sum_\gamma \check{S}_\gamma(x_1, x_2)\,\hat{k}^T_{\leqslant\ell-1}(\gamma)$$

$$\text{with } \check{S}_\gamma(x_1, x_2) = -\nabla^2_{x_1} S_\gamma(x_1, x_2) \tag{4.11}$$

note that

$$\mathcal{S}^T(x_1, x_2) = \sum_{\substack{\gamma\\ \text{restricted}\\ \text{to } j<\ell}} \check{S}_\gamma(x_1, x_2)\,\hat{k}^T(\gamma) + \sum_{\substack{\gamma \text{ with some}\\ \text{cube } \Delta\in\mathcal{D}_j,\ j\geqslant\ell}} \check{S}_\gamma(x_1, x_2)\,\hat{k}^T(\gamma)$$

and similarly for the 4-point fonction.

Definition : for the 2 and 4-point functions $\check{S}$ is given by (4.11) or its equivalent.

for more than 4 points function $\check{S}_\gamma = S_\gamma$

If some part of S contributes to a 2 or 4-point function we want to associate to it the s,t,u dependance of the α or δ corresponding counterterm.

If in formula (4.7), the two and four point diagram were $\check{S}_\Gamma$ then this dependance would be $\prod_{b,\Delta\in\gamma} (s_b)^{\mathcal{S}_b} (t_\Delta)^{\mathcal{B}_\Delta} u_\Delta^{\mathcal{U}_\Delta} / \mathcal{U}_\Delta!$

The equivalent of formula (4.7) is :

$$S(\Omega) = \sum_{\substack{\Pi\in\mathcal{P}(\Omega) \\ \Pi=\{\omega_1,\ldots\}}} \prod_j \sum_{\gamma_1^j\ldots\gamma_n^j} \left\{ \sum_{\substack{\xi^j\in\mathcal{P}(\omega) \\ \xi^j=\{\tau_1\ldots\}}} (\prod_i \check{S}_{\gamma_i^j}(\tau_i))\, \hat{k}^T(\gamma_1,\ldots\gamma_n) \times \frac{Z}{Z_{\sim\bigcup_{i,j}\gamma_i^j}} \right\} Z_{\sim\bigcup_{i,j}\gamma_i^j}$$

where $Z_{\sim\cup\gamma} = Z \Big|_{s,t,u\,=\,0\text{ in }\cup\gamma}$ (the γ can overlap).

The dependance is then given by :

$$\prod_b (s_b)^{\mathcal{S}'_b} \prod_\Delta (t_\Delta)^{\mathcal{B}'_\Delta} (u_\Delta)^{\mathcal{U}'_\Delta} / \mathcal{U}'_\Delta!$$

whith $\mathcal{S}'_b = \sup_{i,j} \{\mathcal{S}_b,\ b\in\gamma_i^j\}$ and similarly for $\mathcal{B}'\Delta$ and $\mathcal{U}'\Delta$.

To separate the connected components it is convenient to expand $Z/Z_{\sim\cup\gamma} = r$ in sum of products of connected functions r^T, each connected graph in the expansion of $S(\Omega)$ has then its corresponding s,t,u dependance given by the γ's contributing to it.

This finishes the presentation of the computation of the counterterms. We have widely used the truncated functions. We have to show that we have qualitatively the same bounds on truncated functions as on non truncated functions.

e) The bounds on the truncated functions.

We consider for example $\hat{k}^T(\gamma_1,\ldots,\gamma_p)$ with (see 4.8) $\tau_i\subset\gamma_i$ and suppose that $x_j \in \tau_i$ is in Δ_j , and let $\eta'(\Delta)$ be the number of variables of $\cup \tau_i$ in Δ, then :

$$|\hat{k}^T(\gamma_1,\ldots,\gamma_p)| \leqslant \prod_{\Delta} (\eta'(\Delta))!)^{O(1)} \exp \sum_i \sum_{\Delta\in\gamma_i} O(1) \tag{4.12}$$

Such a bound fits exactly with our bounds of chapter 3, and we can thus consider the connected components obtained in section (4.d) as connected graphs of the expansion of chapter 2.

To prove (4.12) we use (see [S]) the fact that a truncated function can be written as a derivative of an ad-hoc function :

$$\hat{k}^T(\gamma_1,\ldots,\gamma_n) = \prod_{i=2}^{n} \frac{d}{d\mu_i} \left. \frac{\hat{k}[\gamma_1 \prod_{i=2}^{n}(1+\mu_i\gamma_i)]}{\hat{k}[\prod_{i=2}^{n}(1+\mu_i\gamma_i)]} \right|_{\mu_i = 0 \text{ for all } i} \tag{4.13}$$

where by convention $\hat{k}(\gamma_1 \prod_{i=2}^{n} (1+\mu_i\gamma_i)) =$

$$= \sum_{I\subset\{2,\ldots,n\}} \prod_{i\in I} \mu_i \;\hat{k}(\gamma_1, \{\gamma_i\}_{i\in 1}) \quad \text{see (4.9)}$$

The function of μ in (4.13) is analytic if the μ_i are small enough, roughly if :

$$|\mu_i| < (\prod_{\Delta\in\gamma_i} O(1))^{-1} \tag{4.14}$$

this because [G-J-S] : $|Z_{\sim\cup\gamma_i}/Z| \leqslant \prod_i \prod_{\Delta\in\gamma_i} O(1)$

We compute the r.h.s of (4.13) using the Cauchy formula. It remains to obtain in the region (4.14) a bound on the function of μ . To do this we expand the numerator and the denominator using the expansion of chapter 2. The result is an expansion in terms of graphs, and we have on the connected graphs a bound of type (2.11).

A Kirkwood-Salsburg argument (sketched in [M-S_1] App II) leads then to the bound (4.9). There are factorials in (4.9) because if a given cube is present $\eta(\Delta)$ times we have for each s.c.s. to decide which replica of Δ it contains (if any).

5) THE CONVERGENCE OF THE EXPANSION.

a) The convergence factor due to the renormalization

Each 2-point function gives terms with two gradients acting on the external fields, each 4-point function, terms with one gradient acting on the external fields ; this however if the function has a sufficiently strong decrease. The effect of the renormalization is to transfer "internal" power decrease into external power convergence.

We have thus to verify that the 2 and 4-point functions decrease enough rapidly, provided that the 2 and 4-point subdiagrams are themselves renormalized.

b) The power decrease of the strongly connected functions.

The simplest connected diagrams are the strongly connected ones. A consequence of the factor $\Delta^{-1/2}$ by external field localized in Δ is that a flat (see definition (3.35)) 2-point s.c.s. of momentum M_j is small like $(M_j^{-9})^{3/4}$ if the two points are not coinciding.
Then

$$|G(x,y)| \leqslant M_j^{-27/4} \exp - M_j \text{ dist } (x,y) \leqslant 0(1) \frac{M^{-3/2}}{(1+\text{dist } (x,y))^{2+3+1/4}}$$

which insures the existence of $\widetilde{G}''(p)$ (formula (4.6))
The non flat s.c.s. are even smaller than the flat ones (section 3.f)

c) The convergence of the expansion.

It remains to see that we are able to compute any n point functions (i.e. to sum over all the connected diagrams containing some points), and to show that for the 2 and 4-point function there is enough power decrease.

This is done by inspection ; it is straighforward but lengthy [[M-S$_1$] chapter VI].

We sketch the problem ; a flat s.c.s. contains (roughly) only one size of cube.

A flat s.c.s. G made of cubes Δ has a bound :

$|G| < \exp - \Delta^{-1/3}$diameter of G.

we have :

$\Delta^{-1/2}$ by field localized in Δ.

Moreover there are

2 gradients on the external fields of a 2-point diagram

1 gradient on the external fields of a 4-point diagram

and if a gradient acts on a field localized in Δ, then it is equivalent to a factor $\Delta^{-1/3}$.

Then supposing that there are only A^4 vertices in the theory one has to show (as said above) the convergence of the expansion.

REFERENCES

[B-C-G-N-O-P-S] Benfatto, G., Cassandro, M., Gallavotti, G., Nicolo, F., Presutti, E., Scacciatelli, E., Commun. math. Phys. 59,143 (1978)

[B-F-L-S] Bricmont, J., Fontaine, J.R., Lebowitz, J., Spencer, T., I. Commun. Math. Phys. 78,281 (1980) II. Commun. Math. Phys. 78,363 (1981)

[F_1] Federbush, P., Commun. Math. Phys. 81, 327,341 (1981)

[F_2] Fontaine, J.R., Bounds on the decay of correlations for $\lambda(\nabla\Phi)^4$ models, Université catholique de Louvain (Preprint).

[G-J] Glimm, J., Jaffe, A., Fort. der Physik 21,327 (1973)

[G-J-S] Glimm, J., Jaffe, A., Spencer, T., The cluster expansion, in Lecture Notes in Physics Vol. 25 Springer (1973)

[G-K] Gawedzki, K., Kupiainen, A., Commun. Math. Phys. 82,407 (1981), Commun. Math. Phys. 83,469 (1982) and, Renormalization group for a critical model, I.H.E.S.(Preprint) and their contribution in this book.

[$M-S_1$] Magnen, J., Sénéor, R., The infra-red behaviour of $(\nabla\Phi)^4_3$ Ecole polytechnique (Preprint)

[$M-S_2$] Magnen, J., Sénéor, R., A note on cluster expansions Ann. Inst. Henri Poincaré (to be published)

[S] Spencer, T., Commun. Math. Phys. 44,143 (1975).

ULTRAVIOLET STABILITY IN FIELD THEORY. THE ϕ_3^4 MODEL.

Tadeusz Balaban

1. Preliminary Remarks.

Renormalization group ideas appeared in the early papers on constructive quantum field theory [5,6,7], although they were never stated explicitly. The model $(\phi^4)_d$ was considered in these papers and in the Euclidean approach a generating functional for Schwinger functions in this model is given formally by the expression

$$Z(f) = Z^{-1} \int \prod_x d\phi(x) \exp\{- \int dx[\tfrac{1}{2}\phi(x)(-\Delta\phi)(x) + \tfrac{1}{2}m_0^2\phi^2(x) + \lambda\phi^4(x)] + \int dx f(x)\phi(x)\} \tag{1}$$

where Z is defined by the integral above with f = 0, $x \in R^d$ and ϕ is a function from R^d to R. To give a sense to this expression it is necessary to introduce space and ultraviolet cut-offs, e.g. in the papers mentioned above the expression (1) was replaced by

$$Z_\kappa(\Lambda,f) = Z_\kappa(\Lambda)^{-1} \int d\mu_{C_{m^2}}(\phi) \exp\{- \int_\Lambda dx[\tfrac{1}{2}\delta m^2(\kappa,\lambda)\phi_\kappa^2(x) + \lambda\phi_\kappa^4(x)] + (f,\phi)\} \tag{2}$$

where $d\mu_{C_{m^2}}$ is a Gaussian measure defined on $S'(R^d)$, with covariance $C_{m^2} = (-\Delta + m^2)^{-1}$, $m^2 > 0$, Λ is a compact subset of R^d, $\delta m^2(\kappa,\lambda)$ is some constant (mass renormalization counterterm) and $\tilde{\phi}_\kappa(p) = \zeta_\kappa(p)\tilde{\phi}(p)$, where $\zeta_\kappa(p)$ is a regular function with support in $|p| \leq \kappa$, $\zeta_\kappa(p) \to 1$ as $\kappa \to \infty$. An essential point of constructions in these papers was a division of the set $|p| \leq \kappa$ into a sequence of subsets $\kappa_{r-1} \leq |p| \leq \kappa_r$, $r = 1,2,\dots,n$, $O(1) = \kappa_0 < \kappa_1 < \dots < \kappa_{n-1} < \kappa_n = \kappa$, and calculations of integrals with field $\tilde{\phi}_\kappa$ restricted to each shell separately (e.g. integrating with respect to the restricted fields). These calculations (the so-called "phase space cell expansion") are connected with choices of proper length scales and this feature is basic for all later procedures. This is essentially the scheme of renormalization group method.

The renormalization group appeared more explicitly in the papers [2,3,4]. There an ultraviolet cut-off was introduced with the help of a higher order differential operator, i.e. $-\Delta + m^2$ in (1) was replaced by $(M^2 - m^2)^{-1}(-\Delta + m^2)(-\Delta + M^2)$. Then the corresponding Gaussian measure can be defined on the space of continuous, and even Hölder-continuous, functions and its covariance is equal $C_{m^2} - C_{M^2}$. Let us take for simplicity $m = 1$ and $M = 2^n$, then this covariance can be decomposed into the sum

$$C_1 - C_{2^{2n}} = \sum_{k=1}^{n} \left(C_{2^{2(k-1)}} - C_{2^{2k}}\right) \tag{3}$$

and there is a corresponding decomposition $\phi = \sum_{k=1}^{n} \phi_k$ of the field ϕ into the sum of independent Gaussian fields with covariances $C_{2^{2(k-1)}} - C_{2^{2k}} = C^{(k)}$. The expression (1) is replaced by

$$\begin{aligned} Z_n(\Lambda,f) = \int d\mu_{C^{(1)}}(\phi_1)\cdot\ldots\cdot d\mu_{C^{(n)}}(\phi_n)\exp\{-\int_\Lambda dx[\delta m_n^2(\lambda)(\sum_{k=1}^{n}\phi_k(x))^2 + \\ + \lambda(\sum_{k=1}^{n}\phi_k(x))^4] + (f, \sum_{k=1}^{n}\phi_k) - E_n(\Lambda)\}. \end{aligned} \tag{4}$$

We would like to prove at least the inequality

$$Z_n(\Lambda,f) \le e^{O(1)|\Lambda|} \tag{5}$$

with a constant $O(1)$ independent of n. The integral above is calculated by doing the consecutive integrations with respect to the fields ϕ_n, $\phi_{n-1},\ldots,\phi_1$ and it is enough to calculate sufficiently good approximations of these integrals. "Sufficiently good approximation" means here the cumulant expansion to at least third order in the interaction polynomial. More exactly we have formally

$$\langle e^V\rangle = e^{\langle V\rangle + \frac{1}{2!}\langle V^2\rangle^T + \frac{1}{3!}\langle V^3\rangle^T + \ldots}, \tag{6}$$

where $\langle\cdot\rangle$ is an expectation value with respect to some Gaussian measure, V is a polynomial interaction, and $\langle V^n\rangle^T$ is the truncated expectation value of V^n, i.e. $\langle V^2\rangle^T = \langle V^2\rangle - \langle V\rangle^2$, $\langle V^3\rangle^T = \langle V^3\rangle - 3\langle V^2\rangle\langle V\rangle + 2\langle V\rangle^3,\ldots$. We approximate left side of (6) taking only the first three terms in the formal expansion on the right side. In order to understand why it should be a good approximation let us rescale the field ϕ : $\phi(x) = s^{-\frac{d-2}{2}}\phi'(s^{-1}x)$. This operation transforms a measure with the covariance $C^{(k)} = \left(\frac{(-\Delta + 2^{2(k-1)})(-\Delta + 2^{2k})}{3\cdot 2^{2(k-1)}}\right)^{-1}$ into a measure with the covariance $C'^{(k)} = \left(\frac{(-\Delta + s^2 2^{2(k-1)})(-\Delta + s^2 2^{2k})}{s^2\cdot 3\cdot 2^{2(k-1)}}\right)^{-1}$ and the interaction $\lambda\int_\Lambda dx\phi^4(x)$ into the interaction $\lambda s^{4-d}\int_{s^{-1}\Lambda} dx'\phi'^4(x')$. In the first step, when we integrate with respect to $d\mu_{C'^{(n)}}$, we take $s = 2^{-n}$ and then $C'^{(n)} = \left(\frac{(-\Delta + 2^{-2})(-\Delta + 1)}{3\cdot 2^{-2}}\right)^{-1}$ and the effective coupling constant is equal to $\lambda 2^{-(4-d)n}$. Continuity and exponential decay of the covariance $C'^{(n)}$ easily implies that each term $\langle(\lambda 2^{-(4-d)n}\int_{2^n\Lambda} dx'\phi_n'^4(x'))^m\rangle^T_{C'^{(n)}}$ can be estimated by const. $(\lambda 2^{-(4-d)n})^m|2^n\Lambda| = \text{const. } \lambda^m 2^{-m(4-d)n+dn}|\Lambda|$ and for $-m(4-d) + d < 0$ this estimate is summable in n and so is compatible with (5). This condition for $d = 3$ implies $m > 3$, so we have to consider only the first three terms in the expansion (5), the remaining terms should give rise to an error const. $2^{-\kappa_0 n}|\Lambda|$ with $\kappa_0 > 0$. We can see also the fundamental difficulty in a proof of this estimate: the expression $\langle V^m\rangle^T$ depends on the unbounded field $\sum_{k=1}^{n-1}\phi_k'$. To overcome this difficulty, Gallavotti et al. introduced restrictions on the fields ϕ',ϕ_n', roughly speaking of the form $|\phi'(x)| \le O(1)n^p$ and similar for the field ϕ_n'. More exactly the space of field configurations was divided into a sum of subsets, each subset defined by the condition that the fields are "small", i.e. satisfy the above restrictions, on some domain $R \subset \Lambda$, and are "large" on the complement R^c. Then the interaction V was removed from R^c using some positivity properties of V and the integral

$\langle \chi_R \chi^c_{R^c} \, e^{V(R)} \rangle_{c,(n)}$ was calculated by the cumulant expansion (6) with the first three terms only and the above mentioned error. This procedure was applied for each integration in (4). Summing the errors $\sum_{k=1}^{n} O(1) 2^{-\kappa_0 k} |\Lambda|$ and estimating in a simple way the expression obtained from the cumulant expansion in the last step the inequality (5) was obtained. These ideas form the basis of the procedure which will be explained in the next section. The technical setting is changed in a way that allows generalization to more complicated models with gauge fields; see [1].

2. Lattice approximation, renormalization transformations and basic strategy.

In our approach we will use a regularization in the form of a lattice approximation. It is obtained from (1) by taking some finite subset of the lattice εZ^d instead of R^d and replacing correspondingly derivatives and integrals by finite differences and sums. For simplicity let us consider the d-dimensional torus T_ε identified with the subset of εZ^d:

$$T_\varepsilon = \{x \in \varepsilon Z^d : -L_\mu \le x_\mu < L_\mu, \ \mu = 1,\dots,d\} \tag{8}$$

where L_μ are such nonnegative numbers that $\varepsilon^{-1} L_\mu$ are integers. A distance on T_ε is defined by

$$|x - y| = \max \ \min \{|x_\mu - y_\mu|, \ 2L_\mu - |x_\mu - y_\mu|\} \tag{9}$$

A space of field configurations on T_ε is the set of functions $\phi : T_\varepsilon \to R$. An action of the lattice approximation is defined by

$$S^\varepsilon(\phi) = \sum_{x \in T_\varepsilon} \varepsilon^d [\tfrac{1}{2} \sum_{\mu=1}^{d} (\partial^\varepsilon_\mu \phi)^2(x) + \tfrac{1}{2} m_0^2 \phi^2(x) + \lambda \phi^4(x)] + E \tag{10}$$

where m_0^2, λ, E are real constants, $\lambda > 0$, and

$$(\partial^\varepsilon_\mu \phi)(x) = \varepsilon^{-1}(\phi(x + \varepsilon e_\mu) - \phi(x)), \tag{11}$$

e_μ - a unit vector of μ-th axis. The partition function, or vacuum energy, is given by the integral

$$Z^\varepsilon = \int d\phi e^{-S^\varepsilon(\phi)}, \tag{12}$$

where $d\phi$ is Lebesque measure on the space of field configurations. An expectation value of the function $F(\phi)$ is given by the integral

$$\langle F\rangle^\varepsilon = (Z^\varepsilon)^{-1} \int d\phi e^{-S^\varepsilon(\phi)} F(\phi) \tag{13}$$

and our aim is to prove the existence of a limit of (13) as $\varepsilon \to 0$. It is enough to consider a generating functional of Schwinger functions, or the analytically extended characteristic function of the measure in (13), obtained by taking $F(\phi) = e^{(f,\phi)}$, where (f,ϕ) is the inner product on ε-lattice.

$$(f,g) = \sum_{x\in T_\varepsilon} \varepsilon^d f(x)g(x). \tag{14}$$

Thus we will consider the generating functional

$$Z^\varepsilon(f) = \int d\phi e^{-S^\varepsilon(\phi)+(f,\phi)} \tag{15}$$

and instead of proving the existence of a limit of (15) we will prove the estimates uniform in ε:

$$e^{-E_-(\|f\|)|T_\varepsilon|} \le Z^\varepsilon(f) \le e^{E_+(\|f\|)|T_\varepsilon|}, \tag{16}$$

where $\|\cdot\|$ is some norm, e.g. it can be the L^∞ norm, $|T_\varepsilon| = \sum_{x\in T_\varepsilon} \varepsilon^d$, $E_\pm(\cdot)$ are some continuous functions independent of ε, T_ε. Let us make some comments about the constants m_0^2 and E. We take $m_0^{2\varepsilon} = m^2 + \delta m^2$, where $m^2 > 0$ and δm^2 is a mass renormalization counterterm needed to cancel some divergencies in the theory. It is given by a perturbation expansion up to the order 2 and it is a function of ε, λ, m^2. The constant E is a sum $E = E_0 + E_1$, where E_0 is a normalization constant given by

$$\int d\phi \exp\{- \sum_{x\in T_\varepsilon} \varepsilon^d[\frac{1}{2} \sum_{\mu=1}^{d} (\partial_\mu^\varepsilon\phi)^2(x) + \frac{1}{2}m^2\phi^2(x)]\} = e^{E_0}, \tag{17}$$

and E_1 is a vacuum energy counterterm defined by the perturbation expansion

$$E_1 = \sum_{1\le n\le 3} \frac{1}{n!}\lambda^n \left(\frac{\partial^n}{\partial\lambda^n} \log \int d\phi e^{-S^\varepsilon(\phi)+E}\right)\Big|_{\lambda=0} \tag{18}$$

We will not use the explicit formulas for this renormalization counterterm because we will not consider all the technical details of the method.

Let us now describe very general aspects of this method. Its basic feature is a decomposition of an integration process into a sequence of steps. A one step decomposition is achieved by application of a renormalization group transformation in the form proposed by Wilson and Kadanoff in the papers [9,10]. These transformations, together with the method of Gallavotti et al. are the basic ingredients of the method applied here.

To define the transformation let us divide T_ε into blocks containing L^d points, L is a fixed positive integer, e.g. L = 2 or 3. Each block is determined by a point $y \in T'_\varepsilon = T^{(1)}_{L\varepsilon} = \{y \in L\varepsilon Z^d: \ -L_\mu \le y_\mu < L_\mu,\ \mu = 1,\dots,d\}$ and is defined as the set

$$B(y) = \{x \in T_\varepsilon: \ y_\mu \le x_\mu < y_\mu + L\varepsilon,\ \mu = 1,\dots,d\}. \tag{19}$$

Usually the renormalization transformation is defined as follows: in each block we introduce a new field variable ψ and we integrate over all the configurations ϕ with fixed mean values in the blocks equal to ψ. Here instead of δ-functions $\delta\left(\psi(y)-L^{-d}\sum_{x\in B(y)}\phi(x)\right)$ we take approximations, namely we replace $\delta(\phi)$ by $\sqrt{\frac{c}{2\pi}}\, e^{-\frac{1}{2}c\phi^2}$ with a suitable constant c. This constant depends on a lattice spacing and for ε-lattice we take $c = a(L\varepsilon)^{d-2}$ with a positive constant a, e.g. a = 1. Thus we define a renormalization transformation $T^\varepsilon_{a,L}$ by the formula

$$(T^\varepsilon_{a,L}F)(\psi) = \int d\phi \prod_{y\in T^{(1)}_{L\varepsilon}} \sqrt{\frac{a(L\varepsilon)^{d-2}}{2\pi}}\ \exp\{-\tfrac{1}{2}a(L\varepsilon)^{d-2}(\psi(y) - L^{-d}\sum_{x\in B(y)}\phi(x))^2\}\ F(\phi). \tag{20}$$

This is an integral operator transforming functions F defined on the space of field configurations on the ε-lattice T_ε into functions defined on the space of field configurations on the $L\varepsilon$-lattice $T^{(1)}_{L\varepsilon}$.

It satisfies the fundamental normalization condition

$$\int d\psi (T^{\varepsilon}_{a,L} F)(\psi) = \int d\phi F(\phi). \tag{21}$$

We can now iterate these transformations. We get a sequence of lattices

$$T^{(n)}_{L^n\varepsilon} = \{x \in L^n\varepsilon Z^d: \ -L_\mu \le x_\mu < L_\mu,\ \mu = 1,\dots,d\} \tag{22}$$

and a sequence of functions F_n, each function F_n is defined on the space of field configurations on the $L^n\varepsilon$-lattice $T^{(n)n}_{L^n\varepsilon}$ and satisfies the inductive equation

$$F_{n+1} = T^{L^n\varepsilon}_{a,L} F_n,\ n = 0,1,\dots \qquad F_0 = F, \tag{23}$$

or explicitly

$$F_n = T^{L^{n-1}\varepsilon}_{a,L} \dots T^{\varepsilon}_{a,L} F. \tag{24}$$

We will use (23) as our basic equations, but (24) will be occasionally important also, so let us write a formula for composition of n renormalization transformations in (24):

$$T^{L^{n-1}\varepsilon}_{a,L} \dots T^{\varepsilon}_{a,L} = T^{\varepsilon}_{a_n,L^n}, \tag{25}$$

where $T^{\varepsilon}_{a_n,L^n}$ is defined by (20) with L^n and a_n instead of L and a,

where the numbers a_n satisfy the recursion relations

$$a_{n+1} = \frac{aa_n}{aL^{-2}+a_n}, \qquad a_1 = a. \tag{26}$$

They are uniquely determined by these equations and we have

$$a_n = a\frac{1 - L^{-2}}{1 - L^{-2n}}, \quad a_n \to a_\infty = a(1 - L^{-2}) \text{ as } n \to \infty. \tag{27}$$

For the problem we are interested in the renormalization transformations are applied so many times that we arrive at an almost unit lattice, i.e. $L^n\varepsilon \approx 1$, or $n \approx \log_L \frac{1}{\varepsilon}$. Then an integral of the obtained expression can be estimated easily. Let us make these remarks more precise. We take $F = e^{-S^\varepsilon}$ and apply (23), (24). We get a sequence of densities $e^{-S_n^{L^n\varepsilon}}$, $n = 0,1,\ldots$, $S_0^\varepsilon = S^\varepsilon$. Each density satisfies the condition

$$\int d\psi e^{-S_n^{L^n\varepsilon}(\psi)} = \int d\phi e^{-S^\varepsilon(\phi)} \tag{28}$$

thus it is enough to prove the estimates (16) for the left side of the above equality with $n \approx \log_L \frac{1}{\varepsilon}$. We do not need necessarily the exact action $S_n^{L^n\varepsilon}$, a good approximation is sufficient.

To become acquainted with what the actions look like and what a good approximation is let us consider the first renormalization transformation applied to e^{-S^ε}. We have to calculate the integral

$$\begin{aligned}
&\int d\phi \prod_{y\in T_{L\varepsilon}^{(1)}} \sqrt{\frac{a(L\varepsilon)^{d-2}}{2\pi}} \exp\{-\tfrac{1}{2}a(L\varepsilon)^{d-2}(\psi(y) - L^{-d}\sum_{x\in B(y)}\phi(x))^2\}\cdot \\
&\cdot\exp\{-\tfrac{1}{2}\sum_{x\in T_\varepsilon,\mu}\varepsilon^{d-2}(\phi(x+\varepsilon_\mu) - \phi(x))^2 - \sum_{x\in T_\varepsilon}\varepsilon^d(\tfrac{1}{2}m_0^2\phi^2(x) + \\
&\quad + \lambda\phi^4(x)) + \langle\phi,f\rangle - E\}
\end{aligned} \tag{29}$$

At first let us explain why the integration above is effectively an integration over the fields with momenta localized approximately in the slice $\frac{\pi}{L\varepsilon} \le |p| \le \frac{\pi}{\varepsilon}$. More precisely this means that the range of correlations in (29) is of order $O(L\varepsilon)$ or that "the effective mass" of the field ϕ is of the order $(L\varepsilon)^{-1}$. To see this we rescale the integral (29) from ε-lattice to 1-lattice, i.e. we apply the transformations

$$\begin{aligned}
\phi(\varepsilon x) &= \varepsilon^{-\frac{d-2}{2}}\phi'(x), \qquad x \in T_1 = \varepsilon^{-1}T\varepsilon, \\
\psi(\varepsilon y) &= \varepsilon^{-\frac{d-2}{2}}\psi'(y), \qquad y \in T_L^{(1)} = \varepsilon^{-1}T_{L\varepsilon}^{(1)}.
\end{aligned} \tag{30}$$

Omitting all constants we get

$$\int d\phi' \exp[-\tfrac{1}{2}aL^{d-2} \sum_{y\in T_L^{(1)}} (\psi'(y) - L^{-d}\sum_{x\in B(y)} \phi'(x))^2 - \tfrac{1}{2} \sum_{x\in T_1,\mu} (\phi'(x + e_\mu) - \phi'(x))^2 - \tfrac{1}{2}m_0^2\varepsilon^2 \sum_{x\in T_1} \phi'^2(x) - \lambda\varepsilon^{4-d} \sum_{x\in T_1} \phi'^4(x) + \varepsilon^{\frac{d+2}{2}} (\phi',f) - E] \tag{31}$$

where $y(x)$ is the point of the lattice $T_L^{(1)}$ that $x \in B(y(x))$. The basic quadratic form in ϕ' under the integral (31) is independent of ε and bounded from below:

$$aL^{d-2} \sum_{y\in T_L^{(1)}} (L^{-d} \sum_{x\in B(y)} \phi'(x))^2 + \sum_{x\in T_1,\mu} (\phi'(x + e_\mu) - \phi'(x))^2 \geq \geq \min\{L^{-2},(\tfrac{\pi}{L})^2\} \sum_{x\in T} \phi'^2(x). \tag{32}$$

In estimates we take always $a = 1$. The operator of this form is short-ranged, e.g. its matrix elements are equal to 0 if a distance between points is $> \sqrt{d}\,L$, so (32) implies that the inverse operator is exponentially decaying with a decay of the order $\frac{\pi}{L}$. If we denote this operator by C then we have $|C(x,x')| \leq O(1)\exp\{-O(\frac{\pi}{L})|x - x'|\}$. The situation would be different without the quadratic terms coming from renormalization group transformations. Then the covariance is equal to $(-\Delta)^{-1}(x - x') \approx O(\frac{1}{|x - x'|})$. It has a long range and couples the fields at the points with large distances, so the problem of calculation of the integral (31) is much more difficult and the methods which we are going to describe do not work.

Now let us sketch very shortly a method of calculation of the integral (31). If only the quadratic part of the action would be present in this integral, then we could calculate it making a translation

$$\phi'(x) = \phi''(x) + aL^{-2} \sum_{x'\in T_1} C(x,x')\psi'(y(x')), \tag{33}$$

where the second term on the right side is a minimal configuration of the form. This translation separates the form into two forms, the first is a quadratic form in ψ' and the second is the form (32) in ϕ''. Thus we get a Gaussian integral in ϕ'' with mean 0. We apply the same procedure to the integral (31) and we obtain that it is equal to

$$\exp\{-\frac{1}{2}aL^{d-2}\sum_{y\in T_L^{(1)}}\psi'^2(y)+\frac{1}{2}a^2L^{-4}(\psi',C\psi')\}\ Z\int d\mu_C(\phi'')e^V \tag{34}$$

where $d\mu_C(\phi\ )$ is the Gaussian measure with covariance C,

$$\begin{aligned}V = &-\frac{1}{2}m_0^2\varepsilon^2\sum_{x\in T_1}(\phi''(x)+aL^{-2}(C\psi')(x))^2-\lambda\varepsilon^{4d}\sum_{x\in T_1}(\phi''(x)+\\ &+aL^{-2}(C\psi')(x))^4+\varepsilon^{\frac{d+2}{2}}(\phi''+aL^{-2}C\psi',f)-E,\end{aligned} \tag{35}$$

$$Z=\int d\phi''\exp\{-\frac{1}{2}(\phi'',C_0^{-1}\phi'')\}.$$

Let us now consider the integral in (34). Each term of V has a small parameter ε^{4-d} or ε^2, so this integral can be treated as a small perturbation of the Gaussian integral and the perturbative methods can be applied. We use the cumulant expansion, i.e.

$$\begin{aligned}\int d\mu_C e^V = \langle e^V\rangle = \exp[\langle V\rangle + \frac{1}{2!}(\langle V^2\rangle - \langle V\rangle^2) + \frac{1}{3!}(\langle V^3\rangle - 3\langle V^2\rangle\langle V\rangle +\\ 2\langle V\rangle^3) + \ldots].\end{aligned} \tag{36}$$

The terms of this expression are ordered in a natural way according to the growing powers of ε and in the usual graphical representation are given by connected graphs with the covariance C. Thus terms of an order ≥ 4 can be estimated by $O(\varepsilon^\kappa)|T_1| = O(\varepsilon^{\kappa-d})|T_\varepsilon|$ with $\kappa > d$. If we rescale the terms written explicitly in the cumulant expansion (36) from L-lattice for ψ' field to $L\varepsilon$-lattice, and if we denote them by $P_1^{L\varepsilon}$, then we expect that

$$S_1^{L\varepsilon}(\psi) = \frac{1}{2}(\psi,\Delta_1^{L\varepsilon}\psi) - P_1^{L\varepsilon}(\psi) + O(\varepsilon^{\kappa_0})|T_\varepsilon|, \quad \kappa_0 > 0 \tag{37}$$

where $(\psi,\Delta_1^{L\varepsilon}\psi)$ is the rescaled quadratic form in the exponential function before the integral in (34). It is defined as a result of the renormalization transformation applied to $\exp\{-\frac{1}{2}(\phi,(-\Delta^\varepsilon)\phi)\}$:

$$(T^{\varepsilon}_{a,L}\exp\{-\tfrac{1}{2}(\phi,(-\Delta^{\varepsilon})\phi)\})(\psi) = \text{const.}\exp\{-\tfrac{1}{2}\langle\psi,\Delta_1^{L\varepsilon}\psi\rangle\}, \tag{38}$$

$-\Delta^{\varepsilon}$ is Laplace operator on T_{ε}.

Now the same procedure is applied after each renormalization transformation and we get the following approximate representations:

$$S_n^{L^n\varepsilon}(\phi) = \tfrac{1}{2}(\phi,\Delta_n^{L^n\varepsilon}\phi) - P_n^{L^n\varepsilon}(\phi) + \sum_{m=0}^{n-1} O((L^n\varepsilon)^{\kappa_0})|T_{\varepsilon}|. \tag{39}$$

The operators $\Delta_n^{L^n\varepsilon}$ are obtained applying n times renormalization transformation to the density $\exp\{-\tfrac{1}{2}(\phi,(-\Delta^{\varepsilon})\phi)\}$, hence using (25) we have

$$\left(T^{\varepsilon}_{a_n,L^n}\exp\{-\tfrac{1}{2}(\phi,(-\Delta^{\varepsilon})\phi)\}\right)(\psi) = \text{const.}\exp\{-\tfrac{1}{2}(\psi,\Delta_n^{L^n\varepsilon}\psi)\}. \tag{40}$$

The polynomial $P_n^{L^n\varepsilon}(\phi)$ is defined by the same first three terms in the cumulant expansion (36), only now the integral on the right side is obtained by applying transformation $T^{\varepsilon}_{a_n,L^n}$ instead of $T^{\varepsilon}_{a,L}$.

Continuing the calculations $n(\varepsilon) \approx \log_L \frac{1}{\varepsilon}$ number of times we get a density on $O(1)$-lattice. The expressions $\frac{1}{2}\langle\phi,\Delta_n^{L^n\varepsilon}\phi\rangle - P_n^{L^n\varepsilon}(\phi)$ have some simple positivity properties which allow to get a simple estimate of the integral

$$\int d\psi \exp\{-\tfrac{1}{2}(\psi,\Delta^{O(1)}_{n(\varepsilon)}\psi) + P^{O(1)}_{n(\varepsilon)}(\psi)\} = \exp\{O(1)|T_{\varepsilon}|\}. \tag{41}$$

Gathering together all the constants estimating the accuracy of the approximation in each step and using (41) and (28) we get finally

$$\int d\phi \exp\{-S^{\varepsilon}(\phi)\} = \exp\{\sum_{n=0}^{n(\varepsilon)} O((L^n\varepsilon)^{\kappa_0})|T_{\varepsilon}| + O(1)|T_{\varepsilon}|\} =$$

$$= \exp\{O(1)|T_{\varepsilon}|\sum_{n=0}^{\infty}(L^{-n})^{\kappa_0}\} = \exp\{O(1)|T_{\varepsilon}|\}. \tag{42}$$

Of course the above description of the method was extremely sketchy and some of its fundamental difficulties were not mentioned at all, for example how to estimate the terms with the field ψ' in the cumulant expansion (36). To solve this problem, it is necessary to introduce some restrictions on the fields ϕ',ψ'. This is based on ideas of Gallavotti et al. [2,3,4].

The next section will be devoted to a more detailed analysis of these technical problems.

Let us finish this section with an analysis of the operators $\Delta_n^{L^n \varepsilon}$. It is based on an explicit calculation of the integral in (38). Quite generally if we have a Gaussian density given by a translation invariant, positive operator Δ with a kernel $\Delta(x - x')$, $x,x' \in T_\varepsilon$, then the transformation $T^\varepsilon_{a,L}$ applied to this density gives again a Gaussian density given by an operator Δ' with a kernel $\Delta'(y - y')$, $y,y' \in T^{(1)}_{L\varepsilon}$. The function $\Delta'(y - y')$ can be explicitly calculated in terms of the function $\Delta(x - x')$. Let us introduce a Fourier transform on T_ε:

$$\tilde{f}(p) = \sum_{x \in T_\varepsilon} \varepsilon^d e^{-ip\cdot x} f(x),\ p \in \tilde{T}_\varepsilon, \quad f(x) = (2\pi)^{-d} \sum_{p \in \tilde{T}_\varepsilon} \left(\prod_{\mu=1}^{d} \frac{\pi}{L_\mu}\right) \cdot e^{ix\cdot p} \tilde{f}(p) \tag{43}$$

where $\tilde{T}_\varepsilon$ is a dual torus

$$\tilde{T}_\varepsilon = \{p = (p_1,\dots,p_d):\ p_\mu = \frac{\pi}{L_\mu} n_\mu, n_\mu \text{ is an integer},\ -\frac{L_\mu}{\varepsilon} \le n_\mu < \frac{L_\mu}{\varepsilon},\ \mu = 1,\dots d\}. \tag{44}$$

Of course the corresponding transformations can be defined on each torus, in particular on $T^{(n)}_{L^n \varepsilon}$. Now if $\Delta(p), \Delta'(q)$ are the Fourier transforms of $\Delta(x), \Delta'(y)$, then it can be easily calculated that

$$\Delta'(q) = \frac{1}{\sum_{\ell} |u^\varepsilon_L(q + \ell)|^2 \Delta^{-1}(q + \ell) + a^{-1}(L\varepsilon)^2},\quad q \in \tilde{T}^{(1)}_{L\varepsilon} \tag{45}$$

where

$$u^\varepsilon_L(p) = \prod_{\mu=1}^{d} \frac{\dfrac{e^{-iL\varepsilon p_\mu} - 1}{L\varepsilon}}{\dfrac{e^{-i\varepsilon p_\mu} - 1}{\varepsilon}},\quad \ell = (\ell_1,\dots,\ell_d),\quad \ell_\mu = \frac{2\pi}{L\varepsilon} m_\mu, \tag{46}$$

m_μ is an integer and $-\frac{L - 1}{2} \le m_\mu < \frac{L - 1}{2}$ for L odd, $-\frac{L}{2} \le m_\mu < \frac{L}{2}$ for L even. From these formulas we get an explicit representation of the operators $\Delta_n^{L^n \varepsilon}$ taking $\Delta = -\Delta^\varepsilon, L^n$ and a_n instead of L and a. In each step of the procedure sketched above we use these operators rescaled to the unit scale, so let us write the formulas for these operators

$$\Delta_n(p') = \frac{1}{\sum_{\ell} |u^{\delta}_{L^n}(p' + \ell)|^2(\Delta^{\delta}(p' + \ell))^{-1} + a_n^{-1}},$$

$$\delta = L^{-n}, \quad \Delta^{\delta}(p) = \sum_{\mu=1}^{d} \left|\frac{e^{-i\delta p_\mu} - 1}{\delta}\right|^2, \quad a_n = a\frac{1 - L^{-2}}{1 - L^{-2n}},$$

(47)

$$p' \in T_1^{(n)} = \{p' = (p_1',\ldots,p_d'): \; p_\mu' = \frac{\pi}{L_\mu} L^n \varepsilon n_\mu, \; n_\mu \text{ is an integer}, \; -\frac{L_\mu}{L^n \varepsilon} \le n_\mu < \frac{L_\mu}{L^n \varepsilon}, \mu = 1,\ldots d\}, \; \ell_\mu = 2\pi m_\mu, \; m_\mu \text{ is an integer},$$

$$-\frac{L^n - 1}{2} \le m_\mu < \frac{L^n - 1}{2} \text{ for } L \text{ odd and } -\frac{L^n}{2} \le m_\mu < \frac{L^n}{2} \text{ for } L \text{ even.}$$

Using these formulas we can see easily some uniform estimates of the quadratic forms $(\phi,\Delta n\phi)$. The first one is simply a bound from above

$$\Delta_n(p') \le a_n < 1, \text{ or } (\phi,\Delta_n\phi) \le \|\phi\|^2 = \sum_{x\in T_1^{(n)}} \phi^2(x) \quad (\text{for } a = 1). \tag{48}$$

The second estimate gives a bound from below by Laplace operator on the unit lattice $T_1^{(n)}$. We have

$$|u^{\delta}_{L^n}(p' + \ell)| \le O(1) \prod_{\mu=1}^{d} \frac{|p_\mu'|}{|p_\mu' + \ell_\mu|} \le O(1) \prod_{\mu=1}^{d} \frac{1}{1 + |\ell_\mu|} \tag{49}$$

and denoting $\Delta_0(p') = \sum_{\mu=1}^{d} |e^{-ip_\mu'} - 1|^2$ we have

$$\Delta_n(p') = \frac{\Delta_0(p')}{\sum_{\ell} |u^{\delta}_{L^n}(p' + \ell)|^2 \frac{\Delta_0(p')}{\Delta^{\delta}(p' + \ell)} + a_n^{-1}\Delta_0(p')} \ge \tag{50}$$

$$\ge \frac{\Delta_0(p')}{O(1) \sum_{\ell} \prod_{\mu=1}^{d} \frac{1}{(1 + |\ell_\mu|)^2} \frac{|p'|^2}{|p' + \ell|^2} + 4da_n^{-1}} \ge \gamma_0\Delta_0(p')$$

for some $\gamma_0 > 0$ independent of n and L_μ. This implies

$$(\phi,\Delta_n\phi) \geq \gamma_0\langle\phi,\Delta_0\phi\rangle = \gamma_0 \sum_{x\in T_1^{(n)},\mu} (\phi(x + e_\mu) - \phi(x))^2. \tag{51}$$

This inequality plays an important role because it allows making restrictions on the fields ϕ. It is also important because it implies a uniform bound from below for the quadratic form appearing when the n + 1 renormalization transformation is applied:

$$aL^{d-2} \sum_{y\in T_L^{(n+1)}} (L^{-d} \sum_{x\in B(y)} \phi(x))^2 + \langle\phi,\Delta_n\phi\rangle \geq$$

$$\geq \gamma_0(aL^{d-2} \sum_{y\in T_L^{(n+1)}} (L^{-d} \sum_{x\in B(y)} \phi(x))^2 + \sum_{x\in T_1^{(n)},\mu} (\phi(x + e_\mu) - \tag{52}$$

$$- \phi(x))^2) \geq \gamma_0 \min\{L^{-2},(\tfrac{\pi}{L})^2\} \sum_{x\in T_1^{(n)}} \phi^2(x).$$

This bound in turn implies a uniform exponential decay of all the covariances C_n. If we had taken the whole lattice εZ^d instead of T_ε, then the limit of $\Delta_n(p')$ would exist for $n \to \infty$. For a bounded lattice T_ε we take $n = n(\varepsilon)$ and we ask if the limit of $\Delta_{n(\varepsilon)}$ exists for $\varepsilon \to 0$. For $\varepsilon = L^{-N}$ we have

$$\Delta_{n(\varepsilon)}(p') \underset{N\to\infty}{\to} \Delta_\infty(p') = \frac{1}{\sum_{\ell} \prod_{\mu=1}^{d} \left| \frac{e^{-ip'_\mu} - 1}{p'_\mu + \ell_\mu} \right|^2 \frac{1}{|p' + \ell|^2} + a_\infty^{-1}} \tag{53}$$

$$p' = (p'_1,\ldots,p'_2),\ p'_\mu = \frac{\pi}{L_\mu} n_\mu,\quad - L_\mu \leq n_\mu < L_\mu,$$

$$\ell = 2\pi m,\quad m \in Z^d \text{ and } a_\infty = a(1 - L^{-2}).$$

The operator Δ_∞ for $L_\mu = \infty$ defines a Gaussian density which is a fixed point of the renormalization transformation. This is the basic mechanism behind the uniform estimates described above.

3. Remarks on Technical Problems.

3.1 The Problem of Large Fields.

One of the fundamental problems in the approach is how to estimate the terms of order ≥ 4 in the cumulant expansion (36) by $O(1)\varepsilon^{\kappa_0}|T_\varepsilon|, \kappa_0 > 0$, with a constant $O(1)$ uniform in $n, \varepsilon, T_\varepsilon$. A solution to this problem is obtained by imposing restrictions on the fields. More exactly we decompose the space of field configurations into domains defined by restricting values of fields on subsets of the lattice. The integral (29) or (31) is represented as a sum of integrals and each term in the sum is "calculated" using different procedures for the subsets where the fields are small and where they are large. Now let us describe exactly the restrictions and the procedures in the first step, i.e. for the integral (31). The restrictions are chosen in accordance with positivity properties of the action. There are three basic positive expressions in it which are approximately preserved in all the actions S_n. These are: the quadratic form defining renormalization transformation, gradient terms, $\lambda\varepsilon^{4-d}\phi^4$ terms. Let us denote by T_1^* the set of pairs of nearest neighbour points. We have the following decomposition

$$
\begin{aligned}
1 = \sum_{P\subset T_L^{(1)}} \sum_{Q\subset T_1^*} \sum_{R\subset T_1} & \prod_{y\in P} \chi\Big(\{|\psi(y) - L^{-d}\sum_{x\in B(y)} \phi(x)| > p(\varepsilon)\}\Big)\cdot \\
& \cdot \prod_{y\in P^c} \chi\Big(\{|\psi(y) - L^{-d}\sum_{x\in B(y)} \phi(x)| \leq p(\varepsilon)\}\Big)\cdot \\
& \cdot \prod_{\langle x,x'\rangle\in Q} \chi\Big(\{|\phi(x') - \phi(x)| > p(\varepsilon)\}\Big)\cdot \\
& \cdot \prod_{\langle x,x'\rangle\in Q^c} \chi\Big(\{|\phi(x') - \phi(x)| \leq p(\varepsilon)\}\Big)\cdot \\
& \cdot \prod_{x\in R} \chi\Big(\{|\phi(x)| > \frac{1}{(\lambda\varepsilon^{4-d})^{\frac{1}{4}}}\, p(\varepsilon)\}\Big)\cdot \\
& \cdot \prod_{x\in R^c} \chi\Big(\{|\phi(x)| \leq \frac{1}{(\lambda\varepsilon^{4-d})^{\frac{1}{4}}}\, p(\varepsilon)\}\Big) =
\end{aligned}
\tag{54}
$$

$$= \sum_{P\subset T_L^{(1)}} \sum_{Q\subset \Gamma_1^*} \sum_{R\subset T_1} \chi_P^c \chi_{P^c} \chi_Q^c \chi_{Q^c} \chi_R^c \chi_{R^c},$$

$$p(\varepsilon) = a_0(1 + \log\varepsilon^{-1})^p, \; p > 2. \tag{54}$$

We insert it into the integral (31) and we get the corresponding decomposition of the integral. If any of the following inequalities holds

$$|\psi(y) - L^{-d} \sum_{x\in B(y)} \phi(x)| > p(\varepsilon), \; |\phi(x') - \phi(x)| > p(\varepsilon),$$
$$|\phi(x)| > \frac{1}{(\lambda\varepsilon^{4-d})^{\frac{1}{4}}} p(\varepsilon) \tag{55}$$

then the corresponding exponential factor in (31) satisfies the inequality

$$e^{-(\dots)} \leq e^{-\frac{1}{2}p^2(\varepsilon)} \tag{56}$$

and the number on the right side is very small. It is smaller than arbitrary power ε^N, so it can compensate a factor arising from a very rough estimate of the action in a big neighborhood of a point at which the corresponding inequality (55) holds. On the other hand, a bound on a field or its derivative of the order of $p(\varepsilon)$ is strong enough so that an arbitrarily small power of ε can dominate any polynomial in fields. This is so because $\varepsilon^{\kappa_0} p(\varepsilon)^j \leq 1$ for any fixed κ_0, j. This idea is basic for the procedure described below.

We divide the lattice T_1 into two subsets: Λ is a set of points distant from $P \cup Q \cup R$ more than $r(\varepsilon) = R_0(1 + \log\frac{1}{\varepsilon})^r$, $r > 1$, and its complement Λ^c. Thus $P \cup Q \cup R \subset \Lambda^c$. We define also a subset Λ_2 replacing $r(\varepsilon)$ in the definition of Λ by $2r(\varepsilon)$, thus $\Lambda \supset \Lambda_2$ and a distance between boundaries of these sets is of the order $r(\varepsilon)$.

We begin an estimate of the integral (31) with the decomposition (54) introduced removing the interaction terms from the set Λ_2^c in each term of this decomposition. We use the simple estimate

$$\lambda\varepsilon^{4-d}\phi^4 + \frac{1}{2}m_0^2\varepsilon^2\phi^2 \geq \frac{1}{2}\lambda\varepsilon^{4-d}\phi^4 - O(\varepsilon). \tag{57}$$

which follows from the estimate $m_0^2 = O(\frac{1}{\varepsilon})$ of the counterterm. This implies

$$\chi_R^c e^{-\sum_{x\in\Lambda_2^c}(\frac{1}{2}m_0^2\varepsilon^2\phi^2(x) + \lambda\varepsilon^{4-d}\phi^4(x))} \le e^{-\frac{1}{2}p^4(\varepsilon)|R| + O(\varepsilon)|\Lambda_2^c|}. \quad (58)$$

Defining

$$\sum_{\text{all }\{P,Q,R\}\text{ admissible for }\Lambda} \chi_P^c \chi_{P^c\cap\Lambda^c}\chi_Q^c\chi_{Q^c\cap\Lambda^c}\chi_{R^c\cap\Lambda^c} e^{-\frac{1}{2}p^4(\varepsilon)|R|} = \quad (59)$$

$$= \zeta_{\Lambda^c}$$

and denoting the remaining characteristic functions by χ_Λ we get the following inequality

$$\text{(the integral (31))} \le \sum_{\Lambda}\int d\phi\ \zeta_{\Lambda^c}\chi_\Lambda \exp[-\frac{1}{2}aL^{d-2}\sum_{y\in T_L^{(1)}}(\psi(y) -$$

$$- L^{-d}\sum_{x\in B(y)}\phi(x))^2 - \frac{1}{2}\sum_{<x,x'>\in T_1^*}(\phi(x') - (x))^2 - \quad (60)$$

$$- \sum_{x\in\Lambda_2}(\frac{1}{2}m_0^2\varepsilon^2\phi^2(x) + \lambda\varepsilon^{4-d}\phi^4(x)) + \varepsilon^{\frac{d+2}{2}}(\phi,f) - E + O(\varepsilon)|\Lambda_2^c|].$$

The idea behind this inequality is that, although the error $e^{O(\varepsilon)|\Lambda_2^c|}$ is not small enough, the small factors $e^{-\frac{1}{2}p^2(\varepsilon)}$ coming from positivity estimates (56),(58) produce a very small factor $e^{-O(p(\varepsilon))|\Lambda_2^c|}$ which controls this error and the sum over Λ. The remaining small factors beside those included in ζ_{Λ^c} are connected with terms in basic quadratic form and are obtained when the final Gaussian integration is done.

Now we use the fact that the fields are small on Λ and we calculate a part of each integral on the right side of (60) by conditioning on Λ^c. We get

$$\text{(the right side of (60))} \le \sum_{\Lambda} \int d\phi\, \zeta_\Lambda \exp[-\tfrac{1}{2} a L^{d-2} \sum_{y \in T_L^{(1)}} (\psi(y) -$$

$$- L^{-d} \sum_{x \in B(y)} \phi(x))^2 - \tfrac{1}{2} \sum_{\langle x,x'\rangle} (\phi(x') - \phi(x))^2] \int d\mu_{C_\Lambda}(\phi')\chi_\Lambda \cdot$$

$$\cdot \exp(- \sum_{x \in \Lambda_2} P_x(\phi'(x) + (C_\Lambda \partial_n \Lambda^c \phi)(x) + aL^{-2}(C_\Lambda \psi')(x)) - E) \cdot \qquad (61)$$

$$\cdot\, e^{O(\varepsilon)|\Lambda_2^c|},$$

where $d\mu_{C_\Lambda}$ is a Gaussian measure with the covariance C_Λ, C_Λ is an operator inverse to the operator of the basic quadratic form restricted to Λ, $P_x(\phi) = \frac{1}{2} m_0^2 \varepsilon^2 \phi^2 + \lambda \varepsilon^{4-d} \phi^4 + \varepsilon^{\frac{d+2}{2}} f(x)\phi$, ∂_n is a "normal derivative" to the boundary of Λ, $\psi'(x) = \psi(y(x))$. The covariances C_Λ have similar properties to C, i.e. they are exponentially decaying with a decay rate uniform in Λ and independent of any parameter. Because Λ_2 is far from the boundary of Λ (its distance is $\approx r(\varepsilon)$), we have that the configuration $(C_\Lambda \partial_n \Lambda^c \phi)(x)$ is bounded by $O(\varepsilon^N)$ for arbitrary N and $x \in \Lambda_2$. We can remove it from all the expressions in (61) and the error we make can be estimated by $O(\varepsilon^\kappa)|\Lambda_2| \le O(\varepsilon^{\kappa_0})|T_\varepsilon|$ with $\kappa_0 > 0$.

It can be proved that $(aL^{-2} C_\Lambda \psi')(x) = \psi(y(x)) + O(p(\varepsilon))$, $x \in \chi_2$, and this in turn implies that $\phi'(x) = O(p(\varepsilon))$ for $x \in \Lambda_2$. A very simple analysis based on these facts shows that the characteristic functions χ_Λ in (61) can be bounded by a product of characteristic functions χ_1 giving the restrictions on the field ψ only: $|\psi(y') - \psi(y)| \le O(1)p(\varepsilon)$, $|\psi(y)| \le O(1)\frac{1}{(\lambda\varepsilon^{4-d})^{\frac{1}{4}}} p(\varepsilon)$, $\langle y,y'\rangle \subset \Lambda$, $y \in \Lambda$, and characteristic functions χ' giving the restrictions $|\phi'(x)| \le O(1)p(\varepsilon)$. The second integral on the right side of (61) is estimated by

$$\chi_1 \int d\mu_{C_\Lambda}(\phi')\chi' \exp\Big(- \sum_{x \in \Lambda_2} P_x(\phi'(x) + aL^{-2}(C_\Lambda \psi')(x)) - E\Big) e^{O(\varepsilon^{\kappa_0})|T_\varepsilon|}. \qquad (62)$$

Now we are in a different situation from (36). We have bounds on the fields ϕ',ψ sufficient to prove uniform estimates for the terms in the cumulant expansion (36). In fact we can expand the logarithm of (62) up to fourth order and the remainder terms of this order can be rather easily estimated by $O(\varepsilon^{\kappa})|\Lambda_2| \le O(\varepsilon^{\kappa_0})|T_\varepsilon|$ for $\kappa > d, \kappa_0 > 0$. This follows from a very general lemma proven for lattice systems by Gallavotti et al. [2]. Another proof was given in [8].
The first three terms in the cumulant expansion (36) give a nonlocal interaction polynomial $P_1(\Lambda_2,\psi)$. Thus we have finally the estimate

$$\text{(the integral (31))} \le \sum_{\Lambda} \rho_1(\Lambda,\psi)\exp\{P_1(\Lambda_2,\psi) + O(\varepsilon)|\Lambda_2^c| + \\ + O(\varepsilon^{\kappa_0})|T_\varepsilon|\}, \tag{63}$$

where ρ_1 is given by the first integral on the right side of (61) multiplied by χ_1, or calculating the following conditional integral with conditioning on Λ^c:

$$\rho_1(\Lambda,\psi) = \chi_1 \int d\phi|_{\Lambda^c}\zeta_\Lambda \exp[-\tfrac{1}{2}aL^{d-2} \sum_{y\in T_L^{(1)}\cap\Lambda^c} (\psi(y) - L^{-d}\sum_{x\in B(y)}\phi(x))^2 - \\ - \tfrac{1}{2}(\phi,(-\Delta^D_{\Lambda^c})\phi) + \tfrac{1}{2}(\partial_n\phi, C_\Lambda\partial_n\phi) + aL^{-2}(\partial_n\phi, C_\Lambda\psi')] \tag{64}$$

$$\cdot \exp\{-\tfrac{1}{2}(\psi,\Delta_1(\Lambda)\psi)\},$$

$$(\psi,\Delta_1(\Lambda)\psi) = aL^{d-2}\sum_{y\in T_L^{(1)}\cap\Lambda} \psi^2(y) - a^2L^{-4}(\psi',C_\Lambda\psi').$$

The covariance C_Λ is almost equal to C for arguments restricted to Λ_2, the error is of order ε^N for arbitrary N, so if we are interested in ρ_1 as a function of ψ restricted to Λ_2 we have a representation

$$\rho_1(\Lambda,\psi) = \rho_1'(\psi)\exp\{-\tfrac{1}{2}(\psi|_{\Lambda_2},\Delta_1\psi|\Lambda_2) + O(\varepsilon^{\kappa_0})|T_\varepsilon|\}, \tag{65}$$

in which $\rho_1'(\psi)$ depends on ψ restricted to some neighbourhood of Λ_2^c.
To get e^{-S_1} we rescale all the expressions from L-lattice to the unit lattice for the field ψ.

All the densities e^{-S_n} have the same structure and to calculate the renormalization transformation we apply the same procedure as described above: we introduce the decomposition (54) on a corresponding set Λ instead of $T_1^{(n)}$ and with $p(L^n\varepsilon)$ instead of $p(\varepsilon)$, we estimate the interaction in a neighbourhood of the set where the fields are "large" and we calculate a conditional integral on the set where the fields are "small" applying the cumulant expansion. The estimate of the interaction in the region of large fields is based on the following inequality

$$P_n(\Omega,\phi) \leq -\lambda(L^n\varepsilon)^{4-d} \sum_{x\in\Omega} \phi^4(x) + O((L^n\varepsilon)^{\kappa_0})|\Omega|, \tag{66}$$

holding for the interaction restricted to a subset Ω and fields ϕ restricted by the characteristic function χ_n arising from the last step of the procedure in the same way as the function χ_1 arose in the first step.

This procedure is continued until $n = n(\varepsilon) \approx \log_L \frac{1}{\varepsilon}$, and then the interaction is estimated by $O(1)|T_\varepsilon|$ and a remaining Gaussian integral is calculated giving all the necessary small factors. They control the summations over sets P,Q,R,Λ in all the decompositions.

A proof of the lower bound is simpler because we can always bound an integral from below restricting its range of integration by some characteristic functions. Hence we have to consider only the case with the sets P,Q,R empty.

Finally let us mention that for this particular model it is possible to have a procedure simpler than the one described above, but it does not apply to more complicated models like $(\mathrm{Higgs})_{2,3}$ [1].

3.2 Renormalization.

Another important aspect of the procedure is connected with uniform estimates on the coefficients of the non-local polynomials P_n. A term in this polynomial has the following form

$$(\lambda(L^n\varepsilon)^{4-d})^m \sum_{\substack{y_1,\dots,y_p\in T_1^{(n)} \\ m\leq 3,\ p\leq 8,}} v_{n,m,p}(y_1,\dots,y_p)\phi(y_1)\cdot\ldots\cdot\phi(y_p), \tag{67}$$

and we have to know that the functions $v_{n,m,p}$ are uniformly bounded in n, moreover they have uniform exponential decay. This property is important for all the bounds in the procedure. It is connected with cancellations of divergencies in quantum field theory and it holds only when the parameter m_0^2, or more exactly the parameter δm^2 in the decomposition $m_0^2 = m^2 + \delta m^2$, is chosen in a special way. Let us illustrate how this property can be proven for simple, but very important and typical terms when p = 2 and m = 1,2. The polynomial P_n is obtained by a formal application of the cumulant expansion formula (36) to the expression $T_{a_n,L^n} e^{-S^\delta}, \delta = L^{-n}$. The effective Gaussian integration in this expression is defined by the covariance

$$G_n = (-\Delta^\delta + a_n P_n)^{-1}, \tag{68}$$

where P_n is the projection operator: $(P_n\phi)(x) = \sum_{x' \in B_n(y_n(x))} \delta^d \phi(x')$,

$B_n(y)$ is a block of T_δ defined by a point $y \in T_1^{(n)}$ as in (19), only L replaced by L^n, ε by δ, and $y_n(x)$ is a point in $T_1^{(n)}$ such that $x \in B_n(y_n(x))$. It is easy to calculate the coefficients $v_{n,m,p}$.

Let us denote $g_n(x,y) = a_n \sum_{x' \in B_n(y)} \delta^d G_n(x,x')$, then the term

$v_{n,1,2}(y,y')$ has the form

$$-6\sum_x \delta^d g_n(y,x) G_n(x,x) g_n(x,y'), \tag{69}$$

and an important term in $v_{n,2,2}(y,y')$ has the form

$$\frac{1}{2} 3!\, 4^2 \sum_{x,x'} \delta^{2d} g_n(y,x) G_n^3(x,x') g_n(x',y'). \tag{70}$$

There is also a term with δm^2:

$$-\frac{1}{2} \sum_x \delta^d g_n(y,x) \delta m^2 (L^n \varepsilon)^2 g_n(x,y'). \tag{71}$$

To understand better these expressions we have to know some properties of the functions g_n and G_n. We have the following inequalities

$$|g_n(x,y)|, |\partial g_n(x,y)| \le O(1)e^{-c|x-y|},$$
$$|G_n(x,x')| \le O(1)\frac{e^{-c|x-x'|}}{|x - x'|}, \; x \ne x', \; |G_n(x,x)| = O(\tfrac{1}{\delta}) \text{ for } d = 3. \tag{72}$$

They imply that (69) is linearly divergent in δ, and (70) is logarithmically divergent. We will use the term (71) to cancel these divergencies. The mass renormalization counterterm is given by the formula

$$\delta m^2 = -\lambda 4 \cdot 3C^{\varepsilon}(0) + 3!(4\lambda)^2 \sum_{x' \in T_\varepsilon} \varepsilon^d (C^{\varepsilon}(x - x'))^3 \tag{73}$$

where C^{ε} is a free propagator $C^{\varepsilon} = (-\Delta^{\varepsilon} + m^2)^{-1}$.

Now let us consider for simplicity the case $n = n(\varepsilon)$. Then the covariance C^{ε} is almost equal to G_n in the sense that the difference $C^{\varepsilon} - G_n$ is a bounded function with bounded derivative. It means that we can write

$$\delta m^2 = -\lambda 6G_n(x,x) + \lambda^2 3!4^2 \sum_{x' \in T_\delta} \delta^d G_n^3(x,x') + \delta m_{fin}^2(x) \tag{74}$$

where $\delta m_{fin}^2(x)$ is a uniformly bounded function. The first term on the right side of (74) in (71) cancels exactly (69), for the second term and (70) we get

$$\frac{1}{2}3!4^2 \sum_{x,x'} \delta^{2d} g_n(y,x)G_n^3(x,x\;)(g_n(x',y') - g_n(x,y')) \tag{75}$$

and using (72) we can estimate it by $O(1)e^{-c|y-y'|}$ with some "absolute" constants. Thus we get the required bounds. When $n < n(\varepsilon)$, then we have again the representation (74), but $\delta m_{fin}^2(y)(L^n\varepsilon)^2$ is now uniformly bounded by some positive power of $L^n\varepsilon$.

Similar considerations can be done for all the terms in the polynomial P_n.

Acknowledgements. I would like to thank Prof. A. Jaffe and Dr. J. Imbrie for discussions and suggestions which improved the presentation of this article. This article was written during my current stay at Harvard University and was supported by the NSF Grant PHY79-16812.

References

[1] Balaban, T.: "$(Higgs)_{2,3}$ Field Theory in a Finite Volume. I. A Lower Bound. II. An Upper Bound. III. Renormalization;" to appear in CMP, also Harvard preprints HUTMP 82/B116,B117,B119, and Aarhus University preprint 1981/82 No. 19.

[2] Benfatto, G., Cassandro, M., Gallavotti, G., Nicolò, F., Olivieri, E., Presutti, E., Scacciatelli, E.: "Some Probabilistic Techniques in Field Theory". Commun. Math. Phys. 59, 143-166 (1978).

[3] Benfatto, G., Cassandro, M., Gallavotti, G., Nicolò, F., Olivieri, E., Presutti, E., and Scacciatelli, E.: "Ultraviolet Stability in Euclidean Scalar Field Theories". Commun. Math. Phys. 71, 95-130 (1980).

[4] Benfatto, G., Gallavotti, G., Nicolò, F.: "Elliptic Equations and Gaussian Processes". Journal of Functional Analysis, Vol. 36, 343-400 (1980).

[5] Dimock, J. and Glimm, J.: "Measures on Schwartz Distribution Space and Application to $P(\phi)_2$ Field Theories," Adv. Math. 12, 58-83 (1974).

[6] Glimm, J. and Jaffe, A.: "Field theory models." In: Statistical Mechanics and Quantum Field Theory. C. de Witt and R. Stora, eds., New York: Gordon and Breach.

[7] Glimm, J., Jaffe, A.: "Positivity of the ϕ_3^4 Hamiltonian". Fortschr. Physik 21, 327-376 (1973).

[8] Gawedzki, K., Kupiainen, A.: "A Rigorous Block Spin Approach to Massless Lattice Theories". Commun. Math. Phys. 77, 31-64 (1980).

[9] Ma, S.: "Modern Theory of Critical Phenomena", Benjamin, New York, 1976.

[10] Wilson, K.G., Bell, T.L.: "Finite-lattice approximations to renormalization groups". Physical Review B, Vol. 11, 3431-3445 (1975).

PART II : Dynamical Systems

RENORMALIZATION GROUP ANALYSIS FOR DYNAMICAL SYSTEMS *

P. Collet H. Koch

I. Introduction

During the past few years, an important development has added to our knowledge of dynamical systems. This new development was based on two main ingredients. The first one is the mathematical notion of hyperbolicity. The second one is purely numerical : the use of computers, and more precisely, the possibility of visualizing complicated objects (strange attractors, Julia sets, ...). This is sometimes referred to as experimental mathematics. It turns out that already very simple systems can exhibit very complex behaviour. We shall restrict ourselves to **discrete dynamical systems, obtained by iterating (nonlinear) maps** f from $\mathbb{R}^n$ to $\mathbb{R}^n$. Although this is not much physical, the analysis will be greatly simplified, and a lot of information can be obtained for continuous time evolution, using the method of Poincaré-sections [1]. As a further restriction, let us concentrate only on the long time behaviour of the system, neglecting the transient regime. This leads naturally to the notion of attractors. We shall say that a closed bounded set Ω is an attractor for f, if there is an open neighbourhood U of Ω and a point $X \in \Omega$ such that

(i) $f(U) \subset U$

(ii) $\bigcap_k f^k(U) = \Omega$

(iii) $\{f^k(X) : k \in \mathbb{N}\}$ is dense in Ω (irreducibility)

where $f^k = f \circ f \circ \ldots \circ f$ (k times). If these conditions are satisfied, then every orbit which is trapped by U will stay in U forever, and accumulate to Ω. In order to avoid some confusion, we observe that an attractor is not necessarily connected (examples will be given below), that a transformation f can have several attractors (or none), and that some points may not evolve towards any attractor (or to infinity). For a discussion of the definition of attractor see [39]. The simplest attractors are of the form $\Omega = \{X\}$, where X is a stable fixed point of f, namely

(i) $f(X) = X$

(ii) Spectrum $(Df_X) \subset \{Z \in \mathbb{C} : |Z| \leqslant \zeta\}$ for some $\zeta \in [0,1[$

Df_X denotes the differential (n×n matrix) of f at X. It is easy to show that in some open neighbourhood U of X there is an equivalent metric d such that $d(f(Y),X) \leqslant \sqrt{\zeta}\, d(Y,X)$ for all $Y \in U$. Another example is a stable period, namely $\Omega = \{X, f(X), \ldots f^{m-1}(X)\}$ where X is a stable fixed point of f^m. Note that in both cases the type of the attractor is unchanged under perturbations $f \to f_\alpha = f + O(\alpha)$ for small α, but the attractor itself will in general depend on the parameter α. In a physical experiment such parameters play an important role, as they fix experimental conditions. In a dissipative system, for example, it is necessary to compensate the dissipation by a constant flux of energy, in order to get a nontrivial attractor. In this case α can be viewed as a measure of this energy flux, if all the other external constraints are kept fixed. When α is varied over a large range of values, qualitative changes of the long time behaviour can be observed. For large values of α the system is usually turbulent (more details will be given later).

A change in the topology of the attractor is called a bifurcation. If $\alpha \to f_\alpha$ is a one parameter family of maps on $\mathbb{R}^n$, we call a bifurcation point the value of the parameter at which a bifurcation takes place. As a simple example we consider the Hénon maps

$$f_\alpha(X_1,X_2) = (1-\alpha(1+\beta)^2X_1^2 + X_2, -\beta X_1) \tag{1}$$

on $\mathbb{R}^2$, for fixed $|\beta| < 1$. Dissipation is reflected here by the fact that the volume elements (but not lengths!) in phase space are contracted, $\det((Df)_X) = \beta$. For $\alpha \in [0,2]$ the map f_α has two fixed points $(W_1, -\beta W_1)$ and $(X_1, -\beta X_1)$,

$$W_1(\alpha) = \frac{+1-\sqrt{1+4\alpha}}{2\alpha(1+\beta)}, \quad X_1(\alpha) = \frac{-1+\sqrt{1+4\alpha}}{2\alpha(1+\beta)}. \tag{2}$$

The first one is unstable, and the second one is stable for $\alpha \in [0,3/4[$, but unstable for $\alpha \in]3/4, 2]$. Therefore $\alpha = 3/4$ is a bifurcation point for this family of maps. In order to see what happens with the attractor at $\alpha = 3/4$ we consider the map $f_\alpha^2 = f_\alpha \circ f_\alpha$. The points $(W_1, -\beta W_1)$ and $(X_1, -\beta X_1)$ are of course also fixed points of f_α^2, with the same range of stability (as a function of α) as for f_α. However for $\alpha > 3/4$, two new fixed points appear, namely $(Y_1, \beta Y_1 - \gamma)$ and $(Z_1, \beta Z_1 - \gamma)$ where $\gamma(\alpha) = \beta/\alpha(1+\beta)$ and

$$Y_1(\alpha) = \frac{1+\sqrt{1+4(\alpha-1)}}{2\alpha(1+\beta)}, \quad Z_1(\alpha) = \frac{1-\sqrt{1+4(\alpha-1)}}{2\alpha(1+\beta)} \tag{3}$$

It is easy to check that both of them are stable if $\alpha - 3/4$ is small. This means that at $\alpha = 3/4$ the attractor of f_α bifurcates from a stable fixed point to a stable period two, $\Omega = \{X\} \to \Omega = \{Y,Z\}$.
The same thing might occur for a map $f_\alpha = g_\alpha^m$. In this case the attractor of g_α undergoes a period doubling bifurcation $\Omega = \{X, g_\alpha(X), \ldots, g_\alpha^{m-1}(X)\} \to \Omega = \{Y, g_\alpha(Y), \ldots, g_\alpha^{2m-1}(Y)\}$. Other simple bifurcations are described in [16]. A sequence of bifurcations with a turbulent state reached at the end is called a road to turbulence. We shall concentrate on the period doubling road to turbulence, which is a succession of period doubling bifurcations. This phenomenon was known some time ago (see [36] for example) for maps on the interval $[-1,1]$. As an example we shall describe the observations for the family $f_\alpha(x) = 1 - \alpha x^2$.

There is an increasing sequence $0 = \alpha_0 < \alpha_1 < \alpha_2 < \ldots < \alpha_\infty < 2$ such that f_α has a stable period 2^k for α in the interval $]\alpha_k, \alpha_{k+1}[$ and no other attractor in $[-1, 1]$. The remaining interval $(\alpha_\infty, 2)$ will be discussed later. Note that the number of observed bifurcation points α_k is of course limited by the accuracy of the numerical calculations. The quantitative universal features of the period doubling road to turbulence were discovered by Feigenbaum [19], and independently by Coullet and Tresser [12]. The simplest universality concerns the convergence of the sequence $\{\alpha_k\}$. Namely,

$$\frac{\alpha_{k+1} - \alpha_k}{\alpha_k - \alpha_{k-1}} \to \delta^{-1} \quad \text{as} \quad k \to \infty \tag{4}$$

with $\delta = 4.6692\ldots$ <u>independent of the one parameter family</u> considered. This kind of universality has to be understood in the sense of critical phenomena. For instance there are one parameter families which do not show the sequence of period doubling bifurcations. In this case the theory has nothing to say. For precise statements we refer to the sections IV to VI. In practice, every one parameter family f_α similar to $1-\alpha X^2$ follows the period doubling route to turbulence, and the bifurcation points α_k (which depend on the family f_α) accumulate with the same asymptotic ratio δ. For example, the family $f_\alpha(X)=\cos \alpha X$, $\alpha \in [0,\Pi]$, works perfectly well, as one can verify (up to $k \simeq 10$) by using a programmable **pocket** calculator. Another quantity, which will turn out to be important later on, can also be measured easily. Namely

$$\frac{\xi'_{k+1} - \xi_{k+1}}{\xi'_k - \xi_k} \to \lambda = -.3995\ldots \qquad \text{as} \quad k \to \infty \quad , \tag{5}$$

where ξ_k is the critical point of f_{α_k} ($\xi_k = 0$ in the above examples), and where ξ'_k is the point closest to ξ_k on the periodic orbit of length 2^{k-1} for f_{α_k}. The number λ is universal in the same sense as δ.

In order to justify the expression "road to turbulence", in this context, we shall briefly mention what is observed when α passes beyond α_∞. Considering again the family $f_\alpha(X)=1-\alpha X^2$ as a typical example, a decreasing sequence $2=\tilde{\alpha}_0 > \tilde{\alpha}_1 > \ldots > \tilde{\alpha}_\infty = \alpha_\infty$ can be found (also accumulating with asymptotic ratio δ) such that for $\alpha \in]\tilde{\alpha}_{k+1}, \alpha_k[$ the function f_α induces a cyclic permutation of 2^k subintervals of $[1,1]$. The action of its 2^k-th power on each of these intervals is similar to the action on $[-1,1]$ of some f_α with $\alpha \in]\tilde{\alpha}_1, \tilde{\alpha}_0[$. By varying α in the set $]\tilde{\alpha}_1, \tilde{\alpha}_0[$, a rich variety of **stable** periods is obtained. But for α near 2, most of the functions f_α appear to have no stable periodic orbit (in particular f_2 is conjugate to $X \to 1-2|X|$) and correspondingly, "noisy" periods 2^k are observed in $]\tilde{\alpha}_{k+1}, \tilde{\alpha}_k[$ for parameter values near $\tilde{\alpha}_k$. More details can be found e.g. in [7] and [28] .

The period doubling road to turbulence has also been observed for higher dimensional dissipative systems. This includes the Hénon map discussed above [14] , some systems of differential equations like the nonlinear Duffing and Mathieu equations [27] and the Lorentz model [21] , as well as real physical experiments (see next section). The universal numbers are the same as for one dimensional maps.

Recently, a similar phenomenon has been discovered for one parameter families of area preserving maps of the plane [13]. A typical example is again the family f_α of Hénon maps (1) , but now for $\beta=1$. Note that our previous calculations still apply in this case : f_α has two fixed points for $\alpha \geqslant 0$, and a period two for $\alpha > 3/4$. However, none of them can be stable, since the determinant of $(Df_\alpha)_X$ is equal to one everywhere. Depending on the eigenvalues $e_\pm$ of $(Df_\alpha^m)_X$, a periodic orbit $\{x, f_\alpha(X), \ldots, f_\alpha^{m-1}(X)\}$ of length m is (up to $e_\pm = 1$ or -1) either elliptical ($e_\pm = \exp \pm 2i\varphi$, $\varphi \in \,]0, \pi[$) or hyperbolical ($e_- \in \,]0,1[$, $e_+ = 1/e_-$). So, what happens at $\alpha = 3/4$ is that an elliptic fixed point bifurcates to an elliptic period two. A numerical investigation shows that the period doubling goes on as α is increased . The corresponding bifurcation points α_k define an asymptotic ratio δ as in the dissipative case (4), but its value is different. However δ is again universal in the sense that the same ratio is obtained for many different families of area-preserving maps on the plane [2] , [22] , namely $\delta = 8.721\ldots$. More details can be found in section VI.

II. Some experiments

The first experiment showing a sequence of period doubling bifurcations is due to Libchaber and Maurer [35]. It is a Rayleigh Benard experiment performed with liquid helium. A small container filled with normal liquid helium is heated from below and cooled from above. The container is designed such that only two convective rolls are present (with more than two rolls, other roads to turbulence are observed [25]). The system is analyzed by two bolometers on the top of the container, approximately above the center of each roll. This provides a two dimensional projection of the phase space trajectory of the system when the indication of one bolometer is plotted as a function of the other. Period doubling is observed directly as a bifurcation of a closed periodic orbit of length ℓ to a closed periodic orbit of length 2ℓ . A better way to analyze the signal is to use its power spectrum. In the range of parameters where period doubling bifurcations occur, the spectrum consists of a discrete number of peaks placed at integer multiples of a frequency $\omega/2^k$. Here ω is some basic frequency of the system (carrying most of the spectral energy), and k is the order of period doubling. As the heat flux is increased, crossing a certain threshold value, a new peak appears at $\omega/2^{k+1}$ and its odd multiples. The rate δ at which these thresholds accumulate, as well as the relative magnitude of the peaks in the spectrum (we shall come back to this later on) are roughly the same as for maps on the interval. At the end of the road, a broad band spectrum emerges, indicating that a chaotic state has been reached. A similar experiment was performed by Giglio et al. [23], using water instead of liquid helium. The results are similar to those of Libchaber. We also note that the inverse cascade (peaks of half frequencies emerging from the broad band spectrum when heating is decreased, starting from a turbulent situation) was also observed in both experiments. A curious fact, which is not understood yet, is that the road to turbulence is very sensitive to the history of the experiment. In order to see the period doubling bifurcations, the system has to be prepared in a very special way (essentially a fast heating followed by a cooling, before slowly increasing the heating again to make the measurements). We want to emphasize however that this kind of preparation is very reproducible. Recently, Libchaber et al. [34] have performed a similar experiment with mercury. A constant magnetic field was applied to stabilize the rolls. A more accurate measurement of δ was obtained. It was even possible to observe some of the windows of stable periods inside the chaotic regime.

Other occurences of the period doubling road to turbulence were reported by Lauterborn and Cramer [33] for an acoustical experiment, by Swinney [41] for an oscillating chemical reaction, and by Gibbs et al. [24] who used an optical device. A simple electrical experiment with a good precision is due to Linsay [30] and Perez et al. [38] . Here a varactor diod (nonlinear capacitor) is used in a driven LCR circuit. The amplitude of the sinusoidal excitation serves as a variable external parameter α . By plotting the peak values of the voltage across the diod as a function of α , a diagram is obtained which is very similar to the bifurcation diagram (attractor as a function of α) for the family $f_\alpha(X) = 1 - \alpha X^2$. Some stable windows in the chaotic regime are easily identified. The influence of external noise was also studied. The measured values for various universal numbers are in good agreement with the theory.

An interesting mechanical experiment is due to Croquette and Poitou [11] . A damped compass is placed in an oscillating magnetic field perpendicular to the compass axis. The period doubling phenomenon has been observed, but the universal numbers have not been measured yet. Note that for zero damping we have a Hamiltonian system, so that the universal numbers should be those of the area preserving maps. As for the remaining natural friction, there are two competing aspects. Friction is certainly helpful in finding the periodic orbits, but it also causes a crossover to the dissipative universal values for sufficiently high periods. Whether this is seen or not depends on the quality of the experiment. In order to see one more period doubling without crossover, the accuracy in the measurement of the external parameter has to be increased by a factor $\delta \simeq 8.7$, and the friction has to be lowered by a factor of two.

III. The renormalization transformation

As mentioned before, sequences of period doubling bifurcations are observed not only for maps on the interval, but also for higher dimensional dissipative systems, and for conservative discrete (continuous) systems in two (four) dimensions. Before investigating these different cases in detail, we shall first illustrate the basic concepts, without specifying the space $\mathcal{A}$ of maps considered. To begin with, note that all universal properties are coordinate independent. Therefore they can be reformulated in terms of equivalence classes

$$F = \pi(f) = \{L^{-1} \circ f \circ L : L \in \mathcal{G}\} \tag{6}$$

of maps $f \in \mathcal{A}$, where $\mathcal{G}$ denotes some p-parameter group of transformations of the underlying space $\mathbb{R}^n$. The composition

$$R : F \to F \circ F \tag{7}$$

is well defined in $\mathcal{A}/\mathcal{G}$, namely $\pi(f) \circ \pi(f) = \pi(f \circ f)$. Then for a given family $\alpha \to F_\alpha$, the bifurcation point α_k is the value of α for which the family intersects the manifold $\Sigma_k = R^{-1}(\Sigma_{k-1}) = \ldots = R^{-(k-1)}(\Sigma_1)$, where

$$\Sigma_1 = \{\pi(f) : f \text{ has a fixed point } X \text{ and } Df_X \text{ has an eigenvalue } -1\} . \tag{8}$$

The observation (4) indicates that Σ_∞ exists, and that the spacing between Σ_k and Σ_∞ goes <u>uniformly</u> like δ^{-k} as $k \to \infty$. This is a clear sign that

i) R has a fixed point Φ (in Σ_∞)

ii) DR_Φ has one eigenvalue $\delta > 1$, and the remainder of the spectrum lies strictly inside the unit disc.

iii) Near Φ the geometrical situation in $\mathcal{A}/\mathcal{G}$ is as follows

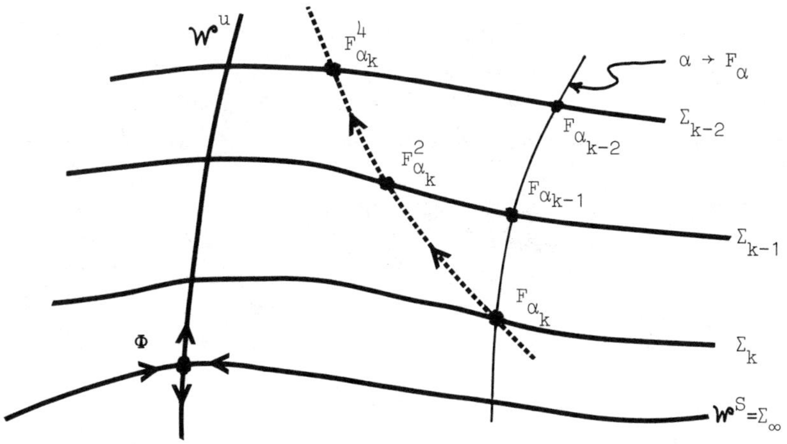

Figure 1 : The action of R

where $\mathcal{W}^u$ is the unstable manifold of dimension one, corresponding to the eigenvalue $\delta > 1$, and where $\mathcal{W}^s$ is the stable manifold of codimension one.

It should now be clear what has to be proved in order to explain (4). Historically, this "renormalization group recipe" was believed to apply to maps on the interval, before the corresponding renormalization transformation was known. The reason is the well known success of the renormalization group in dealing with universality and scaling laws in critical phenomena.

For practical purposes the transformation R might be replaced by a transformation $\mathcal{N}$ on $\mathcal{A}$, satisfying $\pi\mathcal{N} = R\pi$. Such an extension has the form

$$\mathcal{N}: f \to \Lambda_f^{-1} \circ f \circ f \circ \Lambda_f \quad , \quad \Lambda_f \in \mathcal{G} . \tag{9}$$

However Fig. 1 does not automatically apply to $\mathcal{N}$. Even if $\mathcal{N}$ has a fixed point $\varphi \in \pi^{-1}(\Phi)$, the operator $D\mathcal{N}_\varphi$ can have large eigenvalues other than δ, depending on how $\mathcal{N}$ acts on the set $\pi^{-1}(\Phi)$. The corresponding eigenvectors are of the form

$$h = \frac{d}{dt}\,[L(t)^{-1}\,\varphi\,L(t)]_{t=0} \quad , \qquad L(0) = \mathrm{Id}\ . \tag{10}$$

This can be avoided by choosing $f \to \Lambda_f$ in such a way that $\Lambda_f = L^{-1}\circ\Lambda_\varphi$ for $f = L^{-1}\circ\varphi\circ L$, $L \in \mathcal{G}$. Another **possibility** is to determine Λ_f by a normalization condition $\nu(\mathcal{N}(f)) = \nu_0$ for some function $\nu : \mathcal{A} \to \mathbb{R}^p$ with $\nu(\varphi) = \nu_0$. Note that by restricting $\mathcal{N}$ to $\mathcal{A}/\mathcal{G} = \{f \in \mathcal{A} : \nu(f) = \nu_0\}$ we get back **R** .

Besides δ, an infinite number of other universal constants can be measured . This is due to the fact that

$$\mathcal{N}^j(f_{\alpha_{q+j}}) \to \psi_q \qquad \text{as} \qquad j \to \infty \tag{11}$$

where $\psi_q \in W^u \cap \Sigma_q$ is independent of the family considered (there is also the possibility of choosing a different Σ , see e.g. Section V).

In particular there are universal scaling factors $\lambda,\ldots,\mu$ relating a cluster of 2^{q-1} points in the 2^k periodic orbit of f_{α_k} to a similar cluster in the 2^{k+1}-periodic orbit of $f_{\alpha_{k+1}}$, as $k \to \infty$ (an identification is possible because of (11)). To see this, we may choose p coordinates of the 2^{q-1} points under consideration as a normalization function Let $f_j = f_{\alpha_{q+j}}$. Then

$$f_j^{2^j} = \Lambda_j \circ \mathcal{N}^j(f_j) \circ \Lambda_j^{-1} \quad , \tag{12}$$

where $\Lambda_j = \Lambda_{f_j} \circ \Lambda_{\mathcal{N}(f_j)} \circ \cdots \circ \Lambda_{\mathcal{N}^{j-1}(f_j)}$. Since most of the $\mathcal{N}^j(f_j)$ are arbitrarily close to φ as $j \to \infty$, the sequence $L_j = \Lambda_j \circ \Lambda_\varphi^{-j}$ converges to some $L_\infty \in \mathcal{G}$. The coordinates $\nu(f_j^{2^j})$ and $\nu(f_{j+1}^{2^{j+1}})$ are then related by the transformation $L_{j+1} \circ \Lambda_\varphi \circ L_j^{-1}$, which approaches $L_\infty \circ \Lambda_\varphi \circ L_\infty^{-1}$ as $j \to \infty$. Thus $\pi(\Lambda_\varphi)$ is "observable", and in all cases Λ_φ appears to have a stable fixed point ξ . The scaling factors $\lambda,\ldots,\mu$ are the eigenvalues of $\Lambda = D(\Lambda_\varphi)_\xi$.

Finally we note that this also determines a "minimal choice" for $\mathcal{G}$. Namely, suppose that Λ_φ is in a normal form $\Lambda_\varphi = \Lambda \circ R = R \circ \Lambda$. It is easy to see that $L \to \Lambda^{-1} \circ L \circ \Lambda$ and $L \to \Lambda_\varphi^{-1} \circ L \circ \Lambda_\varphi$ have the same unstable manifolds (eigenvalues 1 included) at $L = Id$, which is a finite dimensional group $\mathcal{G}_0$. We may thus take $\mathcal{G} = \mathcal{G}_0$, since it is not necessary to have $\Lambda_f \neq \Lambda_\varphi$ for $f = L^{-1} \circ \varphi \circ L$, if $\Lambda_\varphi^{-j} \circ L \circ \Lambda_\varphi^j \to Id$ as $j \to \infty$.

IV. Some theoretical results in one dimension

According to an earlier remark, we should now investigate families f_α similar to $f_\alpha(X) = 1 - \alpha X^2$. Instead, we shall first generalize the problem in order to be able to explain with some details a simple case. We will simultaneously mention the known results for the original situation. Let us consider a space of functions f such that

i) there is an $\varepsilon > 0$ and a function g, analytic in a complex neighbourhood of $[0,1]$, such that $f(X) = g(|X|^{1+\varepsilon})$

ii) g is monotonously decreasing, $-1 < g(1) < 0$, $g(0) = 1$.

For fixed ε, one parameter families f_α of such functions may present the period doubling phenomenon. The universal numbers are then functions of ε. Note that $\varepsilon = 1$ corresponds to the original situation of even analytic functions.

We shall now investigate the renormalization transformation (9), whose relevance to the problem was pointed out in Section III. A normalization ν is already suggested : we want of course (ii) to be invariant under $\mathcal{N}$. This imposes $\nu(f) = f(0) = 1$. Then $\nu(\mathcal{N}(f)) = 1$ can be achieved by choosing $\Lambda_f(X) = f(1)\, X$, and we get

$$\mathcal{N}(f)(X) = \frac{1}{f(1)}\, f(f(f(1)X)) \qquad \text{(Cvitanowitch)} \qquad (13)$$

The next step in a renormalization group analysis is to find a fixed point for the renormalization transformation. Here we have to solve

$$\varphi(\varphi(\lambda X)) = \lambda\varphi(X) \quad , \quad \lambda = \varphi(1) \quad , \qquad (14)$$

with $\varphi(X) = g(|X|^{1+\varepsilon})$ for some function g satisfying (ii). Note that this equation has a simple solution for $\varepsilon = 0$, namely $\varphi(X) = 1 - |X|$, $\lambda = 0$. We shall see that it is possible to use an ε - expansion, and the various problems of a renormalization group analysis can be solved for small ε without too much effort [6]. The case $\varepsilon = 1$ is much more difficult, and we refer the reader to the corresponding literature.

At this point it is easier to consider λ as an independent parameter, rather than ε . The natural parametrization in terms of ε will be recovered at the end, using the implicit function theorem. For small λ we are now looking for a number $\varepsilon(\lambda)$ and a function g_λ , $g_\lambda(0) = 1$; such that

$$g_\lambda(|g_\lambda(|\lambda|^{1+\varepsilon(\lambda)}t)|^{1+\varepsilon(\lambda)}) - \lambda g_\lambda(t) = 0 \tag{15}$$

for $t \in [0,1]$. This equation is obtained by substituting $\varphi = g_\lambda(|.|^{1+\varepsilon(\lambda)})$ into (14).

We now consider $\lambda \to \varepsilon(\lambda)$ and $\lambda \to g_\lambda$ as functions defined on $\Delta = \{\lambda \in \mathbb{C} : |\lambda| < \lambda_o, |\mathrm{Arg}(-\lambda)| < \theta_o\}$ with values in $\mathbb{C}$ and $\mathcal{B}$ respectively, where $\mathcal{B}$ denotes the Banach space of analytic functions g on $K_\tau = \{t \in \mathbb{C} : |t| < \tau\}$, equipped with the norm $\|g\|_{\mathcal{B}} = \mathrm{Sup}\ \{|g(t)| : t \in K_\tau\}$. The constants $\theta_o \in]0,\pi]$ and $\tau > 1$ are arbitrary, but fixed, and $\lambda_o > 0$ will be chosen later. Let b denote the Banach space of analytic functions $\lambda \in \Delta \to \varepsilon(\lambda) \in \mathbb{C}$ with the norm

$$\|\varepsilon(.)\|_b = \mathop{\mathrm{Sup}}_{\lambda \in \Delta} |\varepsilon(\lambda)| . \left(\frac{1-\log|\lambda|}{|\lambda|}\right) , \tag{16}$$

and let $\mathcal{L}$ denote the Banach space of analytic functions $\lambda \in \Delta \to g_\lambda \in \mathcal{B}$ with the norm

$$\|g_.\|_{\mathcal{L}} = \mathop{\mathrm{Sup}}_{\lambda \in \Delta} \left(\|g_\lambda\|_{\mathcal{B}} \cdot \frac{1}{|\lambda|}\right) . \tag{17}$$

Theorem 1 : There is a positive λ_o such that the equation

$$g_\lambda(g_\lambda((-\lambda)^{1+\varepsilon(\lambda)}\ t)^{1+\varepsilon(\lambda)}) - \lambda g_\lambda(t) = 0 \tag{18}$$

has a unique solution of the form.

$$g_\lambda(t) = 1 - (1-\lambda)t - \lambda t(1-t)\ \ r(\lambda,t) \tag{19}$$

with ε in b and r in $\mathcal{L}$ such that $\|\varepsilon\|_b + \|r\|_{\mathcal{L}} \leq 1$.

Proof : We consider the Banach space $\mathcal{F} = b \times \mathcal{L}$ with the norm $\|(\varepsilon,r)\|_{\mathcal{F}} = \|\varepsilon\|_b + \|r\|_{\mathcal{L}}$. The unit ball in $\mathcal{F}$ is denoted by $\mathcal{F}_1$. For $(\varepsilon,r) \in \mathcal{F}_1$ and $(\lambda,t) \in \Delta \times K_\tau$ we have

$$g_\lambda((-\lambda)^{1+\varepsilon(\lambda)}\ t) = 1 + \lambda t\ \eta_1(\varepsilon,r,\lambda,t),\ |\eta_1(\varepsilon,r,\lambda,t)| = \mathcal{O}(1) \tag{20}$$

Thus $g_\lambda((-\lambda)^{1+\varepsilon(\lambda)}\ t)^{1+\varepsilon(\lambda)})$ is well defined if λ_0 is chosen sufficiently small. Moreover, it is easy to verify that

$$g_\lambda((-\lambda)^{1+\varepsilon(\lambda)}\ t)^{1+\varepsilon(\lambda)} = 1 + \lambda t[(1-\lambda)(-\lambda)^{\varepsilon(\lambda)}(1+\varepsilon(\lambda)) + \lambda g_\lambda(0) + + \lambda\varepsilon(\lambda)\frac{t}{2} + \lambda^2 t r_2(\varepsilon,r;\ \lambda,t)] \tag{20}$$

with $r_2(\varepsilon,r) \in \mathcal{L}$. Therefore $g_\lambda((-\lambda)^{1+\varepsilon(\lambda)}\ t)^{1+\varepsilon(\lambda)} \in K_\tau$, and we can substitute the right hand side of (20) into $r(\lambda,.)$. We get two functions **F** and G such that

$$g_\lambda(g_\lambda((-\lambda)^{1+\varepsilon(\lambda)}\ t)^{1+\varepsilon(\lambda)}) - \lambda g_\lambda(t) = \lambda t\ G(\varepsilon,r;\lambda,t)\ , \tag{21}$$

with $G(\varepsilon,r;\lambda,t) = G(\varepsilon,r;\lambda,1) - \lambda(t-1)\ F(\varepsilon,r;\lambda,t)$. We now have to solve the two equations $G(\varepsilon,r;\lambda,1) = 0$, $F(\varepsilon,r;\lambda,t) = 0$. It is easy to show that they are of the form

$$\begin{aligned} &\varepsilon(\lambda)(1+\log(-\lambda)) - \lambda - \lambda r_5(\varepsilon,r;\lambda) = 0 \\ &r(\lambda) - \frac{1}{2}\varepsilon(\lambda) - \lambda r_4(\varepsilon,r;\lambda,t) \quad = 0 \end{aligned} \tag{22}$$

where $(\varepsilon,r) \to (r_5(\varepsilon,r), r_4(\varepsilon,r))$ is a C^1 map from (a neighbourhood of) $\mathcal{F}_1$ to $\mathcal{F}$. Now (22) is equivalent to the fixed point equation $\mathcal{M}(\varepsilon,r) = (\varepsilon,r)$, with

$$\mathcal{M}(\varepsilon,r;\lambda,t) = (\frac{\lambda}{1+\log(-\lambda)}\ (1+r_5(\varepsilon,r;\lambda)),\ \frac{1}{2}\varepsilon(\lambda) + \lambda r_4(\varepsilon,r;\lambda,t)) \tag{23}$$

But for λ_0 sufficiently small, $\mathcal{M}$ is a **contraction** from $\mathcal{F}_1$ into itself. By the contraction mapping principle, $\mathcal{M}$ has a unique fixed point in $\mathcal{F}_1$, **and therefore** our initial problem has a unique solution under the above hypothesis.

QED

For more details, the reader is referred to [6]. Coming back to the parametrization, it can be shown that $\lambda \to \varepsilon(\lambda)$ can be inverted and that $g_{\lambda(\varepsilon)}(t)$ is an analytic function of t, $\varepsilon \log \varepsilon$, $(\log \varepsilon)^{-1}$

and $(\log \varepsilon)^{-1} \log(-\log \varepsilon)$ for ε small and off the negative real axis.

In the analytic case $\varepsilon = 1$ there are by now three different proofs of the existence of a fixed point φ for $\mathcal{N}$, see [3] , [31] , [32] . All three approaches use, in one way or another, an approximate fixed point φ_0 and the contraction mapping principle. The contractions are obtained in different ways by a clever reformulation of the fixed point equation in order to eliminate the influence of the eigenvalue $\delta > 1$ of $D\mathcal{N}_\varphi$. A very useful fact is that the fixed point φ as well as the eigenvector of $D\mathcal{N}_\varphi$ corresponding to δ are essentially quadratic polynomials. For example

$$\varphi(X) \simeq 1 - 1.5276X^2 + .1048X^4 + .0267X^6 - .0035X^8 + .0001X^{10}$$

One of the proofs is based on a sophisticated computer program which provides good polynomial approximations and rigorous error bounds [31] . More details on this method can be found in Section VI. The analytic properties of φ were investigated in [18] .

In order to establish the correctness of Fig. 1 let us now investigate the differential of $\mathcal{N}$ at φ. From now on we restrict the analysis to real Banach spaces, so that $\lambda < 0 < \varepsilon$. $\mathcal{N}$ acts on $\mathcal{A} = \mathfrak{J}(\mathcal{B})$, where $\mathfrak{J}(g) = g(|X|^{1+\varepsilon})$ and $\|\mathfrak{J}(g)\| = \|g\|$. It is of course equivalent to consider $\mathcal{T} = \mathfrak{J}^{-1} \circ \mathcal{N} \circ \mathfrak{J}$,

$$\mathcal{T}(g)(t) = \lambda^{-1} g(|g(|\lambda|^{1+\varepsilon}\, t)|^{1+\varepsilon}), \qquad \lambda = g(1) \tag{24}$$

A short calculation gives the formula

$$(D\mathcal{T}_g h)(t) = h(y)/\lambda + (1+\varepsilon) g'(y)\, g(z)^\varepsilon\, h(z)/\lambda - [g(t) - (1+\varepsilon) t g'(t)]\, h(1)/\lambda \tag{25}$$

where g is the fixed point of $\mathcal{T}$, and $z = |\lambda|^{1+\varepsilon}\, t$, $y = |g(z)|^{1+\varepsilon}$. In view of the condition (ii), we consider $D\mathcal{T}_g$ on the Banach space $\mathcal{B}_\tau$ of analytic functions h on K_τ which satisfy $h(0) = 0$.

<u>Theorem 2</u> : <u>The spectrum</u> of $D\mathcal{N}_\varphi$ <u>is contained in a disc</u> K_ρ <u>of radius</u> $\rho = \mathcal{O}(\lambda)$ <u>around zero, except for a simple eigenvalue</u> $\delta = 2 - \varepsilon/\lambda + \mathcal{O}(\lambda)$.

Proof : Suppose $-\lambda > 0$ is sufficiently small. We consider $D\mathcal{T}_g$ as a perturbation of the linear operator A ,

$$(Ah)(t) = h(1-\lambda t)/\lambda - (1+\varepsilon)(1-\lambda)\, h(\lambda t)/\lambda - [1+(1-\lambda)\varepsilon t]\, h(1)/\lambda \quad , \tag{26}$$

and it is easy to check that $\|D\mathcal{T}_g - A\| = \mathcal{O}(\lambda)$ on $\mathcal{B}_\tau$. Since A is clearly bounded as an operator from $\mathcal{B}_\tau$ to $\mathcal{B}_{\tau+1}$, and since the injection $\mathcal{B}_{\tau+1} \to \mathcal{B}_\tau$ is compact, we conclude that A is a compact operator on $\mathcal{B}_\tau$. It is now easy to verify that the spectrum of A has the form $\{\delta_i : i = 0,1,2,\dots\}$ with

$$\delta_0 = 1 + (1-\lambda)(1-\frac{\varepsilon}{\lambda} + \varepsilon) \quad , \quad \text{eigenvector} \quad e_0(t) = t \tag{27}$$

$$\delta_i = \lambda^i [1+(-1)^i (1+\varepsilon)(1+\lambda)], \quad \text{eigenvector} \quad e_i(t) = \text{polynomial of degree } i+1 \text{ in } t, i = 1,2,\dots .$$

QED

A similar result (e.g. $\delta = 4.6692\dots$, $\rho > 1$) was obtained by Lanford [31] for the case $\varepsilon = 1$ by using his computer assisted proof. From now on we fix ε and assume that the spectrum of $D\mathcal{N}_\varphi$ is as described above. By using the spectral projection E^u of $D\mathcal{N}_\varphi$ corresponding to the eigenvalue δ , the space $\mathcal{J}(\mathcal{B}_\tau)$ can be written as a sum $\mathcal{A}^u \oplus \mathcal{A}^s$ of an unstable and a stable linear eigenspace for $D\mathcal{N}_\varphi$. We can then apply the general theory of invariant manifolds.

Theorem 3 : [26] : $\mathcal{N}$ has a C^∞ local unstable manifold $\mathcal{W}^u$ of dimension one, tangent at φ to $\varphi + \mathcal{A}^u$. Similarly, $\mathcal{N}$ has a C^∞ local stable manifold $\mathcal{W}^s$ of codimension one, tangent at φ to $\varphi + \mathcal{A}^s$.

Usually, one assumes that $\mathcal{W}^u$ and $\mathcal{W}^s$ are hyperplanes, and that $\mathcal{N}$ can be replaced by $D\mathcal{N}_\varphi$ in the vicinity of φ. It is easy to find a C^∞ change of variables in $\mathcal{A}$ such that the new stable and unstable manifolds become hyperplanes. However the techniques used to reduce a nonlinear map to its linear approximation apply only in very special cases (in particular the map must be invertible). They do not seem to work in our situation. But in fact it is sufficient to linearize the unstable direction, as we shall see below. Let us introduce vectors $h^u \in \mathcal{A}^u$ and $h^s \in \mathcal{A}^s$ as coordinates near φ . We write

$$\varphi + h = \begin{pmatrix} h^u \\ h^s \end{pmatrix} \qquad \text{for} \qquad h = h^u + h^s \tag{28}$$

Theorem 4 : There is a diffeomorphism S, defined in a neighbourhood of φ, such that

$$(S \mathcal{N} S^{-1}) \begin{pmatrix} h^u \\ h^s \end{pmatrix} = \begin{pmatrix} \delta . h^u \\ L(h) . h^s \end{pmatrix} \tag{29}$$

where L(h) is a bounded linear operator on $\mathcal{A}^s$ with $\|L(h)\| \leqslant \rho < 1$ for fixed h in some neighbourhood of zero.

The proof of this theorem relies on the contraction mapping principle. See [6] for more details. In order to complete the renormalization group analysis, we have to show that Fig. 1 is qualitatively correct. Recall that the bifurcation points α_k for a given family $\alpha \to f_\alpha$ are defined by $f_{\alpha_k} \in \Sigma_k$, where $\Sigma_k = \mathcal{N}^{1-k}(\Sigma)$ and $\Sigma = \{ f \in \mathcal{A} : f(X) = X \Rightarrow f'(X) = -1 \}$. It is easy to verify that crossing the (codimension one) manifold Σ_k transversally implies a bifurcation between a stable period 2^k and a stable period 2^{k+1}. In the special case where $\alpha \to f_\alpha$ is a correct parametrization (non zero velocity) of $\mathcal{W}^u$, the transversal intersection with Σ_k follows from

Theorem 5 [6] : Σ crosses $\mathcal{W}^u$ transversally.

Furthermore, this particular family satisfies obviously

$$\frac{\alpha_{k+1} - \alpha_k}{\alpha_k - \alpha_{k-1}} \to \delta^{-1} \qquad \text{as} \qquad k \to \infty \tag{30}$$

where δ is the unstable eigenvalue of $D\mathcal{N}_\varphi$. But by the theorems 4 and 5, the sequence Σ_k accumulates towards $\mathcal{W}^s$ at a uniform rate δ^{-1} in some neighbourhood of φ. Thus (30) holds for every family $\alpha \to f_\alpha$ which satisfies

(i) $\alpha \to f_\alpha$ is a C^1 **curve in** $\mathcal{A}$.

(ii) $\alpha \to f_\alpha$ crosses $\mathcal{W}^s$ transversally near φ (at $\alpha = \alpha_\infty$).

This proves the universality of δ. Given a particular family, the condition (ii) is certainly difficult to check. But if the family passes near φ then it is very "likely" for (ii) to be satisfied. Note also that the property (ii) is stable under C^1 perturbations. Numerically, "near φ " is much less restrictive than in the proofs. But far from φ it is easy to write down families who do not present an infinite period doubling sequence. It is also possible that there are other fixed points with different universal numbers (for the same ε), although this has not been observed yet.

V. Extensions and other universal results

It is now easy to **understand** the other universal properties observed numerically for maps on the interval. All properties of $\mathcal{W}^u$ are universal, since each point $\psi \in \mathcal{W}^u$ can be approached for every family f_α through a sequence $\mathcal{N}^k(f_{\tilde{\alpha}_k})$ with $\tilde{\alpha}_k \to \alpha_\infty$. Namely, if $\tilde{\Sigma}$ is a codimension one manifold intersecting $\mathcal{W}^u$ transversally in ψ, then we can take for $\tilde{\alpha}_k$ the parameter value at which $\alpha \to f_\alpha$ crosses $\tilde{\Sigma}_k = \mathcal{N}^{1-k}(\tilde{\Sigma})$. Note that (if $\varphi \notin \tilde{\Sigma}$) these sequences α_k always accumulate with the same universal ratio δ^{-1}. Let us just pick out a few examples.

The proof of the scaling law (5) is sketched in section III for a more general situation. In our case $\tilde{\Sigma} = \Sigma_1$, and the eigenvalue of Λ_φ at its fixed point is $\lambda = \varphi(1)$. The choice

$$\tilde{\Sigma} = \{f \in \mathcal{B} : f^4(0) = f^3(0)\} \tag{31}$$

is related to the inverse cascade mentionned in the introduction. For $f \in \tilde{\Sigma}_k$ the orbit of zero falls into an unstable period, and this implies turbulence . Crossing $\tilde{\Sigma}_k$ also causes the occurence of a new peak in the power spectrum at $\omega/2^k$ and its odd multiples, where ω is the frequency of the largest peak. Other sequences of $\tilde{\Sigma}_k$ can be found describing the first occurence of a stable period $2^k(2\ell+1)$. This happens inside the chaotic regime (stable windows). For $\alpha = \alpha_\infty$ the dynamical system f_α has a strange attractor (a Cantor set, formally a stable period 2^∞). It is possible to define an invariant ergodic measure, but the transformation is not weakly mixing [6] . Also the power spectrum of f_{α_∞} has some universal properties [20] , [8] , [37] . There is a hierarchy among the peaks of the spectrum. They are located at frequencies of the form $\omega_{k\ell} = \omega\,(2\ell+1)/2^k$ with $|\ell| < 2^{k-1}$, where ω is some basic frequency. If I_q is the power contained in the peaks of frequencies $\omega_{k\ell}$ with k=q, then for large q and j the numbers $[\log I_{q+j} - \log I_q /j]$ lies between 13.5 dB and 14dB.

One extension of the period doubling scenario is motivated by the fact that in experiments only a finite number of bifurcations can be observed. To account for this, the original deterministic system might **be changed** into a stochastic difference equation $X_{k+1} - X_k = (f_\alpha(X_k) - X_k) + \xi_k$. The influence of the external noise ξ was studied in [9] , [10] and [40] . Asymptotically, in order to see one more bifurcation, the noise should be decreased by a factor $K = 6.619\ldots$. The number K^2 is again universal, and it is believed to be largest eigenvalue of $h \to \tilde{h}$ with $\tilde{h}(X) = \lambda^{-2}\, \varphi'(\varphi(\lambda X))^2\, h(X) + \lambda^{-2}\, h(\varphi(\lambda X))$.

As we have already mentioned, the same universal numbers are observed for transformations in higher dimensions. This suggests that there is an extension $E : \mathcal{A}_1 \to \mathcal{A}_n$ of even functions on $\mathbb{R}$ to **maps on** $\mathbb{R}^n$ such that the degrees of freedom in $\mathcal{A}_n \setminus E(\mathcal{A}_1)$ are irrelevant for the doubling transformation (9) on $\mathcal{A}_n$. For simplicity let n=2; the general case is similar. We may assume that a one dimensional linear subspace of $\mathbb{R}^2$ is left invariant by maps in $E(\mathcal{A}_1)$, i.e. that $E(f)(X_1,0) = (f(X_1),0)$. This could be realized by simply taking $E(f)(X_1,X_2) = (f(X_1),0)$. However the renormalization group analysis, as described in Section III, cannot be applied in this case. The reason is that $\mathcal{A}_2/\mathcal{G}$ is not well defined due to the invariance of the maps $f \in E(\mathcal{A}_1)$ under rescaling of the X_2- coordinate. The simplest choice for the extension E, which does not present the same kind of difficulty, seems to be $E(f)(X_1,X_2) = (g(X_1^2 - X_2),\ 0)$ for $f(X) = g(X^2)$. Given the one dimensional fixed point $g(g(\lambda^2X^2)^2) = \lambda g(X^2)$, then

$$\varphi(X_1,X_2) = (g(X_1^2-X_2),\ 0) \tag{32}$$

is the fixed point of any renormalization transformation

$$\mathcal{N}(f) = \Lambda_f^{-1} \circ f \circ f \circ \Lambda_f \tag{33}$$

with $\Lambda_\varphi(X_1,X_2) = (\lambda X_1, \lambda^2 X_2)$. It can be shown that the spectrum of $D\mathcal{N}_\varphi$ coincides outside a disc of radius $\rho < 1$ with the spectrum of $D\mathcal{N}_\varphi$ in one dimension, provided $f \to \Lambda_f$ is chosen correctly (see Section III). The idea of the proof is to show that $D\mathcal{N}_\varphi h = h_1 + h_2$ with $h_1 \in E(\mathcal{A}_1)$ and

h_2 of the form (10) with L arbitrary (not necessarily in $\mathcal{G}$). For details and arbitrary n we refer to [4] . The renormalization group analysis is now easily completed by using the corresponding results in one dimension. This now proves universality for families of maps which (up to a coordinate transformation) cross $\mathcal{W}^s$ transversally near $E(\varphi)$. Note that many of these maps are invertible. Note also that $\alpha \to \mathcal{N}^k(f_\alpha)$ might very well pass near φ for some k, even if the determinant of Df_α is not small. There are however still a lot of families left for other roads to turbulence.

VI. Area preserving maps

Motivated by the findings described in the introduction, we shall investigate here the doubling transformation $\mathcal{N}$: $f \to \Lambda_f^{-1} \circ f \circ f \circ \Lambda_f$, acting on area-preserving maps of the plane, and the corresponding fixed point equation $\mathcal{N}(\varphi) = \varphi$. This should not be surprising, except possibly for the reader who skipped the last three sections. An analysis of the 2^k- periodic orbits (as explained in Section III) indicates that Λ_φ is conjugate to Λ : $(X_1,X_2) \to (\lambda X_1, \mu X_2)$ with the universal scaling factors $\lambda \simeq -.2488$ and $\mu \simeq .06111$ [5] . Other related universal properties have not been investigated yet. As an example, there might be a nested sequence of intervals $]\alpha_k, \tilde{\alpha}_k[$ characterized by the presence of two smooth invariant domains for $f_\alpha^{2^k}$, separated by the (un)stable manifold of a hyperbolic fixed point for $f_\alpha^{2^k}$. This would correspond to the inverse cascade for maps on the interval.

A large class of area preserving diffeomorphisms f possesses a symmetry (see e.g. [15])

$$f \circ T \circ f = T \quad , T \circ T = I_d \ , \det (DT) \equiv -1. \tag{34}$$

In particular, let us consider the family of Hénon maps (1) with $\beta=1$, for convenience expressed in the coordinates $Y_1=X_1+X_2$, $Y_2=X_1-X_2$. Then every member of the family satisfies (34) with

$$T(Y_1,Y_2) = (Y_1, -Y_2) \tag{35}$$

The observation of period doubling for this family $\alpha \to f_\alpha$ has an interesting consequence. Namely, let ν be a normalization such that $\nu(f^{-1}) = \nu(f)$. Such a choice is possible by using coordinates of some period 2^q. Then we choose $f \to \Lambda_f$ by the condition $L(\mathcal{N}(f)) = L(f)$ where $L = L(f)$ is defined by $\nu(L^{-1} \circ f \circ L) = \nu_0$. The sequence $\varphi_k = \mathcal{N}^k(f_\alpha)$ should now converge for $\alpha=\alpha_\infty$. By using that the symmetry (34) is invariant under composition it is easy to see that all scaling transformations Λ_{φ_k} commute with T. Therefore the limit $\varphi = \lim \varphi_k$ satisfies

$$\varphi = \Lambda_{\varphi}^{-1} \circ \varphi \circ \varphi \; \Lambda_{\varphi} \quad , \quad \varphi \circ T \circ \varphi = T \; , \quad \Lambda_{\varphi} \circ T = T \circ \Lambda_{\varphi} \tag{36}$$

We shall now restrict our analysis to maps f satisfying $f \circ T \circ f = T$ with T given by (35), and to scaling transformations Λ_f commuting with T. An argument given in Section III shows that it is sufficient to take Λ_f of the form

$$\Lambda_f(X_1,X_2) = (\lambda(f)X_1 + c(f), \;\; \mu(f)X_2) \; , \; c(\varphi) = 0 \tag{37}$$

with λ, c, μ defined by some normalization condition. According to the renormalization group philosophy the universal scaling factors λ, μ are to be identified with $\lambda(\varphi)$, $\mu(\varphi)$. In fact it can be proved [17] that there is an area preserving map φ satisfying (36), and that

$$|\lambda(\varphi) + .28875| < 2.10^{-5}, \quad |\mu(\varphi) - .061110| < 3.10^{-6}. \tag{38}$$

We shall be more precise later on, and perform first a change of variables $U : S \to f$ in function space . In view of the constraint $\det(Df) \equiv 1$ it is natural to work with a generating function (see also [29]). The condition $f^{-1} = T \circ f \circ T$ imposes a simple symmetry on the graph of $f = (f_1, f_2)$. This symmetry implies

$$\partial_1 S(X_1,Y_1) = \partial_1 S(Y_1,X_1) \tag{39}$$

for the generating function S which satisfies

$$X_2 = - S(f_1(X_1,X_2), X_1), \quad f_2(X_1,X_2) = S(X_1,f_1(X_1,X_2)) \tag{40}$$

Next we define an expression for the new renormalization transformation $\mathcal{T} = U^{-1} \circ \mathcal{N} \circ U$. Let $Z = f \circ \Lambda_f(X)$ and $Y = \Lambda_f^{-1} \circ f(Z)$. Then $Y = \mathcal{T}(S)(X)$, and we get the relations

$$Z_2 = S(\lambda(S)X_1 + c(S), Z_1) = - S(\lambda(S)\, Y_1 + c(S), Z_1), \tag{41}$$

$$\mu(S)\, Y_2 = S(Z_1,\, \lambda(S)\, Y_1 + c(S)) = \mu(S)\, \mathcal{T}(S)(X_1,Y_1), \qquad (42)$$

and a similar equation with $\mu(S)X_2$. By using (41) it is possible to express Z_1 in terms of S,X_1,Y_1. The equation

$$S(\lambda(S)X_1 + c(S),\; Z_1(S;\, X_1,Y_1)) + S(\lambda(S)Y_1 + c(S),\; Z_1(S;\, X_1,Y_1)) = 0 \qquad (43)$$

defines a symmetric function $Z_1(S)$. Inserting this Z_1 into (42) gives

$$\mathcal{T}(S)(X_1,Y_1) = \frac{1}{\mu(S)}\; S(Z_1(S;\, X_1,\, Y_1),\, \lambda(S)\, Y_1 + c(S)). \qquad (44)$$

For simplicity we now set $c(.) \equiv 0$. Let $S(1,0) = 1$ and $\partial_2 S(1,0) = \frac{1}{5}$. Then we determine λ, μ by imposing the same normalization conditions on $\mathcal{T}(S)$. A short calculation shows that this leads to the equations $Z_1(S;\, 1,0) = 1$ and

$$S(\lambda(S),\, 1) + S(0,1) = 0, \qquad (45)$$

$$\mu(S) = \lambda(S) + 5\partial_1\, S(1,0)\; \partial_1 Z_1(S;\, 1,0) \qquad (46)$$

Given now a fixed point S^* of $\mathcal{T}$, a fixed point $\varphi = U(S^*)$ for $\mathcal{N}$ can be reconstructed by using (40). Let $\mathcal{B}''$ denote the Banach **space** of analytic functions

$$S(X_1,Y_1) = \Sigma C_{ij} \left(\frac{X_1 - .5}{1.6}\right)^i \left(\frac{Y_1 - .5}{1.6}\right)^j \;,\quad C_{ij} \in \mathbb{R}, \qquad (47)$$

equipped with the norm $\|S\|_1 = \Sigma |C_{ij}|$, and let $\mathcal{B}'$ denote the affine subspace defined by the above normalization conditions and by (39).

<u>Theorem 6</u> [17] : $\mathcal{T}$ <u>is a</u> C^∞ <u>map from an open domain</u> $\mathfrak{D}_{\mathcal{T}} \subset \mathcal{B}'$ <u>into</u> $\mathcal{B}'$ <u>and has a unique fixed point</u> S^* <u>in</u> $\mathfrak{D}_{\mathcal{T}}$. <u>The scaling factors</u> $\lambda(S^*)$ <u>and</u> $\mu(S^*)$ <u>satisfy the bound (38). Furthermore the spectrum of</u> $D\mathcal{T}_{S^*}$ <u>is contained in a disc</u> $K_\rho = \{\, \xi \in \mathbb{C} : |\xi| < \rho \,\}$ <u>of radius</u> $\rho = .85$, <u>except for two simple eigenvalues</u>, $\lambda(S^*)^{-1}$ <u>and</u> $\delta \in [\,8.2,\, 9.2\,]$.

Note that the eigenvalue λ^{-1} only arises since there **is** no compensation for translations in the 1-direction. This eigenvalue disappears if $C(\cdot) \equiv 0$ is replaced by an additional normalization $S(0,1) = S^*(0,1)$.

e shall now outline the proof of the existence of S^*. The other tatements are proved similarly.

Our first step is the construction of a contraction $\mathcal{M}$, which has he same fixed point as $\mathcal{T}$. The definition of $\mathcal{M}$ is inspired by the fact hat, using the language of section V, the sequence $\mathcal{N}^k(f_{\alpha_k})$ converges o φ if $\tilde{\Sigma}$ crosses $\mathcal{W}^u$ transversally in φ. Let us consider the manifold

$$\tilde{\Sigma} = \{S \in \mathcal{B}' : \mathcal{T}(S)(0,1) = S(0,1),\ \partial_2\mathcal{T}(S)(0,1) = \partial_2 S(0,1)\} \tag{48}$$

it passes through the fixed point S^* and is chosen of codimension two, since we expect $D\mathcal{T}_{S^*}$ to have two unstable directions. To every $S \in \mathcal{B}'$ we now associate a two-parameter family $\{S_\alpha\}$, where

$$S_\alpha(X_1, Y_1) = S(X_1,Y_1) + a(X_1 - 1) + bY_1^2 , \qquad \alpha = (a,b) \tag{49}$$

Then the transformation $\mathcal{M}$: $S \to (\{S_\alpha\} \cap \tilde{\Sigma})$ has the desired property with two eigenvalues zero replacing λ^{-1} and δ. It is convenient to project everything onto the linear subspace $\mathcal{B} \subset \mathcal{B}''$ of functions σ which satisfy $\partial_1\sigma(X_1,Y_1) = \partial_1\sigma(Y_1,X_1)$ and $\sigma(1,0) = \sigma(0,1) = \partial_2\sigma(1,0) = \partial_2\sigma(0,1) = 0$. Let $\tau(X_1,Y_1) = (X_1+Y_1-1)/5$. Then our final transformation is

$$\mathcal{M}(\sigma) = \mathcal{T}((\sigma + \tau)_{\alpha_1}) - \tau_{\alpha_1} \tag{50}$$

where α_1 is the value of the parameter α at which the family $\alpha \to (\sigma+\tau)_\alpha$ crosses $\tilde{\Sigma}$. Note that $S^* = (\sigma^*+\tau)_{\alpha_1}$ is a fixed point for $\mathcal{T}$ if $\mathcal{M}(\sigma^*) = \sigma^*$.

The existence of a fixed point for $\mathcal{M}$ is derived from the two bounds

$$\|\mathcal{M}(\sigma_0) - \sigma_0\| < 1.3\ 10^{-8} , \tag{51}$$

$$\|D\mathcal{M}_\sigma\| < .55 \qquad \text{for} \qquad \|\sigma - \sigma_0\| < 5.10^{-8} \tag{52}$$

where σ_0 is a given polynomial of degree 20 and where

$$\|\sigma\| = |c_{02}| + \sum_{i+j \geq 3} |c_{ij}|$$

These estimates ensure that a ball of radius $r = 5.10^{-8}$ around σ_o is contracted into itself by $\mathcal{M}$. Following ideas of Lanford [31], we prove the inequalities (51) and (52), simply speaking, by reducing them to about a hundred billion inequalities of the form $a+b < c$ or $a.b < c$. These relations are **then** shown to be consistent by constructing a set of numbers satisfying them. This can be done on a computer, if the rounding errors are taken into account correctly.

The approximate fixed point σ_o is obtained by **writing** a computer program which manipulates polynomials of two variables, truncating everything at a given degree d. In this approximation $\mathcal{M} = \mathcal{M}^{\text{trunc.}}$ is a nonlinear map on the space R^N of Taylor coefficients C_{ij}, and its differential is a matrix. Then one simply iterates $\mathcal{M}^{\text{trunc.}}$ a sufficient number of times, starting e.g. with $\sigma \equiv 0$. In a subsequent step every number C is replaced by an interval $\overline{C} = [C_-, C_+]$, and interval operations $\overline{a}\overline{+}\overline{b}$, $\overline{a} \overline{*} \overline{b}$, ... are introduced in such a way that $a+b \subset \overline{a}\overline{+}\overline{b}$, $a.b \subset \overline{a} \overline{*} \overline{b}$, ... for every $a \in \overline{a}$ and $b \in \overline{b}$. Within the polynomial representation, a neighbourhood $\overline{\sigma}$ around σ_o can then be represented as a vector of N intervals $\overline{C}_{ij} = [C_{ij}-r, C_{ij} + r]$ centered at the coefficients of σ_o, and the set of operators $D\mathcal{M}_\sigma^{\text{trunc.}}$, for $\sigma \in \overline{\sigma}$, can be viewed as a matrix of intervals. Such a program provides no strict equalities, but only inclusion relations. It is now clear how (51) and (52) could be proved rigorously for $\mathcal{M} = \mathcal{M}^{\text{trunc.}}$. But note that the incorporation of an unspecified function into N identical coefficients $[-r,r]$ leads to bad estimates. Therefore we introduce a new interval $\overline{r} = [0,r]$ representing the ball $R = \{ h \in \mathcal{B}'' : \|h\|_1 \leq r \}$. Similarly the set E of functions of degree $> d$ and with norm $\leq e$ is represented by the interval $\overline{e} = [0,e]$. The basic object in the final program is then the pointwise sum of R,E and a set of polynomials, implemented as a vector $\overline{\sigma} = (\overline{r}, \overline{e}, \{\overline{C}_{ij}\})$ of 2(N+2) numbers. The previous program has to be complemented by error-estimating parts. This includes the change of domains for derivatives and compositions of $\overline{\sigma}$'s, the accumulation of truncation errors into $\overline{e}$, bound for implicitely defined quantities,... . More details can be found in [17].

References

[1] V. Arnold, Chapitres supplémentaires de la théorie des équations différentielles ordinaires, Editions MIR, Moscou 1980.

[2] G. Benettin, C. Cercignani, L. Galgani, A.Giorgilli, Lett. Nuovo Cimento 28, 1(1980).

[3] M. Campanino, H. Epstein, Comm. Math. Phys. 79, 261 (1980).

[4] P. Collet, J-P. Eckmann H. Koch, J. Stat. Phys. 25, 1 (1981).

[5] P. Collet, J-P.Eckmann, H. Koch, Physica 3D, 457 (1981).

[6] P. Collet, J-P. Eckmann, O.E. Lanford, Commun. Math. Phys. 76, 211 (1980).

[7] P. Collet, J-P. Eckmann, Commun. Math. Phys. 73, 115 (1981).

[8] P. Collet, J-P. Eckmann, L. Thomas, Commun. Math. Phys. 81, 261 (1981).

[9] J.P. Crutchfield, M. Nauenberg, J. Rudnick, Phys. Rev. Lett. 46, 933 (1980).

[10] J.P. Crutchfield, J.D. Farmer, B.A. Hubermann, Preprint, University of California at Santa Cruz (1981).

[11] V. Croquette, C. Poitou, J. Phys. Lett. (Paris) 42, 537 (1981).

[12] P. Coullet, J. Tresser, C.R. Acad. Sci. 287, 577 (1978).

[13] B. Derrida, Y. Pomeau, Phys. Lett. 80A, 217 (1980).

[14] B. Derrida, A. Gervois, Y. Pomeau, J. Phys. A12, 269 (1979).

[15] R. De Vogelaere, In "Contributions to the theory of Nonlinear Oscillations", Vol. IV, S. Lefschetz ed. Princeton University Press, Princeton 1958.

[16] J-P. Eckmann, Rev. Mod. Phys. 53, 643 (1981).

[17] J-P. Eckmann, H. Koch, P. Wittwer, Preprint UGVA-DPT/04-345, University of Geneva 1982.

[18] H. Epstein, J. Lascoux, Commun. Math. Phys. 81, 437 (1981).

[19] M.J. Feigenbaum, J. Stat. Phys. 19, 25 (1978); 21, 669 (1979).

[20] M.J. Feigenbaum, Commun. Math. Phys. 77, 65 (1980).

[21] V. Franceschini, C. Tebaldi, J. Stat. Phys. 21, 707 (1979).

[22] J.M. Greene, R.S. Mackay, F. Vivaldi, M. Feigenbaum, Physica 3D, 468 (1981).

[23] M. Giglio, S. Muzzati, U. Perini, Phys. Rev. Lett. 47, 243 (1981).

[24] H.M. Gibbs, F.A. Hopf, O.L. Kaplan, R.L. Shoemaker, Phys. Rev. Lett. 46, 474 (1981).

[25] G. Gollub in Proceedings of Les Houches Summer School 1981, R. Helleman and G. Ioss editors, to appear.

[26] M. Hirsch, C. Pugh, M. Shub, Invariant manifolds . Lecture notes in Mathematics, Vol. 583, Springer Verlag Berlin, Heidelberg, New York (1977).

[27] A. Ito, Prog. Theor. Phys. 61, 815 (1979).

[28] M.V. Jakobson, Commun. Math. Phys. 81, 39 (1981).

[29] L.P. Kadanoff, M. Widom, Physica D, to appear.

[30] P.S. Linsay, Phys. Rev. Lett. 47, 1349 (1981).

[31] O.E. Lanford III, Bull. Amer. Math. Soc. 6, (New Series) 427 (1982).

[32] O.E. Lanford III, To appear.

[33] W. Lautenborg, E. **Cramer,** Phys. Rev. Lett. 47, 1445 (1981).

[34] A. Libchaber, C. Laroche, S. Fauve, J. Phys. Lett. (Paris) 43, 211 (1982).

[35] A. Libchaber, J. Maurer , J. Phys. (Paris) 41, C3, 51 (1980).

[36] R.M. May, Nature 261, 459 (1976).

[37] M. **Nauenberg**, J. **Rudnick**, Preprint, University of California at Santa Cruz (1981).

[38] J. Perez, J. Testa, C. Jeffries, Phys. Rev. Lett. 48, 714 (1982).

[39] D. Ruelle, Commun. Math. Phys. 82, 137 (1981).

[40] B. Shraimam, C.E. **Wayne,** P.C. Martin, Phys. Rev. Lett. 46, 935 (1981).

[41] H.L. Swinney, J.C. Roux, Topology of chaos in chemical reaction. In "nonlinear Phenomena in Chemical Dynamics", C. Vidal and A. Pacault editors. Springer Verlag, Berlin-Heidelberg-New York 1981.

P. Collet

Physique Théorique
Ecole Polytechnique
91128 Palaiseau-France

H. Koch

Lyman Laboratory
Harvard University
Cambridge, Mass. 02138 USA

*Work supported by NSF Grant PHY 79-16812

BOWEN'S FORMULA FOR THE HAUSDORFF DIMENSION OF SELF-SIMILAR SETS

David RUELLE

1. Introduction and statement of results.

Geometric self-similarity of a set J (with a metric) means that the microscopic structure of the set (in any neighborhood of any point) can be magnified to ressemble the macroscopic structure of the set. The Hausdorff dimension t of J is then a natural notion, as stressed by Mandelbrot [5]. Bowen's formula expresses t in terms of concepts of statistical mechanics, and can be used to show that the dimension t is not an integer (see Bowen [3]) or that it depends smoothly on parameters (see Ruelle [7]).

The definition of Hausdorff dimension is as follows (see Billingsley [1]). For $\alpha > 0$ and $\delta > 0$ let

$$\ell_\alpha(J,\delta) = \inf \sum_i (\mathrm{diam}\ B_i)^\alpha$$

where the infimum is over all coverings of S by countably many balls B_i with diameter $\leq \delta$. Define also

$$\ell_\alpha(J) = \lim_{\delta\to 0} \ell_\alpha(J,\delta)$$

The Hausdorff dimension $t = \dim J$ is the infimum of those α such that $\ell_\alpha = 0$. If $\beta < t$, then $\ell_\beta = \infty$. When $\ell_t(J)$ is neither 0 nor ∞ , the t-Hausdorff measure is defined by the set function $S \mapsto \ell_t(S)$ for $S \subset J$.

We shall assume that J is a compact subset of a Riemann manifold M , and is invariant under an expanding map f . More precisely, we shall assume that $f : M \to M$ is a map of class $C^{1+\varepsilon}$ *) and that J has an open neighborhood V in M with the following properties

(a) There exist $C > 0$ and $\alpha > 1$ such that **)

$$\|(T_x f^n)u\| \geq C\alpha^n \|u\|$$

for all $x \in J$, $u \in T_x M$, $n \geq 1$.

(b) $J = \{x \in V : f^n x \in V \text{ for all } n > 0\}$

(c) f is topologically mixing on J, i.e., for every non-empty open set $\mathcal{O}$ intersecting J there is $n > 0$ such that $f^n \mathcal{O} \supset J$.

We say that J is a (mixing) repeller for f. From (b) and (c) it follows that $fJ = J$. Note that f is in general many to one.

We shall express self-similarity of J by the fact that the map f is conformal, i.e., the tangent map $T_x f$ is (for each x) a scalar times an isometry.

Theorem. Let J be a repeller for a map $f : V \to M$. We assume that f is conformal with respect to some continuous Riemann metric, and of class $C^{1+\varepsilon}$ $(\varepsilon > 0)$. If we write

$$\varphi(x) = -\log \|T_x f\|$$

the Hausdorff dimension t of J is defined by Bowen's formula

$$P(t\varphi) = 0$$

where P is the pressure functional (see below). Furthermore the t-Hausdorff measure ν on J is equivalent to the Gibbs measure ρ corresponding to $t\varphi$.

The main application of the above theorem is to the case where M

---

*) This means that f has derivatives which are Hölder continuous of exponent $\varepsilon > 0$

**) Here $T_x f$ denotes the tangent map to f at x (the derivative at x if M is $\mathbb{R}^N$).

is the Riemann sphere, f a rational function, and J the Julia set *) (assumed to be hyperbolic) **). Bowen's original application was to a discontinuous f (see [3]); his proof can readily be adapted to the present situation.

If $\psi : J \to \mathbb{R}$ is a continuous function, the pressure $P(\psi)$ may be defined by

$$P(\psi) = \max \{h(\sigma) + \sigma(\psi) : \sigma \text{ is an } f\text{-invariant probability measure}\}$$

where $h(\sigma)$ is the entropy (Kolmogorov-Sinai invariant) of σ with respect to f. The functional P is convex. If $\varphi : J \to \mathbb{R}$ is Hölder continuous, there is a unique Radon measure ρ on J such that

$$P(\varphi+\psi) - P(\varphi) \geq \rho(\psi)$$

for all ψ, and ρ is an f-invariant probability measure, which we call Gibbs measure associated with φ.

The above notions of pressure and Gibbs measure reduce to the similar concepts for a one-dimensional lattice spin system, thanks to the existence of Markov partitions of J. A Markov partition $\{S_\alpha\}$ is a finite collection of closed non-empty subsets of J such that $\cup S_\alpha = J$ and $\operatorname{int} S_\alpha$ is dense in S_α (int denotes the interior in J). Furthermore,

(i) $\operatorname{int} S_\alpha \cap \operatorname{int} S_\beta = \emptyset$ if $\alpha \neq \beta$

(ii) each $f S_\alpha$ is a union of sets S_β.

A sequence $(S_{\alpha_i})_{i \geq 0}$, where $f S_{\alpha_i} \supset S_{\alpha_{i+1}}$ for each i, determines a unique point $x \in J$ such that $f^i x \in S_{\alpha_i}$ for all i. The sequences (S_{α_i}) may be considered as configurations of a spin system in one dimension, and the pressure and Gibbs measure are then identified with familiar concepts of statistical mechanics. For more details, the reader must be referred to Bowen [2] or Ruelle [6].

---

*) See Brolin [4].

**) A formal calculation in [7] shows that when $f(z) = z^q + \lambda$, then $t = 1 + \frac{|\lambda^2|}{4 \log q}$ + higher order in λ.

2. Proof of the theorem.

Let $\{S_\alpha\}$ be a Markov partition of J into small subsets. We call K the maximum number of S_β which intersect any S_α :

$$K = \max_\alpha \operatorname{card} \{S_\beta : S_\alpha \cap S_\beta \neq \emptyset\} .$$

Let $\tilde{S}_\alpha$ be a small open neighborhood of S_α in V, for each α, such that $\tilde{S}_\alpha \cap \tilde{S}_\beta = \emptyset$ whenever $S_\alpha \cap S_\beta = \emptyset$. We assume that for all α the diameter of $\tilde{S}_\alpha$ is $< \Delta$, and that $\tilde{S}_\alpha$ contains the δ-neighborhood of S_α $(0 < \delta < \Delta)$. If $\xi_0, \xi_1, \ldots, \xi_n$ is an admissible sequence of elements of the Markov partition, i.e., $f\xi_{j-1} \supset \xi_j$ for $j = 1,\ldots,n$, we define

$$E(\xi_0,\ldots,\xi_n) = \bigcap_{j=0}^{n} f^{-j}\xi_j$$

$$\tilde{E}(\xi_0,\ldots,\xi_n) = \bigcap_{j=0}^{n} f^{-j}\tilde{\xi}_j$$

The sets $\tilde{E}(\xi_0,\ldots,\xi_n)$ which intersect a given $\tilde{E}(\xi_0^*,\ldots,\xi_n^*)$ are determined successively as follows :

(a) choose ξ_n such that $\xi_n \cap \xi_n^* \neq \emptyset$

(b) ξ_j is uniquely determined for $k = n-1,\ldots,1, 0$ by

$$\Big[\bigcap_{j=k}^{n} f^{-(j-k)}\xi_j\Big] \cap \Big[\bigcap_{j=k}^{n} f^{-(j-k)}\xi_j^*\Big] \neq \emptyset .$$

In particular the sets $\tilde{E}(\xi_0,\ldots,\xi_n)$ which intersect $\tilde{E}(\xi_0^*,\ldots,\xi_n^*)$ correspond precisely to the sets $E(\xi_0,\ldots,\xi_n)$ which intersect $E(\xi_0^*,\ldots,\xi_n^*)$, and there are at most K of those. We also see that, if Δ has been taken sufficiently small, there are $\beta \in (0,1)$ and $G > 0$ (β and G independent of n, $\xi_0^*,\ldots,\xi_n^*$) such that

$$\operatorname{dist}(\xi,\xi^*) \leq G\beta^n \text{ if } \tilde{\xi} \in \tilde{E}(\xi_0,\ldots,\xi_n) \text{ and } \xi^* \in \tilde{E}(\xi_0^*,\ldots,\xi_n^*) \qquad (1)$$

[use part (a) of the definition of a repeller]. In particular, $\operatorname{diam} \tilde{E}(\xi_0^*,\ldots,\xi_n^*) \leq G\beta^n$.

Let $F_{\xi_0,\ldots,\xi_n} : \tilde{\xi}_n \mapsto \tilde{E}(\xi_0,\ldots,\xi_n)$ be the inverse of the restriction of f^n to $\tilde{E}(\xi_0,\ldots,\xi_n)$. If $x \in \tilde{\xi}_n$ we have, since f is conformal,

$$\log \|F'_{\xi_0,\dots,\xi_n}(x)\| = \sum_{k=0}^{n-1} \log \|(f^{-1})'(F_{\xi_{k+1},\dots,\xi_n}(x))\|$$

$$= -\sum_{k=0}^{n-1} \log\| f'(F_{\xi_k,\dots,\xi_n} x)\|$$

$$= \sum_{k=0}^{n-1} \varphi(F_{\xi_k,\dots,\xi_n} x) \tag{2}$$

where we have denoted the tangent map by a dash. If $\widetilde{E}(\xi_0,\dots,\xi_n) \cap \widetilde{E}(\xi_0^*,\dots,\xi_n^*) \neq \emptyset$, and $x \in \widetilde{\xi}_n$, $x^* \in \widetilde{\xi}_n^*$ we have thus, using (1),

$$|\log \|F'_{\xi_0,\dots,\xi_n}(x)\| - \log \|F'_{\xi_0^*,\dots,\xi_n^*}(x^*)\| |$$

$$\leq C_\varepsilon \sum_{k=0}^{n-1} (G\beta^{n-k})^\varepsilon < \frac{C_\varepsilon G^\varepsilon}{1-\beta^\varepsilon} = D \tag{3}$$

where C_ε is the ε-Hölder norm of φ . In particular, <u>if</u> $x^* \in \xi_n$, <u>the ball of radius</u> $e^{-D}\delta \| F'_{\xi_0^*,\dots,\xi_n^*}(x^*)\|$ <u>centered at</u> $F_{\xi_0^*,\dots,\xi_n^*} x^*$ <u>is entirely contained in</u> $\widetilde{E}(\xi_0^*,\dots,\xi_n^*)$ *) .

The Gibbs measure ρ corresponding to $t\varphi$ is determined (since $P(t\varphi) = 0$) by the fact that there is a constant γ such that **)

$$|\log \rho(E(\xi_0,\dots,\xi_n)) - \sum_{k=0}^{n-1} t\varphi (F_{\xi_k,\dots,\xi_n} x)| < \gamma \tag{4}$$

where γ is independent of n , $E(\xi_0,\dots,\xi_n)$, and $x \in \xi_n$. Using (2) and (4) we have, for each $E(\xi_0,\dots,\xi_n)$, the following estimate of the t-Hausdorff measure ν :

$$\nu(E(\xi_0,\dots,\xi_n))$$

---

*) We assume here for simplicity that $\varphi < 0$.

**) See Bowen [2] or Ruelle [6] .

$$\leq \lim_{p\to\infty} \sum_{\xi_{n+1}\ldots\xi_{n+p}} (\operatorname{diam} \tilde{E}(\xi_o,\ldots,\xi_{n+p}))^t$$

$$\leq \lim_{p\to\infty} \sum_{\xi_{n+1}\ldots\xi_{n+p}} (2\Delta e^D \| F'_{\xi_o,\ldots,\xi_{n+p}}(f^p x) \|)^t$$

$$\leq (2\Delta e^D)^t \lim_{p\to\infty} \sum_{\xi_{n+1}\ldots\xi_{n+p}} \exp \sum_{k=0}^{n+p-1} t\varphi(F_{\xi_k,\ldots,\xi_{n+p}} f^p x)$$

$$\leq (2\Delta e^D)^t \, e^\gamma \rho(E(\xi_o,\ldots,\xi_n)) .$$

This shows that ν is absolutely continuous with respect to ρ .

On the other hand $\nu(E(\xi_o,\ldots,\xi_n))$ is the infimum of $\sum_{j=1}^{\infty} (\operatorname{diam} U_j)^t$ for an open cover $\{U_j\}$ of $E(\xi_o,\ldots,\xi_n)$ when $\operatorname{diam} U_j \to 0$. For each j take $y_j \in E(\xi_o,\ldots,\xi_n) \cap U_j$, and notice that $E(\xi_o,\ldots,\xi_n)$ is covered by the balls $B_{y_j}(\operatorname{diam} U_j)$. For each j let n_j be the smallest integer such that if $y_j \in E(\xi^*_o,\ldots,\xi^*_{n_j+1})$ then

$$e^{-D}\delta \,\| F'_{\xi^*_o,\ldots,\xi^*_{n_j+1}}(f^{n_j+1} y_j)\| \leq \operatorname{diam} U_j . \tag{5}$$

(We may assume that $\operatorname{diam} U_j$ is small, and therefore $n_j > n$, $\xi^*_o = \xi_o,\ldots,\xi^*_n = \xi_n$, the further ξ_k depend on j). By assumption

$$e^{-D}\delta \,\| F'_{\xi^*_o,\ldots,\xi^*_{n_j}}(f^{n_j} y_j)\| > \operatorname{diam} U_j .$$

Therefore, the set $E(\xi_o,\ldots,\xi_n)$ is covered by the $\tilde{E}(\xi^*_o,\ldots,\xi^*_{n_j})$ and, using (5) and (2), we see that

$$\sum_{j=1}^{\infty} (\operatorname{diam} U_j)^t$$

$$\geq e^{-Dt}\delta^t \sum_{j=1}^{\infty} \exp t \sum_{k=0}^{n_j} \varphi(F_{\xi^*_k,\ldots,\xi^*_{n_j+1}} f^{n_j+1} y_j)$$

$$\geq e^{-Dt\,-Et}\,\delta^t \sum_{j=1}^{\infty} \exp t \sum_{k=0}^{n_j-1} \varphi(F_{\xi^*_k,\ldots,\xi^*_{n_j}} f^{n_j} y_i)$$

where E is an upper bound to $|\varphi(x)|$. We recall that each

$\tilde{E}(\xi^*_o,\ldots,\xi^*_{n_j})$ intersects at most K sets $E(\xi_o,\ldots,\xi_{n_j})$. Redistributing the contribution of the index j among those, and using (2) and (3) we find

$$\sum_{j=1}^{\infty} (\text{diam } U_j)^t \geq K^{-1} e^{-2Dt -Et} \delta^t \sum_{\lambda} \exp t \sum_{k=0}^{n_\lambda - 1} \varphi(F_{\xi^\lambda_k,\ldots,\xi^\lambda_{n_\lambda}} x_\lambda)$$

where the $E(\xi^\lambda_o,\ldots,\xi^\lambda_{n_\lambda})$ cover $E(\xi_o,\ldots,\xi_n)$. So, finally, using (4), we obtain

$$\nu(E(\xi_o,\ldots,\xi_n)) \geq K^{-1} e^{-2Dt -Et} \delta^t e^{-\gamma} (E(\xi_o,\ldots,\xi_n))$$

This shows that ρ is absolutely continuous with respect to ν , completing the proof of the proposition.

REFERENCES.

[1] P. Billingsley. Ergodic theory and information. John Wiley New York, 1965.

[2] R. Bowen. Equilibrium states and the ergodic theory of Anosov diffeomorphisms. Lecture Notes in Math. N° 470. Springer, Berlin, 1975.

[3] R. Bowen. Hausdorff dimension of quasi-circles. Publ. math. IHES 50, 11 - 26 (1976).

[4] H. Brolin. Invariant sets under iteration of rational functions. Arkiv för Mat. 6, 106 - 144 (1965).

[5] B. Mandelbrot. Fractals : form, chance, and dimension. W.H. Freeman, San Francisco, 1977.

[6] D. Ruelle. Thermodynamic formalism. Addison-Wesley, Reading, 1978.

[7] D. Ruelle. Repellers for real analytic maps. Ergodic Theory and dynamical systems. To appear.

Perturbation Theory for

Classical Hamiltonian Systems.

Giovanni Gallavotti*
I.H.E.S., Bures-sur-Yvette, France 91440
Math. Dept., Princeton University
Princeton, New Jersey 08544
U.S.A.

Abstract: the relation between the Kolmogorov-Arnold-Moser theory of the non resonant motions in nearly integrable Hamiltonian systems and the renormalization group methods is pointed out. It is followed by a very detailed proof of a version of the KAM theorem based on dimensional estimates (in which no attention is paid to obtaining best constants).

* On leave from Universitá di Roma. The author gratefully acknowledges the Stiftung Volkswagenwerk for partial financial support, through I.H.E.S. Permanent address: Istituto Matematico, Universitá di Roma, Moro 5, 00185 Roma, Italy.

§1. Introduction

The theory of the perturbations of classical Hamiltonian systems predicts the existence of some nontrivial prime integrals for the motions of systems close to integrable ones: therefore such systems are still not ergodic, contrary to the hope, widespread until recently that this would not be the case.

Here we propose an analysis of such prime integrals with special attention to their smoothness properties. Although such an analysis is certainly implicit in the basic papers of the KAM-theory, [1], [2], and [3], it was not explicitly done in them and it is certainly worth it to be brought to the attention of the physicists as it is there that, in my view, the physical meaning of the results appears more clearly.

The purpose of these lectures is mainly to give a complete treatment of the KAM theory for anisochronous systems in the special case of analytic systems. The analytic case can be treated with very few calculations basing all the hard estimates on "dimensional bounds" (i.e. bounds in which a derivative of a holomorphic function at a point is bounded by its supremum in a region divided by the distance of the point to the boundary of the same region). It seems of some interest to collect a consistent treatment of the problem, along these intuitive lines, not referring to previous literature (which is the usual obstacle encountered by those who study this subject for the first time). I hope that in this way I shall make clear that the KAM theory can be understood basically without any calculations by a physicist familiar with classical perturbation theory (reviewed in §2). The fact that for various reasons this theory has become forgotten by most theoretical physicists can be the only explanation of the fact that such a beautiful and rich subject has so far remained in the area of pure mathematics and mathematical physics

I will therefore follow [5] quite closely, providing all the details of the statements which in [5] refer to the existing literature and discussing with some care the motivations of the various steps. Occasionally the discussion is simplified with respect to [5], often improving it and only seldom getting slightly worse bounds at the advantage of a simpler analysis.

The results on smoothness were derived explicitly and independently in [4] and [5]: in [4] the attention is devoted to the treatment of the differentiable case which is much harder since, now, the dimensional bounds do not directly help.

Another purpose of these notes is to stress the connection between the KAM theory and the general class of theories (and methods) which in the last decade has become known among physicists as the "renormalization group" methods.

The "renormalization group" can be loosely described as follows: one starts with the aim of studying some properties of a certain function H and then performs a change of variables transforming the initial problem into an identical problem in terms of a new function $H' = KH$. The transformation K is chosen such that after a few iterations (or at least in the limit $n \to \infty$) $H^{(n)} = K^n H$ becomes treatable by some other techniques.

In the successive changes of coordinates there is some loss of information due to the fact that the changes of coordinates will not, as a rule, be one to one or everywhere defined (otherwise the problem could not really become simpler in the new coordinates).

The KAM theory, described below independently of the previous literature, provides a good example of how the above program can be enforced in a particular case. For a general overview of the renormalization group in other areas see [6] and for some mathematically rigorous versions and applications see [7].

Let us set up some notation. We consider a classical system with ℓ degrees of freedom which is integrable and, therefore, can be described in action-angle variables $(\underline{A},\underline{\phi}) \in V \times T^\ell$, where V = {open sphere of radius r in R^ℓ}* and T^ℓ = {ℓ-dimensional torus} = {$[0,2\pi]^\ell$} with opposite sides identified , by a Hamiltonian which

---

*we shall use the distance $|\underline{x}| = \sum_{i=1}^{d} |x_i|$ in R^d: so a sphere is in fact a square if $d = 2$, an octahedron for $d = 3$, and, its volume is $2^d r^d (d!)^{-1}$, if r is its radius.

depends only on $\underline{A}$:

$$H_0(\underline{A},\underline{\phi}) = h_0(\underline{A}) \tag{1.1}$$

which we call the "free Hamiltonian".

If the free system is perturbed via an analytic perturbation f_0 ("interaction between the normal modes of (1.1)") we shall consider the Hamiltonian motions ("flow") on $V{\times}T^{\ell}$ associated with the Hamiltonian

$$H_0(\underline{A},\underline{\phi}) = h_0(\underline{A}) + f_0(\underline{A},\underline{\phi}). \tag{1.2}$$

The functions h_0 and f_0 have many features which are irrelevant for the analysis and we shall describe the relevant ones in terms of few parameters:

1) a parameter E_0 measuring the size of h_0 (or, better, of it's gradient: recall that the motion is described in terms of the gradient of the Hamiltonian).

2) a parameter ε_0 measuring the size of f_0 (or, better, of it's gradient).

3) a parameter η_0 measuring the anisochrony of the quasiperiodic motions of the free Hamiltonian.

4) a parameter ρ_0 measuring the analyticity radius in the action variables ("radial analyticity").

5) a parameter ξ_0 measuring the analyticity in the angle variables ("toral analyticity").

To introduce such parameters unambiguously we need some notation.

First notice that f_0 can be written as a Fourier's series:

$$f_0(\underline{A},\underline{\phi}) = \sum_{\underline{\gamma}\in Z^{\ell}} f_{0\underline{\gamma}}(\underline{A})\, e^{i\,\underline{\gamma}\cdot\underline{\phi}} \tag{1.3}$$

which suggests introducing the variables

$$\underline{z} = (z_1,\dots,z_{\ell}) = (e^{i\phi_1},\dots,e^{i\phi_{\ell}}) \tag{1.4}$$

and the notation

$$\underline{z}^{\underline{\gamma}} = \prod_{i=1}^{\ell} z_i^{\gamma_i}\,. \tag{1.5}$$

This allows us to regard f_0 as defined on $C^{2\ell}$. Henceforth we identify T^ℓ with the set of the points of the form (1.4) in C^ℓ. By requiring that f_0 is analytic we shall mean that f_0 is extendible to a holomorphic function on $C^{2\ell}$ in the vicinity of VxT^ℓ, more precisely to the set

$$W(\rho_0,\xi_0;V) = \bigcup_{\underline{A}\in V} C(\rho_0,\xi_0;\underline{A})$$

$$C(\rho_0,\xi_0;\underline{A}) = \{(\underline{A}',\underline{z})\,|\,(\underline{A}',\underline{z}) \in C^{2\ell}, e^{-\xi_0} < |z_j| < e^{\xi_0}, \qquad (1.6)$$

$$|A'_j - A_j| < \rho_0,\ \forall\ j = 1,\dots,\ell\}$$

(the multiannulus $C(\rho_0,\xi_0;\underline{A})$ will be our standard complex neighborhood of VxT^ℓ). The parameters $\rho_0 > 0$, $\xi_0 > 0$ are a natural measure of the radial and of the toral analyticity of f_0. We assume for simplicity that $\rho_0 < r$.

To introduce E_0, η_0, ε_0 we define the length of a vector $\xi \in C^\ell$ or of an $\ell x\ell$-matrix M:

$$|\underline{\xi}| = \sum_{i=1}^{\ell} |\xi_i|$$

$$|M| = \sum_{i,j=1}^{\ell} |M_{ij}| \qquad (1.7)$$

Furthermore given h_0, analytic in $W(\rho_0,\xi_0;V)$, we define its "free pulsations"

$$\underline{\omega}_0(\underline{A}) = \frac{\partial h_0}{\partial \underline{A}}(\underline{A}) \qquad (1.8)$$

where $\frac{\partial}{\partial \underline{A}} = (\frac{\partial}{\partial A_1},\dots,\frac{\partial}{\partial A_\ell})$ is the gradient.

Then E_0, η_0, ε_0 will be the bounds:

$$\sup \left|\frac{\partial h_0}{\partial \underline{A}}(\underline{A})\right| \le E_0$$

$$\sup \left|\left(\frac{\partial^2 h_0}{\partial \underline{A}\partial \underline{A}}(\underline{A})\right)^{-1}\right| \le \eta_0 \qquad (1.9)$$

$$\sup\left(\left|\frac{\partial f_0}{\partial \underline{A}}(\underline{A},\underline{z})\right| + \rho_0^{-1}\left|\frac{\partial f_0}{\partial \underline{\phi}}(\underline{A},\underline{z})\right|\right) \le \varepsilon_0$$

where $\frac{\partial}{\partial \underline{\phi}} \equiv (-iz_1 \frac{\partial}{\partial z_1}, \ldots, -iz_\ell \frac{\partial}{\partial z_\ell})$ and $\frac{\partial^2 h_0}{\partial \underline{A} \partial \underline{A}}$ is the matrix

$$\frac{\partial^2 h_0}{\partial A_i \partial A_j} \equiv \frac{\partial \omega_{0i}}{\partial A_j} .$$

We have called η_0 the "anisochrony parameter" because it is $+\infty$ both for the harmonic oscillators and for any system in which one pulsation is a function of the others.

In (1.9) E_0, η_0 and ε_0 are defined as bounds to the respective ℓ.h.s. rather than as the ℓ.h.s. themselves just for convenience. In the applications one usually knows only bounds to the ℓ.h.s. .

Finally the factor ρ_0^{-1} in the last equation of (1.9) is included to fix the physical dimensions. It has the same dimension of $\underline{A}$, so that ε_0 has the same dimension as E_0 .

Clearly the free system has ℓ independent analytic prime integrals (the $\underline{A}$'s themselves).

It is however well known by now that it will not be possible [8] to define prime integrals for the perturbed system in the whole phase space $V \times T^\ell$: not even in the regions where one could naively expect them to exist, i.e. in the whole phase space "up to boundary effects" due to the fact that near $\partial V \times T^\ell$ the perturbed trajectories will in general have a tendency to leave $V \times T^\ell$ so that the equations of motion are no longer defined).

The main result of the KAM-theory is that nevertheless the prime integrals may be defined and turn out to be smooth on a large subset $\Gamma(f_0)$ filling $V \times T^\ell$ up to a small fraction λ of its volume, provided ε_0 is small enough:

$$\mathrm{vol}\,\Gamma(f_0) \geq (1-\lambda)\,\mathrm{vol}(V \times T^\ell) . \tag{1.10}$$

There is a simple way to find out what kind of restriction on the smallness of f_0 is to be expected. We have so far characterized our system by the following parameters: r, ρ_0, ξ_0, E_0, η_0, ε_0 and we hope to give a procedure for building $\Gamma(f_0)$ which only depends on the above parameters and on λ , not on other structural properties of h_0, f_0 .

With the above parameters we can form the dimensionless following parameters

$$\frac{\varepsilon_0}{E_0}\,,\ \lambda\,,\ (\eta_0\rho_0^{-1}E_0)^{-\ell}\,,\ \xi_0,\ \frac{\rho_0}{r} \tag{1.11}$$

and for simplicity we shall henceforth suppose $\rho_0 < r$, $\varepsilon_0 < E_0$ (a case to which the general case could anyway always be reduced) so that all the parameters (1.11) are ≤ 1: notice that in any case $E_0\eta_0\rho_0^{-1} \ge 1$.*

Hence, assuming that there is a theorem saying that (1.10) follows under suitable restrictions on ε_0, η_0, E_0, ρ_0, ξ_0, r, λ, it appears that such restrictions must take the form:

$$\frac{\varepsilon_0}{E_0} < \overline{G}(\lambda,(\eta_0E_0\rho_0^{-1})^{-\ell},\xi_0) \tag{1.12}$$

where $\overline{G}$ is a suitable univeral function (which depends only on ℓ) such that $\overline{G}\longrightarrow 0$ as $\lambda\longrightarrow 0$, $\xi_0\longrightarrow 0$ or $(\eta_0E_0\rho_0^{-1})^{-1}\longrightarrow 0$: it is independent of r/ρ_0 because if the theorem is true for (say) $r' = 2\rho_0$, then one can cover V, the sphere of radius r, with spheres of radius r' (recall that our spheres are, actually, polyhedra) and then deduce the theorem for the larger sphere.

If near zero $\overline{G}(x,y,z)$ has the form $\overline{G}(x,y,z) = x^\alpha y^\beta z^\gamma g_0(x,y,z)$ with $\overline{g}$ continuous at the origin then (1.12) can be replaced by

$$\frac{\varepsilon_0}{E_0} < \overline{G}\,\lambda^\alpha(\eta_0\rho_0^{-1}E_0)^{-\beta}\xi_0^{\ \gamma} \tag{1.13}$$

where $\overline{G},\alpha,\beta,\gamma$ are universal constants depending only upon ℓ, the number of degrees of freedom. In the coming sections we shall prove in detail the following proposition which we still wish to call KAM theorem since it only involves trivial additional arguments compared to the ones that can be found in the classical papers [1], [2], [3].

---

*in fact call $M_0(\underline{A}) = \frac{\partial^2 h_0}{\partial\underline{A}\partial\underline{A}}(\underline{A})$ and let $\underline{A}\in V$: notice that $1 = \sum_{j=1}^{\ell} M_0(\underline{A})^{-1}_{ij}\, M_0(\underline{A})_{ji} \le \sum_{j=1}^{\ell}|M_0(\underline{A})^{-1}_{ij}|\,|\frac{\partial\omega_{0j}}{\partial A_i}(\underline{A})| \le \sum_{j=1}^{\ell}|M_0(\underline{A})^{-1}_{ij}|\frac{E_0}{\rho_0} < \eta_0E_0\rho_0^{-1}$ having bounded $\frac{\partial\omega_0}{\partial\underline{A}}(\underline{A})$, $\underline{A}\in V$, dimensionally as $\frac{E_0}{\rho_0}$.

Proposition. Given $\ell > 0$ one can find four positive constants $G > 1$, α, β, $\gamma > 0$ such that if

$$\frac{\varepsilon_0}{E_0} < G\,\lambda^{\alpha}(\eta_0 E_0 \rho_0{}^{-1})^{-\beta} \xi_0{}^{\gamma} \tag{1.14}$$

then it is possible to find a subset $\Gamma(f_0) \subset V x T^{\ell}$ and ℓ functions $A_1', \ldots, A_\ell' \in C^{\infty}(VxT^{\ell})$ such that $\Gamma(f_0)$ is invariant under the perturbed flow and:

i)
$$\text{vol}\,\Gamma(f_0) \geq (1-\lambda)\,\text{vol}(VxT^{\ell}) \tag{1.15}$$

ii) $A_1', \ldots, A_\ell'$ are prime integrals for the perturbed motions starting in $\Gamma(f_0)$.

iii) Any other function $\underline{A} \in C^{\infty}(VxT^{\ell})$ which is a prime integral on $\Gamma(f_0)$ is, on $\Gamma(f_0)$, a function of $A_1', \ldots, A_\ell'$.

iv) $A_1', \ldots, A_\ell'$ are "independent" on $\Gamma(f_0)$ i.e. their Jacobian determinant with respect to the $A_1, \ldots, A_\ell$ does not vanish.

Actually the proof gives much more than what is stated in the above proposition. It is useful to state a few other propositions which are proved simultaneously:

Proposition. 1) The set $\Gamma(f_0)$ can be chosen to consist of invariant tori on which non resonant quasiperiodic motions take place with pulsations $\underline{\omega}$ verifying, $\forall\, \underline{\gamma} \in Z^{\ell}$, $\underline{\gamma} \neq \underline{0}$:

$$|\underline{\omega} \cdot \underline{\gamma}|^{-1} \leq G_1\, \lambda^{-1}(E_0 \eta_0 \rho_0{}^{-1})^{\ell}\, |\underline{\gamma}|^{\ell+\frac{1}{2}} \tag{1.16}$$

where G_1 is a suitable universal constant, i.e. depends only on ℓ.

2) There exist ℓ T^{ℓ}-valued $C^{\infty}(VxT^{\ell})$-functions $(\phi_1', \ldots, \phi_\ell')$ such that the change of variables $(\underline{A}, \underline{\phi}) \longleftrightarrow (\underline{A}', \underline{\phi}')$ is one to one and canonical on $\Gamma(f_0)$. This means that there is a $C^{\infty}(VxT^{\ell})$ - function Φ such that if $\underline{A}' = \underline{A}'(\underline{A}, \underline{\phi})$, $\underline{\phi}' = \underline{\phi}'(\underline{A}, \underline{\phi})$, $(\underline{A}, \underline{\phi}) \in \Gamma(f_0)$ then $(\underline{A}', \underline{\phi}') \in VxT^{\ell}$ and

$$\begin{aligned} \underline{A} &= \underline{A}' + \frac{\partial \Phi}{\partial \phi}(\underline{A}', \underline{\phi}) \\ \underline{\phi}' &= \underline{\phi} + \frac{\partial \Phi}{\partial \underline{A}}(\underline{A}', \underline{\phi}) \end{aligned} \tag{1.17}$$

3) As $\underline{\phi}'$ varies in T^{ℓ}, (1.17) gives the (implicit) parametric equations of the torus in $\Gamma(f_0)$ on which the integrals $\underline{A}'$ take given values.

(4) On $\Gamma(f_0)$ ϕ solves the Hamilton-Jacobi equation: i.e.

$$h_0(\underline{A}'+\frac{\partial\Phi}{\partial\underline{\phi}}(\underline{A}',\underline{\phi})) + f_0(\underline{A}'+\frac{\partial\Phi}{\partial\underline{\phi}}(\underline{A}',\underline{\phi}),\underline{\phi}) = \{\underline{\phi}\text{-independent function}\} \tag{1.18}$$

if $\underline{A}' = \underline{A}'(\underline{A},\underline{\phi})$, $(\underline{A},\underline{\phi}) \in \Gamma(f)$.

Finally we can say something about the relations between the free tori and the perturbed ones.

Proposition. 1) The function Φ can be bounded as

$$|\frac{\partial\Phi}{\partial\underline{A}'}(\underline{A}',\underline{\phi})| + \frac{1}{\rho_0}|\frac{\partial\Phi}{\partial\underline{\phi}}(\underline{A}',\underline{\phi})| \le G_2(\frac{\varepsilon_0}{E_0})^{\alpha_1}\lambda^{-\alpha_2}(E_0\rho_0{}^{-1}\eta_0)^{-\alpha_3}\xi_0{}^{-\alpha_4} \tag{1.19}$$

with G_2, α_1, α_2, α_3, α_4 suitable positive universal constants: i.e. the perturbed tori are close to the free ones, when they exist.

2) There is a C^{∞}-function on V: $\underline{A}' \longrightarrow \underline{\omega}(\underline{A}')$ which gives the pulsations of the quasiperiodic motion taking place on a perturbed torus on which the integrals $\underline{A}'$ take given values. Also, $\forall \underline{A}' \in V$;

$$|\underline{\omega}(\underline{A}')-\underline{\omega}_0(\underline{A}')| < G_3(\frac{\varepsilon_0}{E_0})^{\beta_1}(E_0\rho_0{}^{-1}\eta_0)^{\beta_2}\xi_0{}^{-\beta_3}\lambda^{-\beta_4} \tag{1.20}$$

where G_3, $\beta_1, \dots, \beta_4$ are positive universal constants.

3) Finally, as $\underline{A}'$ varies over the prime integral values on $\Gamma(f_0)$ the range of $\underline{\omega}_0(\tilde{\underline{A}}')$ covers the set of values of $\underline{\omega}_0(\tilde{\underline{A}})$ with $\tilde{\underline{A}}$ V such that $\forall|\underline{\gamma}| \neq 0$, $\underline{\gamma}$ Z^{ℓ}:

$$|\underline{\omega}_0(\tilde{\underline{A}})\cdot\underline{\gamma}|^{-1} \le G_4\,\lambda^{-1}(E_0\eta_0\rho_0{}^{-1})^{\ell}\,|\underline{\gamma}|^{\ell-\frac{1}{2}} \tag{1.21}$$

with a positive universal constant G_4.

It would be nice to have good, if not optimal, estimates for the stants G, α, β, γ, $\cdots$. Our estimates however are pretty bad particularly concerning the ℓ-dependence.

However it should be stressed that this does not mean that the estimates, as they are presented here, at least up to (3.34), are without interest for applications. On the contrary they are often either already very sharp or they can be made so with not too much extra effort. The reason they look lousy, probably even if carefully performed, is to be found in the fact that they are universal. In the applications, however, f_0 (as well as h_0) will always be very special and the theory should be applied "cum grano salis": namely one should follow the pattern of the proof in the first few renormalizations, possibly with the help of a computer, making use of the special features of the model under consideration until the calculations become untreatable and, at this point, apply the general result with the universal constants.

In this way in some special $\ell=2$ cases where tests have been made, [9], one can prove stability of certain irrational motions with surprising accuracy without excessive efforts. We hope that, in view of the last comment, the reader will not laugh seeing the following values for G, γ, β, γ as they come out of these notes:

$$G = (\ell+31)!^{31} \; , \; \alpha = 11 \; , \; \beta = 5 + 22\ell \; , \; \gamma = 30(\ell+\tfrac{1}{3}). \qquad (1.22)$$

I believe their main features will not be changed by better estimates. The dependence of G on $\ell^{a\ell}$, $a > 0$, the linear dependence of β and γ on ℓ and the nice fact that α does not depend on ℓ seem to be necessary features of the results. The values 31, 11, 5, 22, 30, $\frac{1}{3}$ can, however, be greatly improved. For a discussion of some "best estimates" see H. Russmann [11].

In §2 we give an account of the classical perturbation theory and in §3 we give the proof of roposition 1, in all imaginable detail. The reader will have, afterwards, no trouble in extracting from it the proof of the statements of the other propositions which are not explicitely proved there. The only statement which needs some heavy extra work is (1.21). Its proof requires essentially going through the whole proof of Proposition 1 keeping track at each step of the set of values taken by the sequence of functions which approximate $\underline{\omega}$ and which are constructed in the proof. This could have been done in the proof of proposition 1 at little extra cost. We did not do so because it seemed to us that the gain in generality would be paid by excessive formal complications on a side issue.

As mentioned above the main technical tool in the proofs are the dimensional estimates. To avoid misunderstanding, let me explicitely state what I mean by this.

Let f be holomorphic in $W \subset C^{2\ell}$ and suppose that we wish to estimate $\frac{\partial^{\underline{a}+\underline{b}} f}{\partial \underline{A}^{\underline{a}} \partial \underline{z}^{\underline{b}}}(\underline{A},\underline{z})$ where $\underline{a} = (a_1,\dots,a_\ell)$, $\underline{b} = (b_1,\dots,b_\ell) \in Z_+^\ell$. Suppose that we know that $(\underline{A},\underline{z})$ is "well inside W" i.e. suppose that for some $\rho > 0$, $\delta > 0$, $\xi > 0$

$$(\underline{A},\underline{z}) \in C(\rho,\xi-\delta;\underline{A}) \subset C(\rho,\xi;\underline{A}) \subset W \tag{1.23}$$

using the notation (1.6). Then

$$\begin{aligned}\left|\frac{\partial^{\underline{a}+\underline{b}} f}{\partial \underline{A}^{\underline{a}} \partial \underline{z}^{\underline{b}}}(\underline{A},\underline{z})\right| &\le (\sup|f|) \frac{\underline{a}!\ \underline{b}!}{\rho^{|\underline{a}|}(e^{-(\xi-\delta)}-e^{-\xi})^{|\underline{b}|}} \\ &\le (\sup|f|) \frac{\underline{a}!\ \underline{b}!\ e^{\xi|\underline{b}|}}{\rho^{|\underline{a}|}\ \delta^{|\underline{b}|}}\end{aligned} \tag{1.24}$$

where $\underline{a}! = \prod_i (a_i!)$, $\underline{b}! = \prod_i (b_i)!$ and the supremum is taken over $C(\rho,\xi;\underline{A})$.

(1.2) follows immediately from Cauchy's theorem on the multiannulus $C(\rho,\xi;\underline{A})$ and for $|\underline{a}| + |\underline{b}| = 1$, we read it "the derivative of f is bounded by its maximum divided by the distance to the boundary of the holomorphy domain".

Another consequence of the analyticity of f in $C(\rho,\xi,\underline{A})$ is that its Laurent coefficients in the $\underline{z}$-variables at fixed $\underline{A}'$, $|A_i'-A_i| < \rho$, are bounded by

$$|f_{\underline{\gamma}}(\underline{A}')| \le (\sup|f|)\ e^{-\xi|\underline{\gamma}|} \tag{1.25}$$

for $\forall\, \underline{\gamma} \in Z^\ell$.

§2. The classical theory of perturbations by the Hamilton-Jacobi method.

The classical theory of perturbations deals with a perturbation whose form is, for some $N < \infty$:

$$f_0(\underline{A},\underline{\phi}) = \sum_{\substack{\underline{\gamma}\in Z^\ell \\ |\underline{\gamma}|\le N}} f_{0\underline{\gamma}}(\underline{A})\, e^{i\underline{\gamma}\cdot\underline{\phi}} \tag{2.1}$$

i.e. with interactions which are simply trigonometric polynomials in the angle variable (think of a double pendulum).

For a better understanding of the structure of the theory we did not put the usual ε in front of the sum on the ℓ.h.s. of (2.1) and we chose to measure f_0 by the supreme ε_0 of

$$\left|\frac{\partial f_0}{\partial \underline{A}}(\underline{A},\underline{z})\right| + \frac{1}{\rho_0}\left|\frac{\partial f_0}{\partial \underline{\phi}}(\underline{A},\underline{z})\right| \tag{2.2}$$

as $(\underline{A},\underline{z})$ vary in $W(\rho_0,\xi_0;V)$, using the notation (1.6) and assuming that f_0 is holomorphic in $(\underline{A},\underline{\phi})$ (i.e. in $\underline{A}$ as the ϕ-holomorphy is trivial from (2.1) and $N < \infty$). So the reader should mentally think that f_0, in (2.1), is small of "$O(\varepsilon_0)$".

Classical perturbation theory tries to find a canonical map $C^{(0)}: (\underline{A}',\underline{\phi}') \longrightarrow (\underline{A},\underline{\phi})$ defined on $\tilde{V}\times T^\ell$, $\tilde{V} \subset V$, with values in $R^\ell \times T^\ell$ which will be used to describe the motions that take place in $C^{(0)}(\tilde{V}\times T^\ell) \cap (V\times T^\ell)$ via a new Hamiltonian $H_1(\underline{A}',\underline{\phi}') = H_0(C^{(0)}(\underline{A}',\underline{\phi}'))$. The choice of $C^{(0)}$ has to be made according to the following criterion: H_1 should have the form

$$H_1(\underline{A}',\underline{\phi}') \equiv H_0(C^{(0)}(\underline{A}',\underline{\phi}')) = h_1(\underline{A}') + f_1(\underline{A}',\underline{\phi}') \tag{2.3}$$

with f_1 of $O(\varepsilon_0^2)$. The search for $C^{(0)}$ can be reduced to the search for its generating function

$$\overline{\Phi}_0(\underline{A}',\underline{\phi}) = \underline{A}'\cdot\underline{\phi} + \Phi_0(\underline{A}',\underline{\phi}) \tag{2.4}$$

related to $C^{(0)}$ by

$$\underline{A} = \underline{A}' + \frac{\partial \Phi_0}{\partial \underline{\phi}} (\underline{A}',\underline{\phi})$$
$$\underline{\phi}' = \underline{\phi} + \frac{\partial \Phi_0}{\partial \underline{A}'} (\underline{A}',\underline{\phi}) \tag{2.5}$$

establishing the defining relation between $(\underline{A}',\underline{\phi}')$ and $(\underline{A},\underline{\phi}) = C^{(0)}(\underline{A}',\underline{\phi}')$. With (2.5) in mind one immediately finds an equation for Φ_0

$$h_0(\underline{A}'+\frac{\partial \Phi_0}{\partial \underline{\phi}}(\underline{A}',\underline{\phi})) + f_0(\underline{A}'+\frac{\partial \Phi_0}{\partial \underline{\phi}}(\underline{A}',\underline{\phi}),\underline{\phi})$$
$$\doteq \{\underline{\phi}\text{-independent function up to } O(\varepsilon_0^{\,2})\} \quad . \tag{2.6}$$

Since one reasonably expects that Φ_0 should be of $O(\varepsilon_0)$ one expands (2.6) neglecting $O(\varepsilon_0^{\,2})$, bearing in mind that f_0 is already of $O(\varepsilon_0)$. The easily found result is

$$\underline{\omega}_0(\underline{A}') \cdot \frac{\partial \Phi_0}{\partial \underline{\phi}} (\underline{A}',\underline{\phi}) + f_0(\underline{A}',\underline{\phi}) \doteq \{\underline{\phi}\text{-independent}\} \tag{2.7}$$

which, averaging over $\underline{\phi}$, says that the r.h.s. of (2.2) must be $f_{\underline{00}}(\underline{A}')$, i.e. the average of f_0 over the angles, and (2.7) becomes:

$$\underline{\omega}_0(\underline{A}') \cdot \frac{\partial \Phi_0}{\partial \underline{\phi}} (\underline{A}',\underline{\phi}) + f_0(\underline{A}',\underline{\phi}) - f_{\underline{00}}(\underline{A}') = 0 \tag{2.8}$$

which should determine Φ_0.

It is clear that if Φ_0 exists and if it is really of $O(\varepsilon_0)$ and if the canonical transformation (2.6) can really be defined (i.e. if one can solve the implicit functions problems arising in extracting from (2.5) the map $C^{(0)}$ and its inverse $\tilde{C}^{(0)}$) then

$$H_0(C^{(0)}(\underline{A}',\underline{\phi}')) = h_0(\underline{A}') + f_{\underline{00}}(A') + O(\varepsilon_0^2). \tag{2.9}$$

The preliminary problem is, of course, that of understanding whether the "first order" Hamilton-Jacobi equation (2.8) admits a solution.

If it exists the solution is given clearly by

$$\Phi(\underline{A}',\underline{\phi}) = \sum_{\substack{\underline{\gamma}\in Z^{\ell} \\ 0<|\underline{\gamma}|\le N}} \frac{f_{0\underline{\gamma}}(\underline{A})\, e^{i\underline{\gamma}\cdot\underline{\phi}}}{-i\,\underline{\omega}_0(\underline{A}')\cdot\underline{\gamma}} \tag{2.10}$$

which shows where the problem really is. In fact the denominator in (2.10) could vanish somewhere inside V.

In formal perturbation theory one discards this possibility by restricting consideration to regions V so small that $\underline{\omega}_0(\underline{A}')\cdot\underline{\gamma}\neq 0$, $\forall\,\underline{A}'\in V$, $\forall\, 0<|\underline{\gamma}|\le N$. There are some interesting cases when this cannot be done within a reasonable approximation (harmonic oscillators, celestial mechanics, etc.) In such cases some other ideas are necessary in order to transform the problem into one that could be attacked along the above lines, with good approximations. However, we do not need to develop these matters as they will not show up here.

For our future discussions it is useful to try to see what happens if V is not such that $\underline{\omega}_0(\underline{A}')\cdot\underline{\gamma}\neq 0$, $\forall\,\underline{A}'\in V$, $\forall\, 0<|\underline{\gamma}|\le N$. This can be more easily done if the notion of "resonance" is introduced. A resonance of order (C,N) for h_0 is defined as a torus $\{\underline{A}\}\times T^{\ell}$ such that

$$|\underline{\omega}_0(\underline{A})\cdot\underline{\gamma}|^{-1} \le C|\underline{\gamma}|^{\ell} \quad \forall\, 0<|\underline{\gamma}|\le N \tag{2.11}$$

is not verified.

Clearly if $|\underline{\omega}_0(\underline{A})\cdot\underline{\gamma}| = 0$, $\underline{\gamma}\neq 0$, then the torus $\{\underline{A}\}\times T^{\ell}$ is a resonant torus of order $(C,|\underline{\gamma}|)$, $\forall\, C>0$.

In the future we shall identify the tori $\{\underline{A}\}\times T^{\ell}$ in $V\times T^{\ell}$ with the points $\underline{A}\in V$ and so we shall also say that $\underline{A}$ is a (C,N)-resonance for h_0 if (2.11) does not hold.

It is useful to stress that if the free system is "anisochronous" (see §1), i.e. $\eta_0 < +\infty$, then the resonant tori of order (C,∞) are rare in phase space.

More precisely if we call $V_{C,N}\subset V$ the subset of V consisting of the resonances of order (C,N) and if $V_C \equiv V_{C,\infty} = \bigcup_{N=0}^{\infty} V_{C,N}$ then, for $S_1 = 4^{\ell}\ell^{3/2}\sum_{\underline{\gamma}\neq\underline{0}} |\underline{\gamma}|^{-\ell-1}$:

$$\mathrm{vol}\,V_C \le S_1(\eta_0 E_0\rho_0^{-1})^{2\ell}(E_0 C)^{-1}\,\mathrm{vol}\,V \xrightarrow[C\to\infty]{} 0 \,. \tag{2.12}$$

It is worth while to digress and derive this simple estimate explicitely

$$\text{vol } V_C = \int_{V_C} d\underline{A} \leqq T \int_{\underline{\omega}_0(V_C)} d\underline{\omega} \max\left|\det\left(\frac{\partial\underline{\omega}}{\partial\underline{A}}\right)^{-1}\right| \leq T\, \eta_0^{\ell} \int_{\underline{\omega}_0(V_C)} d\underline{\omega} \tag{2.13}$$

where T is the maximum multiplicity of the map $\underline{A} \longrightarrow \underline{\omega}(\underline{A})$, $\underline{A} \in V$.

Clearly this map is locally 1 - to - 1, as $\eta_0 < +\infty$, and in fact if $\underline{A}$, $\tilde{\underline{A}} \in V$, $|\underline{A}-\tilde{\underline{A}}| < \hat{r}$ we have, using $|M\underline{x}| \geq |M^{-1}|^{-1}|\underline{x}|$:

$$\begin{aligned}|\underline{\omega}_0(\underline{A})-\underline{\omega}_0(\tilde{\underline{A}})| &\equiv |M_0(\tilde{\underline{A}})\cdot(A-\tilde{\underline{A}})+\underline{\omega}_0(\underline{A})-\underline{\omega}_0(\tilde{\underline{A}})-M_0(\tilde{\underline{A}})(\underline{A}-\tilde{\underline{A}})| \\ &\geq \eta_0^{-1}|\underline{A}-\tilde{\underline{A}}| - \tfrac{1}{2}\sup\left|\frac{\partial^2\underline{\omega}_0}{\partial A_i\,\partial A_k}(\underline{A}')\right| |\underline{A}-\tilde{\underline{A}}|^2 \\ &\geq \eta_0^{-1}\, |\underline{A}-\tilde{\underline{A}}|\, \left(1-\tfrac{1}{2}\,\eta_0\, \frac{E_0}{\rho_0^2}\,\hat{r}\right) > 0\end{aligned} \tag{2.14}$$

if $\hat{r} = \frac{1}{2}\,\rho_0(\eta_0 E_0 \rho_0^{-1})^{-1}$. We see that the multiplicity of $\underline{\omega}_0(\underline{A})$ cannot exceed the number of spheres of radius $\hat{r}$ necessary to cover V : i.e.

$$T < \left(\frac{2r}{\hat{r}}\right)^{\ell} \tag{2.15}$$

(recall that our spheres are polyhedra and that r needs not be a multiple of $\hat{r}$: this explains the factor 2).

So (making use of $\frac{1}{\sqrt{\ell}}\,|\underline{\gamma}| \leq (\sum_i \gamma_i^2)^{\frac{1}{2}} \leq |\underline{\gamma}|$):

$$\begin{aligned}\text{vol } V_C &< \left(4(\eta_0 E_0 \rho_0^{-1})\frac{r}{\rho_0}\right)^{\ell}\, \eta_0^{\ell} \int_{\underline{\omega}_0(V_C)} d\underline{\omega} \\ &\leq \left(4(\eta_0 E_0 \rho_0^{-1})\frac{r}{\rho_0}\right)^{\ell}\, \eta_0^{\ell} \sum_{\underline{\gamma}\neq\underline{0}} \int_{\substack{|\underline{\omega}\cdot\underline{\gamma}| < C^{-1}|\underline{\gamma}|^{-\ell}\sqrt{\ell} \\ |\underline{\omega}| < E_0}} d\underline{\omega} \\ &\leq \frac{(2E_0)^{\ell-1}}{(\ell-1)!}\, \left(4(\eta_0 E_0 \rho_0^{-1})\frac{r}{\rho_0}\right)^{\ell}\, \eta_0^{\ell}\, \frac{2\sqrt{\ell}}{C}\, \left(\sum_{\underline{\gamma}\neq\underline{0}} |\underline{\gamma}|^{-\ell-1}\right) \\ &\leq \ell^{3/2}\, 4^{\ell}\, \frac{(\eta_0 E_0 \rho_0^{-1})^{2\ell}}{E_0 C}\, \left(\sum_{\underline{\gamma}\neq 0} |\underline{\gamma}|^{-\ell-1}\right) \text{Vol } V\end{aligned} \tag{2.16}$$

Returning to the problem of perturbation theory: In a region where there are resonances one can try to proceed if one gives up the hope of describing what happens near the resonating tori and describes only the motions which evolve far from these tori.

This can be done by defining the canonical transformation and, beforehand, its generating function Φ_0 on a set $V^{(0)}xT^{\ell} \subset VxT^{\ell}$ leaving out the resonating tori of order (C,N) for some acceptable $C > 0$. Also one should not try to describe the motions too close to the boundary ∂VxT^{ℓ} as the perturbed motions will have the property that they reach the boundary ∂VxT^{ℓ} in a finite time, if they start too close to it. So a natural choice of $V^{(0)}$ would seem to be the following

$$\overline{V}^{(0)} = \{\underline{A} \mid \text{distance of } \underline{A} \text{ from } \partial V > \tilde{\rho}_0 \text{ and } \underline{A} \text{ does not resonate to order } (C,N)\} \tag{2.17}$$

where $\tilde{\rho}_0$ is some positive number.

Clearly we can define Φ_0 on $V^{(0)}xT^{\ell}$, as the denominators in (2.11) do not vanish on $\overline{V}^{(0)}$: however $\overline{V}^{(0)}$ is a rather badly shaped set with a very wiggly boundary. Since we must consider derivatives of Φ_0 it is better to slightly enlarge $\overline{V}^{(0)}$ into

$$V^{(0)} = \bigcup_{\underline{A}\in\overline{V}^{(0)}} S(\underline{A},\frac{\tilde{\rho}_0}{2}) \tag{2.18}$$

where $S(\underline{A},x) = \{\text{open sphere of radius } x \text{ and center } \underline{A}\}$.

Of course we shall try to hold $\tilde{\rho}_0$ as large as possible but demanding that the resonances in $V^{(0)}$ should still be not too strong, say at most of the order $(2C,N)$. This last requirement severely restricts $\tilde{\rho}_0$ and allows us to find an expression for it: in fact let $\underline{A} \in V^{(0)}$; then there is $\underline{A}_0 \in \overline{V}^{(0)}$ such that $|\underline{A}-\underline{A}_0| > \frac{\tilde{\rho}_0}{2}$. We can then evaluate the order of resonance of $\underline{A}$:

$$\begin{aligned} |\underline{\omega}_0(\underline{A})\cdot\underline{\gamma}|^{-1} &< |\underline{\omega}_0(\underline{A}_0)\cdot\underline{\gamma}|^{-1}\,\left|1-\frac{(\underline{\omega}_0(\underline{A})-\underline{\omega}_0(\underline{A}_0))\cdot\underline{\gamma}}{\underline{\omega}_0(\underline{A}_0)\cdot\underline{\gamma}}\right|^{-1} \le \\ &\le C|\underline{\gamma}|^{\ell}\,(1-C|\underline{\gamma}|^{\ell+1}|\underline{\omega}_0(\underline{A})-\underline{\omega}_0(\underline{A}_0)|)^{-1} \le \\ &\le C|\underline{\gamma}|^{\ell}\,(1-C\ell N^{\ell+1}\frac{E_0}{\rho_0}\frac{\tilde{\rho}_0}{2})^{-1} \le 2C\,|\underline{\gamma}|^{\ell} \end{aligned} \tag{2.19}$$

if

$$\tilde{\rho}_0 = \frac{1}{2}\rho_0(\ell\, C\, N^{\ell+1} E_0)^{-1} \tag{2.20}$$

where in (2.19) we have estimated $|\underline{\omega}_0(\underline{A})_k - \underline{\omega}_0(\underline{A}_0)_k| \equiv |(\underline{A}-\underline{A}_0)\cdot\frac{\partial\underline{\omega}_{0k}}{\partial A}(\underline{\tilde{A}})| \leq$ $\leq |\underline{A}-\underline{A}_0|\,\frac{E_0}{\rho_0}$, by a dimensional estimate of the type of (1.24) using the (complex) holomorphy of $\underline{\omega}_0$. Having chosen $\tilde{\rho}_0$ in this way we still have to check that Φ_0 is actually of order ε_0 : i.e. we have to find bounds for Φ_0.

This can be done easily by remarking that the assumed analyticity of f_0 together with the bounds (1.24), (1.25) ("Laurent's theorem") imply:

$$\begin{aligned} |\underline{\gamma}|\,|f_{0\underline{\gamma}}(\underline{A})| &< \varepsilon_0\rho_0\, e^{-\xi_0|\underline{\gamma}|} \\ |\frac{\partial f_{0\underline{\gamma}}}{\partial \underline{A}}(\underline{A})| &< \varepsilon_0\, e^{-\xi_0|\underline{\gamma}|} \end{aligned} \tag{2.21}$$

Using (2.21) and the bounds

$$\begin{aligned} |\underline{\omega}_0(\underline{A})\cdot\underline{\gamma}|^{-1} &< 2C|\underline{\gamma}|^{\ell} \\ |\frac{\partial(\underline{\omega}_0(A)\cdot\underline{\gamma})^{-1}}{\partial\underline{A}}| \equiv |\underline{\omega}_0(\underline{A})\cdot\gamma|^{-2}\,|\frac{\partial\underline{\omega}}{\partial\underline{A}}(\underline{A})\cdot\underline{\gamma}| &\leq 4C^2|\underline{\gamma}|^{2\ell+1}\frac{E_0}{\rho_0} \end{aligned} \tag{2.22}$$

valid for $\underline{A}$ $V^{(0)}$ we can easily find a bound for Φ_0:

$$|\frac{\partial\Phi_0}{\partial A}| + \frac{1}{\tilde{\rho}_0}|\frac{\partial\Phi_0}{\partial\underline{\phi}}| \leq F\,\varepsilon_0\, C(E_0C)^2\,\xi_0^{-2\ell-1}\,N^{\ell+1} \tag{2.23}$$

$\forall(\underline{A},\underline{\phi}) \in V^{(0)}xT^{\ell}$, if $F > 0$ is suitably chosen. (2.23) shows that Φ_0 is actually of $O(\varepsilon_0)$ in the same sense as f_0 and that one can try to invert (2.6).

This inversion can be done very easily if one notices that Φ_0 is actually defined and holomorphic in the larger, complex, region $W(\tilde{\rho}_0,\xi_0;V^{(0)})$. This however requires further discussion. We postpone this analysis and terminate the discussion of classical perturbation theory by adding a few last comments. 1) (2.23) shows that even if we restrict our Φ_0 to a small domain avoiding the resonances we cannot derive that Φ_0 is of order ε_0 uniformly in the number of frequencies appearing in f_0, (2.1).

2) Because of 1) we must expect that even if we solve the implicit function problem of defining $C^{(0)}$, $\tilde{C}^{(0)}$, the function f_1 will also turn out to be of $O(\varepsilon_0^2 N^{\ell+1})$, rather than $O(\varepsilon_0^2)$ uniformly in N.

3) The comments 1), 2) show that to solve the problem of studying a general perturbation one will have to first truncate the Fourier's series of f_0 to $|\underline{\gamma}| \leq N_0$ for some N_0 and then apply the methods of classical perturbation theory to remove the perturbation, or better its truncated part, to $O(\varepsilon_0^2 N_0^{\ell+1})$. Naturally N_0 will have to be chosen so that the part of f_0 which is truncated is of $O(\varepsilon_0^2 N_0^{\ell+1})$. or $O(\varepsilon_0^2)$, too. This leads to an ε_0 dependence of N_0.

Since the coefficients of f_0 decay exponentially as $e^{-\xi_0|\underline{\gamma}|}$ by the analyticity hopotheses, one expects to have to choose $N_0 \sim \frac{1}{\xi_0} \log(\frac{\varepsilon_0}{E_0})^{-1}$. The main point is that $O(\varepsilon_0^2 N_0^{\ell+1})$ will still be much smaller than $O(\varepsilon_0)$ and almost of $O(\varepsilon_0^2)$: one can then hope that this is, after all, more than sufficient if we consider an iteration of the procedure leading to Hamiltonians with smaller and smaller interactions, which should allow us to eliminate entirely the perturbation in the region of phase space which remains after the successive removal of the resonating tori.

4) In order that this region be not empty one has to choose the parameter C measuring the resonances large to start with; $C = C_0$, so that the amount of phase space volume thrown out (see (2.12)) is small. At the successive iterations C has to grow, say at the k-th iteration the resonance limit will be measures by some C_k. By (2.12), it is natural to choose $C_k = C_0(1+k)^2$ so that $\sum_k C_k^{-1} < \infty$. With this choice of C_k we may hope to loose, in the end, a phase space volume of the order C_0^{-1} only: hence eventually C_0 will have to be taken as a function of λ.

§3. The iterative renormalization

The idea of the proof of proposition 1 is to construct a procedure K allowing the analysis of the motions taking place "away from the resonances" by representing them in coordinates in which the perturbation is made much smaller than in the original variables.

The algorithm is based on classical perturbation theory and consists of the following seven steps.

Given two holomorphic functions h_0, f_0 on $W(\rho_0,\xi_0;V)$, as in §1, whose size is measured by the parameters E_0, η_0, ε_0 introduced in §1 and given λ we shall find it more convenient to proceed in the discussion by first trying to show that the region $\Gamma(f_0)$ consists of invariant tori on which the perturbed motion is quasiperiodic with pulsations which do not resonate to order (C,∞) and then checking that if C is suitably chosen as a function of λ then $\Gamma(f_0)$ has volume $\geq (1-\lambda)\mathrm{vol}(V\times T^{\ell})$ (see that last comment to §2).

So instead of λ we give for the time being $C \equiv C_0 > 0$.

1) One first divides f_0 into a large wavelength part $f_0^{[\leq N_0]}$ and a short wavelength part $f_0^{[>N_0]}$:

$$f_0^{[\leq N_0]}(\underline{A}.\underline{z}) = \sum_{|\underline{\gamma}|\leq N_0} f_{0\underline{\gamma}}(\underline{A})\, \underline{z}^{\underline{\gamma}}$$
$$f_0^{[>N_0]}(\underline{A},\underline{z}) = \sum_{|\underline{\gamma}|>N_0} f_{0\underline{\gamma}}(\underline{A})\, \underline{z}^{\underline{\gamma}} \tag{3.1}$$

both holomorphic in $W(\rho_0,\xi_0;V)$.

The value of N_0 will be chosen so that $f_0^{[>N_0]}$ is $O(\varepsilon_0^2)$. The plan is to change variables by classical perturbation theory to put the Hamiltonian

$$h_0(\underline{A}) + f_0^{[\leq N_0]}(\underline{A}.\underline{z}) \tag{3.2}$$

into the form

$$h_1(\underline{A}) + O(\varepsilon_0^2). \tag{3.3}$$

If such a plan succeds we clearly achieve at the same time (by the same change of variables) writing $h_0(\underline{A}) + f_0(\underline{A},\underline{z})$ into the form (3.3) with a different $O(\varepsilon_0^2)$ (see comment 3) at the end of §2).

So N_0 is determined by requiring

$$\left|\frac{\partial f_0^{[>N_0]}}{\partial \underline{A}}(\underline{A},\underline{z})\right| + \frac{1}{\rho_0}\left|\frac{\partial f_0^{[>N_0]}}{\partial \underline{\phi}}(\underline{A},\underline{z})\right| < B_1\, C_0\, \varepsilon_0^2 \tag{3.4}$$

for $(\underline{A},\underline{z}) \in W(\rho_0,\xi_0-\delta_0;V)$ where B_1 is some constant and $\delta_0 > 0$.

The reason why we write the r.h.s. of (3.4) as $C_0\varepsilon_0^2$ rather than simply as ε_0^2 is due to our desire to keep all the constants dimensionless so that they will be "universal"; i.e. they will only depend on the dimension ℓ of the phase space. This attention to the dimensions of the constants is essential if one does not wish to be overwhelmed by a massive number of constants which depend not only on ℓ but also on "everything else".

In my opinion writing the proof in dimensionless form, as we shall be doing (see also [2], [3]), is an essential step towards understanding it in a simpleminded way.

Using C_0 to fix the dimension in (3.4) is, however, an arbitrary choice. One could use to the same effect e.g. E_0^{-1} without affecting the rest of the proof.

Before continuing let us suppose that $C_0\varepsilon_0 < 1$, as this will eventually have to be required.

To compute N_0 one still needs to choose δ_0. First notice that it is clearly necessary that δ_0 is positive. If we wish to estimate in a simple way the remainder term of the Laurent series, of a holomorphic function, the only way is to bound its coefficients in terms of the supremum of the function in the annulus of definition (see (1.21) and then to replace the powers of the variables by their maxima in the region where the estimate is needed. Clearly if one looks in this way for a bound in the <u>whole</u> annulus of definition one finds $+\infty$.

In our case the analyticity in $W(\rho_0\xi_0;V)$ and the bounds (1.9),

(1.25) allow to infer that

$$|\underline{\gamma}||f_{0\underline{\gamma}}(\underline{A})| \le e^{-\xi_0|\underline{\gamma}|}\varepsilon_0\rho_0$$
$$\left|\frac{\partial f_{0\underline{\gamma}}}{\partial \underline{A}}(\underline{A})\right| \le e^{-\xi_0|\underline{\gamma}|}\varepsilon_0 \tag{3.5}$$

and that the maximum of $|\underline{z}^{\underline{\gamma}}|$ in the region $W(\rho_0,\xi_0-\delta_0;V)$ is $e^{(\xi_0-\delta_0)|\underline{\gamma}|}$.

Therefore we must try to estimate $f_0^{[> N_0]}$ in a region like $W(\rho_0,\xi_0-\delta_0;V)$ with $\delta_0 > 0$. Of course δ_0 must be $< \xi_0$. Since we shall have to reduce the domain of "toral analyticity" to make later estimates we shall take $4\delta_0 < \xi_0$ instead. Furthermore, since in the j-th step of the induction we shall, always for the same reason, need a further reduction of the anayticity region by $4\delta_j$, we shall fix beforehand the sequence δ_j, $j = 0, 1, \dots,$ as

$$\delta_j = \xi_0/16(1+j)^2. \tag{3.6}$$

The sequence of these "analyticity loss paramameters" is rather arbitrary provided $\sum_{j=1}^{\infty} 4\delta_j < \xi_0$ and provided δ_j does not approach zero too fast as $j \longrightarrow \infty$, as it will emerge later.

Having chosen δ_0 we can immediately perform the calculation of N_0.

In fact using (3.5) we can estimate the second term of (3.2) in $W(\rho_0,\xi_0-\delta_0;V)$, as

$$\sum_{|\underline{\gamma}|>N_0} \ell\,\varepsilon_0\, e^{-\xi_0|\underline{\gamma}|}\, e^{(\xi_0-\delta_0)|\underline{\gamma}|} + \rho_0^{-1}\sum_{|\underline{\gamma}|>N_0} \ell\,\varepsilon_0\rho_0\, e^{-\xi_0|\underline{\gamma}|}\, e^{(\xi_0-\delta_0)|\underline{\gamma}|}$$

$$= 2\,\varepsilon_0\,\ell \sum_{|\underline{\gamma}|>N_0} e^{-\delta_0|\underline{\gamma}|} \le 2\ell\,\varepsilon_0\, e^{-\frac{\delta_0}{2}N_0} \sum_{\underline{\gamma}} e^{-\frac{\delta_0}{2}|\underline{\gamma}|} \tag{3.7}$$

$$\le 2\ell\,\varepsilon_0\, e^{-\frac{\delta_0}{2}N_0}\left(\frac{1+e^{-\frac{\delta_0}{2}}}{1-e^{-\frac{\delta_0}{2}}}\right)^{\ell} \le B_1\,\varepsilon_0\, e^{-\frac{\delta_0}{2}N_0}\,\delta_0^{-\ell}$$

with $B_1 = 2\ell 4^{\ell} = 2^{2\ell+1}\ell$.

Then (3.7) tells us that we must choose

$$N_0 = \frac{2}{\delta_0} \log(C_0 \varepsilon_0 \delta_0{}^{\ell})^{-1} > 1 \quad . \tag{3.8}$$

This completes the first step of the algorithm.

2) Next we fix the region $V^{(0)} \times T^{\ell}$ in which the new canonical coordinates (yet to be defined) will vary.

The region $V^{(0)}$ will be obtained roughly speaking by omitting from V a band of width $\sim \tilde{\rho}_0$ both near the boundary ∂V as well as near all the resonances of order (C_0, N_0).

More precisely we set:

$$\overline{V}^{(0)} = \{\underline{A} | \underline{A} \in V, \text{ distance of } \underline{A} \text{ from } \partial V > \tilde{\rho}_0\}$$

$$\cap \{\text{non resonant points of order } (C_0, N_0), \text{ for } h_0\} \equiv \tag{3.9}$$

$$\equiv \{\underline{A} | \underline{A} \in V, \ d(\underline{A}, \partial V) > \tilde{\rho}_0, \ |\underline{\omega}_0(\underline{A}) \cdot \underline{\nu}|^{-1} \le C_0 |\underline{\nu}|^{\ell}, \ \forall\, 0 < |\underline{\nu}| \le N_0\}$$

and

$$V^{(0)} = \bigcup_{\underline{A} \in \overline{V}^{(0)}} S(\underline{A}, \tilde{\rho}_0 / 2) \tag{3.10}$$

The reason for the double definition (3.9), (3.10) has been explained in some detail in §2.

The problem is then choosing $\tilde{\rho}_0$. This choice was also discussed in §2. There, however, we were working in real phase space whereas here it is convenient to work in its complex extension in order to be able to systematically establish easy dimensional bounds.

So we determine $\tilde{\rho}_0$ by imposing that (see §2) in the whole complex region $W(\tilde{\rho}_0, \xi_0; V^{(0)})$

$$|\underline{\omega}_0(\underline{A}) \cdot \underline{\nu}|^{-1} \le 2C_0 \ |\underline{\nu}|^{\ell}. \tag{3.11}$$

This leads to a slight modification of (2.20):

$$\tilde{\rho}_0 = \rho_0 (2B_2 C_0 E_0 N_0{}^{\ell+1})^{-1}$$
$$B_2 = 2^3 \ell \tag{3.12}$$

(see appendix A for details).

With this choice $\tilde{\rho}_0$ is $<\rho_0/2$ since to continue the discussion we must suppose that $V^{(0)} \neq \emptyset$: we have $\bar{V}^{(0)} \neq \emptyset$. Hence for some $\underline{A} \in \bar{V}^{(0)}$, $|\underline{\omega}_0(\underline{A})\cdot\underline{\gamma}|^{-1} \leq C_0|\underline{\gamma}|^{\ell}$ and, taking $\underline{\gamma} = (0, .., 1, ..0)$, also $|\underline{\omega}_0(\underline{A})_i| \geq C_0^{-1}$, i.e. $E_0 > C_0^{-1}$.

It is clear that

$$V^{(0)} \subset V \ , \ W(\tilde{\rho}_0,\xi_0;V^{(0)}) \subset W(\tilde{\rho}_0,\xi_0;V).$$

The reader may wonder why one removes the resonances of the free Hamiltonian from what is to become the phase space of the first renormalized Hamiltonian. It would perhaps be more natural to take out the resonances of $h_1(\underline{A}) = h_0(\underline{A}) + f_{00}(\underline{A})$. However it will turn out that the region that we are throwing away is so large that it makes no difference whether we define the new phase space as the old one minus the resonances of the "old" (h_0) or of the "new" (h_1) free Hamiltonian.

3) We define the generating function Φ_0 of the change of variables in the whole complex region $W(\tilde{\rho}_0,\xi_0;V^{(0)})$.

We are, again, so careful in defining our functions in complex domains to save calculations. It is very easy to estimate derivatives of a function at a point $(\underline{A},\underline{z})$ inside some domain W via dimensional estimates like (1.24).

The definition of Φ_0, motivated by perturbation theory, is

$$\Phi_0(\underline{A}',\underline{z}) = \sum_{0<|\underline{\gamma}|\leq N_0} \frac{f_{0\underline{\gamma}}(\underline{A}')\underline{z}^{\underline{\gamma}}}{-i\underline{\omega}_0(\underline{A}')\cdot\underline{\gamma}} \ . \tag{3.14}$$

This function is easily bounded, by (3.5), in $W(\tilde{\rho}_0,\xi_0-\delta_0;V^{(0)})$:

$$\left|\frac{\partial\Phi_0}{\partial\underline{A}'}(\underline{A}',\underline{z})\right| \leq B_3'\varepsilon_0 C_0 E_0 C_0 \delta_0^{-3\ell} \tag{3.15}$$

$$\left|\frac{\partial\Phi_0}{\partial\underline{\phi}}(\underline{A}',\underline{z})\right| \leq B_3''\varepsilon_0 \rho_0 C_0 \delta_0^{-2\ell} \tag{3.16}$$

with

$$B_3' = \ell!\ 2^{3\ell+1}\ell \qquad B_3'' = 2^{4(\ell+1)}(2\ell)!\,\ell.$$

Since it is straightforward, the calculation is relegated to appendix B.

4) Next we define the map of $W(\tilde{\rho}_0, \xi_0 - \delta_0; V^{(0)})$ into $C^{2\ell}$: $(\underline{A}', \underline{z}) \longrightarrow (\underline{A}, \underline{z}')$ as

$$\underline{A} = \underline{A}' + \frac{\partial \Phi_0}{\partial \underline{\phi}} (\underline{A}', \underline{z})$$

$$\underline{z}' = \underline{z} \exp i \frac{\partial \Phi_0}{\partial \underline{A}'} (\underline{A}', \underline{z}) \tag{3.17}$$

where, to compactify the notation, we set $\underline{w}\, e^{\underline{\mu}} = (w_1 e^{\mu_1}, \ldots, w_\ell\, e^{\mu_\ell})$, $\forall\, \underline{w}, \underline{\mu} \in C^\ell$.

5) We try to invert the second line of (3.17) to express $\underline{z}$ in terms of $(\underline{A}', \underline{z}')$ at fixed $\underline{A}'$ and the first line of (3.17) to express $\underline{A}'$ in terms of $(\underline{A}, \underline{z})$. We look for inversions of the respective form

$$\underline{z} = \underline{z}' \exp i\, \underline{\Delta}(\underline{A}', \underline{z}')$$

$$\underline{A}' = \underline{A} + \underline{\Xi}'(\underline{A}, \underline{z}) \quad . \tag{3.18}$$

To do this we have to use an implicit function theorem.

Clearly the condition for this to be possible will have to look like:

$$\frac{\partial \Phi_0}{\partial \underline{\phi}} (\underline{A}', \underline{z}) \qquad \text{small (compared to } \tilde{\rho}_0)$$

$$\frac{\partial^2 \Phi_0}{\partial \underline{A}' \partial \underline{\phi}} (\underline{A}', \underline{z}) \qquad \text{small (compared to 1)} \tag{3.19}$$

to invert the first, or

$$\frac{\partial^2 \Phi_0}{\partial \underline{\phi} \partial \underline{A}'} (\underline{A}', \underline{z}) \qquad \text{small (compared to 1)} \tag{3.20}$$

to invert the second equation of (3.17).

However, for the usual reasons, we cannot expect to be able to bound the derivatives of Φ_0 on its holomorphy domain in a useful way. So we shall invert the second equation in

$$W(\tilde{\rho}_0,\xi_0-2\delta_0;V^{(0)}) \tag{3.21}$$

and the first in

$$W(\frac{\tilde{\rho}_0}{2},\xi_0-\delta_0;V^{(0)}) \tag{3.22}$$

giving up a little holomorphy as much as appears sufficient for the later developments (the philosophy being always to give up as little holomorphy as possible).

In these domains we can easily dimensionally estimate the derivatives (3.19), (3.20) using (3.15) and (3.16):

$$\begin{aligned} \left|\frac{\partial\Phi_0}{\partial\underline{\phi}}(\underline{A}',\underline{z})\right| &\le B_3'' \varepsilon_0\rho_0 C_0\delta_0^{-2\ell} \\ \left|\frac{\partial^2\Phi_0}{\partial\underline{A}'\partial\underline{\phi}}(\underline{A}',\underline{z})\right| &\le B_3'' \varepsilon_0\rho_0 C_0\delta_0^{-2\ell}\frac{1}{\frac{\tilde{\rho}_0}{2}} \end{aligned} \tag{3.23}$$

and

$$\left|\frac{\partial^2\Phi_0}{\partial\underline{\phi}\partial\underline{A}'}(\underline{A}',\underline{z})\right| \le B_3' \varepsilon_0 C_0 E_0 C_0\delta_0^{-3\ell}\frac{1}{\delta_0} \; . \tag{3.24}$$

Recalling the expression $\tilde{\rho}_0 = \rho_0/2B_2E_0C_0N_0^{\ell+1}$ we see that (3.19), and (3.20) follow by (3.23), (3.24), from a single condition of the type:

$$B_4\,\varepsilon_0 C_0 E_0 C_0 N_0^{\ell+1}\delta_0^{-3\ell-1} < 1 \; . \tag{3.25}$$

So, if we impose (3.25) with B_4 large enough, it should be possible to define functions $\underline{\Xi}'$, $\underline{\Delta}$, holomorphic in the regions (3.22), (3.21) respectively and inverting the first or the second equations of (3.17).

We also see from (3.17) that the following identities should be satisfied:

$$\begin{aligned} \underline{\Xi}'(\underline{A},\underline{z}) &= -\frac{\partial\Phi_0}{\partial\underline{\phi}}(\underline{A}',\underline{z}) \\ \underline{\Delta}(\underline{A}',\underline{z}') &= \frac{\partial\Phi_0}{\partial\underline{A}'}(\underline{A}',\underline{z}) \end{aligned} \tag{3.26}$$

and from (3.23), (3.15), (3.16), (3.25)

$$|\underline{\Xi}'| \le B_3'' \varepsilon_0\rho_0 C_0\delta_0^{-2\ell} < \frac{\tilde\rho_0}{\delta_0} \qquad \text{in } W(\frac{\tilde\rho_0}{2},\xi_0-\delta_0;V^{(0)})$$

$$|\underline{\Delta}| \le B_3' \varepsilon_0 C_0 E_0 C_0\delta_0^{-3\ell} < \delta_0 \qquad \text{in } W(\frac{\tilde\rho_0}{2},\xi_0-2\delta_0;V^{(0)}) \tag{3.27}$$

where the inequality on the r.h.s. is a consequence (3.25) (possibly increasing there the factor B_4).

The (3.27) show that it would make sense to consider

$$\underline{\Delta}'(\underline{A},\underline{z}) = \frac{\partial\Phi_0}{\partial\underline{A}'}(\underline{A}+\underline{\Xi}'(\underline{A},\underline{z}),\underline{z}) \text{ in } W(\frac{\tilde\rho_0}{2},\xi_0-\delta_0;V^{(0)})$$

$$\underline{\Xi}(\underline{A}',\underline{z}') = \frac{\partial\Phi_0}{\partial\underline{\phi}}(\underline{A}',\underline{z}' \exp i\,\underline{\Delta}(\underline{A}',\underline{z}')) \text{ in } W(\frac{\tilde\rho_0}{2},\xi_0-2\delta_0;V^{(0)}) \tag{3.28}$$

since Φ_0 is defined in $W(\tilde\rho_0,\xi_0;V^{(0)})$.

This makes it possible to define

$$C^{(0)}:\ (\underline{A}',\underline{z}') \longrightarrow (\underline{A},\underline{z}) \quad \text{as} \quad \begin{cases} \underline{A} = \underline{A}' + \underline{\Xi}(\underline{A}',\underline{z}') \\ \underline{z} = \underline{z}' \exp i\,\underline{\Delta}(\underline{A}',\underline{z}') \end{cases}$$

$$\tilde{C}^{(0)}:\ (\underline{A},\underline{z}) \longrightarrow (\underline{A}',\underline{z}') \text{ as } \begin{cases} \underline{A}' = \underline{A} + \underline{\Xi}'(\underline{A},\underline{z}) \\ \underline{z}' = \underline{z} \exp i\,\underline{\Delta}'(\underline{A},\underline{z}) \end{cases}.$$

By (3.26), (3.27), (3.16), the functions $\underline{\Delta},\underline{\Delta}',\underline{\Xi},\underline{\Xi}'$ will verify

$$-\underline{\Delta}(\underline{A}',\underline{z}') = \frac{\partial\Phi_0}{\partial\underline{\phi}}(\underline{A}',\underline{z}) = \underline{\Delta}'(\underline{A},\underline{z})$$

$$-\underline{\Xi}(\underline{A}'.\underline{z}') = \frac{\partial\Phi_0}{\partial\underline{\phi}}(\underline{A}',\underline{z}) = \underline{\Xi}'(\underline{A},\underline{z}) \tag{3.30}$$

and

$$|\underline{\Delta}|\ ,\ |\underline{\Delta}'| \le B_3' \varepsilon_0 C_0 E_0 C_0\delta_0^{-3\ell} < \delta_0$$

$$|\underline{\Xi}|\ ,\ |\underline{\Xi}'| \le B_3'' \varepsilon_0\rho_0 C_0\delta_0^{-2\ell} < \tilde\rho_0/8 \tag{3.31}$$

in their respective domains of definition.

Furthermore (3.30), (3.31) show that $\mathcal{C}^{(0)}$, $\tilde{\mathcal{C}}^{(0)}$ will be "real" on $V^{(0)}xT^{\ell}$ and will have a common domain of definition

$$\begin{aligned} \mathcal{C}^{(0)}: \quad & W(\frac{\tilde{\rho}_0}{4},\xi_0-3\delta_0;V^{(0)}) \longrightarrow W(\frac{\tilde{\rho}_0}{2},\xi_0-2\delta_0;V^{(0)}) \\ \tilde{\mathcal{C}}^{(0)}: \quad & W(\frac{\tilde{\rho}_0}{4},\xi_0-3\delta_0;V^{(0)}) \longrightarrow W(\frac{\tilde{\rho}_0}{2},\xi_0-2\delta_0;V^{(0)}) \end{aligned} \qquad (3.32)$$

on which their composition is defined and, of course, is the identity map $\tilde{\mathcal{C}}^{(0)}\mathcal{C}^{(0)} = \mathcal{C}^{(0)}\tilde{\mathcal{C}}^{(0)}$ = identity on $W(\frac{\tilde{\rho}_0}{4},\xi_0-3\delta_0;V^{(0)})$.

By construction $\mathcal{C}^{(0)}$ and $\tilde{\mathcal{C}}^{(0)}$ will be canonical on $V^{(0)}xT^{\ell}$.

All the above constructions rest on the statement on the implicit functions used in connection with (3.25).

Clearly, the above statement was guessed on purely dimensional grounds. Therefore it should not only be true but also quite simple to prove under our holomorphy assumptions (and "false" otherwise).

This is actually so and is discussed in appendix C where it is shown that one can take

$$B_4 = 2^{4(\ell+3)}(2\ell+2)!. \qquad (3.33)$$

Let us summarize the results of the above discussion. Under the condition

$$B_4(\varepsilon_0 C_0)(E_0 C_0)N_0^{\ell+1}\delta_0^{-2\ell} < 1 \qquad (3.34)$$

one does actually define a pair of canonical maps $\mathcal{C}^{(0)}$ and $\tilde{\mathcal{C}}^{(0)}$

$$\begin{aligned} \mathcal{C}^{(0)}(\underline{A}',\underline{z}') &= (\underline{A}'+\underline{\Xi}'(\underline{A}',\underline{z}'),\ \underline{z}'\exp i\,\underline{\Delta}(\underline{A}',\underline{z}')) \quad \text{on } W(\tilde{\rho}_0,\xi_0-2\delta_0;V^{(0)}) \\ \tilde{\mathcal{C}}^{(0)}(\underline{A},\underline{z}) &= (\underline{A}+\underline{\Xi}'(\underline{A},\underline{z}),\ \underline{z}\exp i\,\underline{\Delta}'(\underline{A},\underline{z})) \quad \text{on } W(\frac{\tilde{\rho}_0}{2},\xi_0-2\delta_0;V^{(0)}) \end{aligned} \qquad (3.35)$$

with $\underline{\Xi}$, $\underline{\Delta}$ holomorphic in $W(\tilde{\rho}_0,\xi_0-2\delta_0;V^{(0)})$ and $\underline{\Xi}'$, $\underline{\Delta}'$ holomorphic in $W(\frac{\tilde{\rho}_0}{2},\xi_0-\delta_0;V^{(0)})$, verifying the bounds:

$$\begin{aligned} |\underline{\Delta}|,|\underline{\Delta}'| &\le B_3'\ \varepsilon_0 C_0 E_0 C_0 \delta_0^{-3\ell} < \delta_0 \\ |\underline{\Xi}|,|\underline{\Xi}'| &\le B_3''\ \varepsilon_0 C_0 \rho_0 \delta_0^{-2\ell} < \frac{\tilde{\rho}_0}{8} \end{aligned} \qquad (3.36)$$

in the respective domains of definition. Furthermore (3.32) holds.

Having constructed the canonical transformations $\mathcal{C}^{(0)}, \tilde{\mathcal{C}}^{(0)}$, when (3.34) holds, we can go to the next step.

6) Assuming the validity of (3.34) we can study in the new variables $(\underline{A}',\underline{\phi}')$ the Hamiltonian flow which, in the old variables, takes place in $\mathcal{C}^{(0)}(V^{(0)}xT^{\ell}) \subset VxT^{\ell}$.

In the new variables the Hamiltonian is:

$$H_1(\underline{A}',\underline{z}') = h_0(\underline{A}'+\Xi(\underline{A}',\underline{z}')) + f_0(\underline{A}'+\Xi(\underline{A}',\underline{z}'),\ \underline{z}'\, e^{i\underline{\Delta}(\underline{A}',\underline{z}')}) \tag{3.37}$$

which, as suggested by classical perturbation theory, should be written as $H_1 = h_1 + f_1$ with $f_1 = H_1 - h_1$ and

$$h_1(\underline{A}') = h_0(\underline{A}') + f_{\underline{00}}(\underline{A}') \tag{3.38}$$

and, on the basis of perturbation theory, we expect f_1 to be $\sim O(\varepsilon_0^2 N_0^{\ell+1})$.

The functions h_1 and f_1 are defined on $W(\tilde{\rho}_0,\xi_0-2\delta_0;\underline{A}_0)$. However for the purpose of iterating the algorithm we shall measure their size in a smaller region $W(\rho_1,\xi_1;V^{(0)})$ with

$$\rho_1 = \frac{\tilde{\rho}_0}{8} = \rho_0/16\ B_2\ E_0 C_0 N_0^{\ell+1}$$
$$\xi_1 = \xi_0 - 4\delta_0 \tag{3.39}$$

and, for later reference, we remark that

$$\mathcal{C}^{(0)}W(\rho_1,\xi_1;V^{(0)}) \subset W(\rho_0,\xi_0;V)$$
$$\mathcal{C}^{(0)}(V^{(0)}xT^{\ell}) \subset VxT^{\ell} \quad . \tag{3.40}$$

The above sentence means that we shall look for estimates E_1, η_1, ε_1 of

$$\sup\left|\frac{\partial h_1}{\partial \underline{A}'}\right| ,\ \sup\left|\left(\frac{\partial^2 h_1}{\partial \underline{A}'\partial \underline{A}'}\right)^{-1}\right| ,\ \sup\left(\left|\frac{\partial f_1}{\partial \underline{A}'}\right| + \rho_1^{-1}\left|\frac{\partial f_1}{\partial \underline{\phi}'}\right|\right) \tag{3.41}$$

valid in the whole domain $W(\rho_1,\xi_1;V^{(0)})$ of H_1.

We expect, as can easily be seen by dimensional considerations, estimates of the type

$$\sup\left|\frac{\partial h_1}{\partial \underline{A}'}\right| \le E_1 = E_0 + \varepsilon_0 \qquad \text{in } W(\rho_0,\xi_0;V) \tag{3.42}$$

$$\sup\left|\left(\frac{\partial h_1}{\partial \underline{A}'\partial \underline{A}'}\right)^{-1}\right| \le \bar{\eta}_1 = \eta_0\,(1+\bar{B}_5\eta_0\varepsilon_0\rho_0^{-1}) \qquad \text{in } W(\tfrac{\rho_0}{2},\xi_0;V) \tag{3.43}$$

$$\sup\left(\left|\frac{\partial f_1}{\partial \underline{A}'}\right|+\rho_1^{-1}\left|\frac{\partial f_1}{\partial \underline{\phi}'}\right|\right) \le \bar{\varepsilon}_1 = \bar{B}_6\,\varepsilon_0^{\,2}C_0(E_0C_0)^2N_0^{\,\ell+1}\delta_0^{-\bar{x}} \qquad \text{in } W(\rho_1,\xi_1;V^{(0)}) \tag{3.44}$$

if (3.34) holds and if

$$\bar{B}_5\,\eta_0\,\varepsilon_0\rho_0^{-1} < 1. \tag{3.45}$$

The above inequalities can in fact easily be proved by dimensional estimates (relegated to appendix D,E). We find

$$\bar{B}_5 = 4\ell^2$$

$$\bar{B}_6 = 2^{-2}\,2^{10(\ell+1)}\,\ell(\ell!)^2 \qquad \bar{x} = 4\ell+1. \tag{3.46}$$

For the purposes of a simpler discussion of the iterations it is convenient to replace the bound $\bar{\varepsilon}_1$ by the weaker bound ε_1 (see appendix E):

$$\varepsilon_1 = B_6(\varepsilon_0C_0\delta_0^{\,\ell})^{\frac{3}{2}}\;E_0^{\,2}C_0\;\delta_0^{-7\ell-2}$$
$$B_6 = 2^{14\ell+12}\,(\ell+1)!^4 \quad . \tag{3.47}$$

It will be convenient to replace the conditions (3.45), (3.46) by a single condition. To do this we use $E_0C_0 \ge 1$ to replace (3.45) by the stronger condition $\bar{B}_5\,\varepsilon_0C_0\eta_0E_0\rho_0^{-1} < 1$ which, to simplify the bound for η_1 (see below), we replace by the even stronger $\bar{B}_5\sqrt{C_0\varepsilon_0}\,(\eta_0E_0\rho_0^{-1}) < 1$ (since $\varepsilon_0C_0 < 1$, as assumed at the beginning). Then we combine (3.34) and the latter inequality into a stronger one:

$$B_4\bar{B}_5\;\varepsilon_0C_0E_0C_0N_0^{\,\ell+1}\;\delta_0^{-2\ell}(\eta_0E_0\rho_0^{-1})^2 < 1. \tag{3.48}$$

If (3.48) holds and since $\bar{B}_5\eta_0\varepsilon_0\rho_0^{-1} < \sqrt{C_0\varepsilon_0}$ we can bound $\bar{\eta}_1$ by a simpler formula:

$$\bar{\eta}_1 \le \eta_0 (1 + \sqrt{C_0\varepsilon_0}) \ . \qquad (3.49)$$

The above estimates end step 6.

7) The first six steps allow us to define a new phase space $V^{(0)} x T^\ell$ and a Hamiltonian $h_1 + f_1$ holomorphic in $W(\rho_1, \xi_1; V^{(0)})$ characterized by the parameters ε_1, η_1, E_1 where

$$\begin{aligned}
\xi_1 &= \xi_0 - 4\delta_0 \\
\rho_1 &= \rho_0/B_8\, E_0 C_0 \delta_0^{-\ell-1} (\log(\varepsilon_0 C_0 \delta_0^{+\ell})^{-1})^{\ell+1} \\
E_1 &= E_0 + \varepsilon_0 \qquad (3.50)\\
\eta_1 &= \eta_0(1 + \sqrt{C_0\varepsilon_0}) \\
\varepsilon_1 &= B_6(\varepsilon_0 C_0 \delta_0^{\ell})^{3/2}\, E_0^2 C_0 \delta_0^{-7\ell-2}
\end{aligned}$$

provided

$$B_7 \varepsilon_0 C_0 E_0 C_0 N_0^{\ell+1} \delta_0^{-2\ell} (\eta_0 E_0 \rho_0^{-1})^2 < 1 \qquad (3.51)$$

with $B_6 = 2^{14\ell+12}(\ell+1)!^4$, $B_7 = B_4\bar{B}_5 = 2^{6\ell+5}(\ell+1)!^2$, $B_8 = \ell 2^{\ell+8} = 2^4 B_2 2^{\ell+1}$.

We now wish to iterate the procedure.

We only have to explain how $V^{(1)}$ will be constructed and, later, how $V^{(k)}$ will be constructed for $k \ge 1$.

The idea will simply be to set:

$$\bar{V}^{(k)} = \{\underline{A} | \underline{A} \in V^{(k-1)}, \text{distance of } \underline{A} \text{ from } \partial V^{(k-1)} > \frac{\tilde{\rho}_k}{2}\} \cap \qquad (3.52)$$

$$\cap \ \{\text{non resonant points of order } (C_k, N_k) \text{ for } h_k\}$$

and (see (3.10)):

$$V^{(k)} = \bigcup_{\underline{A}\in\overline{V}^{(k)}} S(\underline{A},\frac{\tilde{\rho}_k}{2}) \qquad (3.53)$$

where $N_k = 2\delta_k^{-1} \log(\varepsilon_k C_k \delta_k^{+\ell})^{-1}$.

Then the whole argument can be iterated leading to the relations*

$$\begin{aligned}
\xi_{k+1} &= \xi_k - 4\delta_k \\
E_{k+1} &= E_k + \varepsilon_k \\
\eta_{k+1} &= \eta_k\,(1+\sqrt{\varepsilon_k C_k}) \qquad (3.54)\\
\rho_{k+1} &= \rho_k / B_6\, E_k C_k \delta_k^{-\ell-1} (\log \mu_k^{-1})^{\ell+1} \\
\mu_{k+1} &= B_6\, \mu_k^{3/2} (E_k C_k)^2\, \mu_k^{-6\ell-2}\, \frac{C_{k+1}}{C_k} (\frac{k+1}{k+2})^2
\end{aligned}$$

where $\mu_k \equiv \varepsilon_k C_k \delta_k^{\ell}$, provided

$$B_7 \mu_k E_k C_k N_k^{\ell+1} \delta_k^{-3\ell} (\eta_k E_k \rho_k^{-1})^2 < 1 \,. \qquad (3.55)$$

The value of C_k is still free. However for the reasons explained in comment 4) at the end of §2 we shall choose

$$C_k = C_0 (1+k)^2. \qquad (3.56)$$

The induction algorithm is now ready and we can iterate it indefinitely as long as (3.55) holds.

The study of the sequence $(\xi_k, \rho_k, E_k, \eta_k, \mu_k)$ generated by (3.54) leads, after some straightforward manipulations (see appendix F) to

---

*using the explicit form of δ_k to write $\frac{\delta_{k+1}}{\delta k} = (\frac{k+1}{k+2})^2$.

the result:

$$\mu_k \geq (\mu_0)^{(3/2)^k}$$

$$\mu_k \leq (B_9 (E_0 C_0)^4 \xi_0^{-4(3\ell+1)} \mu_0)^{(3/2)^k}$$

$$\left(\frac{\rho_k}{\rho_0}\right)^{-1} \leq (B_{10} \xi_0^{-\ell-1} E_0 C_0 (\log \mu_0^{-1})^{\ell+1})^k \cdot [(k+1)!^2 (\tfrac{3}{2})^{\frac{k(k+1)}{2}}]^{\ell+1} \tag{3.57}$$

$$E_k \leq 4\, E_0$$

$$\eta_k \leq 4\, \eta_0$$

provided the condition

$$B_{11}\ \varepsilon_0 C_0 \xi_0^{\ell} (E_0 C_0)^{10} (E_0 \eta_0 \rho_0^{-1})^5 \xi_0^{-10(3\ell+1)} < 1 \tag{3.58}$$

holds.

The constants are $B_9 = 2^{100\ell+68}(\ell+1)!^8$, $B_{10} = 2^{5\ell+14}\ell$, $B_{11} = (\ell+20)!^{20}$

If (3.58) holds it is furthermore true that (3.55) holds $\forall\, k \geq 0$. So (3.58) is the condition permitting infinitely many iterations.

§4. Consequences of the infinite renormalizability of the Hamiltonian.

1) Having established the possibility of infinitely many iterations of the algorithm described in §3 we shall suppose that (3.58) is valid and we can therefore consider the sequences $C^{(0)}, C^{(1)}, \dots, \tilde{C}^{(0)}, \tilde{C}^{(1)}, \dots$ of canonical maps associated with the iterations.

We write

$$C^{(n)}(\underline{A}',\underline{z}') = \begin{cases} \underline{A}' + \underline{\Xi}_{(n)}(\underline{A}',\underline{z}') \\ \underline{z}' \exp i\, \underline{\Delta}_{(n)}(\underline{A}',\underline{z}') \end{cases}$$
$$\tilde{C}^{(n)}(\underline{A}',\underline{z}') = \begin{cases} \underline{A} + \underline{\Xi}'_{(n)}(\underline{A},\underline{z}) \\ \underline{z}\ \exp i\, \underline{\Delta}'_{(n)}(\underline{A},\underline{z}) \end{cases} \quad . \tag{4.1}$$

Our next task will be to choose convenient domains for considering the compositions $C^{(0)} \dots C^{(n-1)} \equiv C_{n-1}$ and $\tilde{C}^{(n-1)} \dots \tilde{C}^{(0)} \equiv \tilde{C}_{(n-1)}$.

The following choice for the domain W_n of $\tilde{C}_{(n-1)}$ is natural:

$$W_n = C^{(0)} \dots C^{(n-1)}(W(\rho_n,\xi_n;V^{(n-1)})) \tag{4.2}$$

where one should notice that $W(\rho_n,\xi_n;V^{(n-1)})$ is a natural domain for $C^{(n-1)}$.

Then $\tilde{C}_{(n-1)}$ is well defined on W_n since, $\forall\, n \geq 0$:

$$C^{(n)} W(\rho_{n+1},\xi_{n+1};V^{(n)}) \subset W(\tfrac{\tilde{\rho}_n}{4},\xi_n - 3\delta_n;V^{(n)}) \subset W(\rho_n,\xi_n;V^{(n-1)}) \tag{4.3}$$

by (3.36) and by the definition of $V^{(n)}$ (see (3.51), (3.52)).

In fact (4.3) implies that

$$W_n \subset W_{n-1} \subset \dots \subset W_1 = C^{(0)}(W(\rho_1,\xi_1;V^{(0)})) \tag{4.4}$$

which can be used inductively to see that, on W_n, $\tilde{C}_{(n-1)} = \tilde{C}^{(n-1)} \dots$

$\tilde{C}^{(0)}$ is well defined.

Also, if we define

$$\Gamma_n = C^{(0)} \ldots C^{(n-1)}(V^{(n-1)} x T^{\ell}) \tag{4.5}$$

it follows from the second line of (3.40) that

$$\Gamma_n \supset \Gamma_{n-1} \ . \tag{4.6}$$

For later use we recall that the maps $C^{(n)}$, $\tilde{C}^{(n)}$ are well defined and holomorphic on the sets

$$W(\frac{\tilde{\rho}_n}{2}, \xi_n - 2\delta_n ; V^{(n)}) \tag{4.7}$$

and there (see (3.36)):

$$\begin{aligned} |\underline{\Xi}_{(n)}| \ , \ |\underline{\Xi}'_{(n)}| &\le B''_3 \ \varepsilon_n C_n \rho_n \delta_n^{-3\ell} < \frac{\tilde{\rho}_n}{8} \\ |\underline{\Delta}_{(n)}| \ , \ |\underline{\Delta}'_{(n)}| &\le B'_3 \ \varepsilon_n C_n \rho_n \delta_n^{-2\ell} < \delta_n \end{aligned} \tag{4.8}$$

2) Having been able to define the canonical maps $\tilde{C}_{(n-1)}$ on W_n and the maps $C_{(n-1)}$ on $W(\rho_n, \xi_n ; V^{(n-1)})$ we set

$$W_\infty = \bigcap_{n=1}^{\infty} W_n \ , \ \Gamma_\infty = \bigcap_{n=1}^{\infty} \Gamma_n \tag{4.9}$$

We wish to show that the limits

$$(\underline{A}_\infty(\underline{A},\underline{z}), \underline{z}_\infty(\underline{A},\underline{z})) = \lim_{n \to \infty} \tilde{C}^{(n-1)} \ldots \tilde{C}^{(0)}(\underline{A},\underline{z}) \tag{4.10}$$

exist for all $(\underline{A},\underline{z}) \in W_\infty$.

In fact $\tilde{C}^{(n)}$ differs from the identity map by a quantity (see (4.1)) that can be easily estimated dimensionally by remarking that because of (4.3), (4.4) we are interested only in its values on a subset of $W(\frac{\tilde{\rho}_n}{4}, \xi_n - 3\delta_n ; V^{(n)})$.

Since $\underline{\Xi}'_{(n)}$, $\underline{\Delta}'_{(n)}$ are holomorphic in $W(\frac{\tilde{\rho}_n}{2}, \xi_n - 2\delta_n ; V^{(n)})$ (see (4.7), (4.8)) we get by a dimensional estimate:

$$\left|\frac{\partial^{\underline{k}+\underline{h}}}{\partial \underline{A}^{\underline{k}}\partial \underline{z}^{\underline{h}}}(\pi_1\tilde{\mathcal{C}}^{(n)}-\pi_1)\right| \equiv \left|\frac{\partial^{\underline{k}+\underline{h}}}{\partial \underline{A}^{\underline{k}}\partial \underline{z}^{\underline{h}}}\ \underline{\Xi}'_{(n)}\right|$$

$$\le B_3'' \varepsilon_n C_n \rho_n \delta_n^{-2\ell} \frac{\underline{k}!}{(\frac{\tilde{\rho}_n}{4})^{|\underline{k}|}} \frac{\underline{h}!}{\delta_n^{|\underline{k}|}} (2e)^{|\underline{h}|} \tag{4.11}$$

$$\left|\frac{\partial^{\underline{k}+\underline{h}}}{\partial \underline{A}^{\underline{k}}\partial \underline{z}^{\underline{h}}}(\pi_2\tilde{\mathcal{C}}^{(n)}-\pi_2)\right| \equiv \left|\frac{\partial^{\underline{k}+\underline{h}}}{\partial \underline{A}^{\underline{k}}\partial \underline{z}^{\underline{h}}}(\underline{z}\, e^{i\underline{\Delta}'_{(n)}}-\underline{z})\right|$$

$$\le e^2 B_3' \varepsilon_n C_n E_n C_n \delta_n^{-3\ell} \frac{\underline{k}!}{(\frac{\tilde{\rho}_n}{4})^{|\underline{k}|}} \frac{\underline{h}!}{\delta_n^{|\underline{h}|}} (\frac{e}{2})^{|\underline{h}|} .$$

on $W(\frac{\tilde{\rho}_n}{4}, \xi_n - 3\delta_n; V^{(n)})$.

The convergence of the sequences

$$(\underline{A}_n(\underline{A},\underline{z}), \underline{z}_n(\underline{A},\underline{z})) = \tilde{\mathcal{C}}^{(n-1)} \cdots \tilde{\mathcal{C}}^{(0)}(\underline{A},\underline{z}) \tag{4.12}$$

on W_∞ is clearly guaranteed by (4.11) and (3.57) (which implies that the r.h.s. of (4.11) converges to zero faster than any exponential as $n \longrightarrow \infty$).

(4.11) shows more. In fact, the chain rule of differentiation implies that the functions $\underline{A}_n$, $\underline{z}_n$ in (4.12) have derivatives on W_n which are bounded there by:

$$\left|\frac{\partial^{\underline{a}+\underline{b}}\underline{A}_n}{\partial \underline{A}^{\underline{a}}\partial \underline{z}^{\underline{b}}}(\underline{A},\underline{z})\right| < B_1(|\underline{a}|,|\underline{b}|)\, \rho_0^{-|\underline{a}|} \quad \forall (\underline{A},\underline{z}) \in W_n$$

$$\left|\frac{\partial^{\underline{a}+\underline{b}}\underline{z}_n}{\partial \underline{A}^{\underline{a}}\partial \underline{z}^{\underline{b}}}(\underline{A},\underline{z})\right| < B_2(|\underline{a}|,|\underline{b}|)\, \rho_0^{-|\underline{a}|} \quad \forall (\underline{A},\underline{z}) \in W_n \tag{4.13}$$

where B_1, B_2 depend on ℓ, ε_0, E_0, C_0, ξ_0, ρ_0, η_0 but not on n.

Furthermore (4.11) shows that the derivatives appearing on the ℓ.h.s. of (4.13) converge, as $n \longrightarrow \infty$, to some limit if $(\underline{A},\underline{z}) \in W_\infty$. This allows us to define some functions on W_∞ which will be naturally denoted by

$$\frac{\partial^{|\underline{a}|+|\underline{b}|}\underline{A}_\infty}{\partial \underline{A}^{\underline{a}}\partial \underline{z}^{\underline{b}}}, \quad \frac{\partial^{|\underline{a}|+|\underline{b}|}\underline{z}_\infty}{\partial \underline{A}^{\underline{a}}\partial \underline{z}^{\underline{b}}} \tag{4.14}$$

and again the convergence is faster than $(\frac{\rho_n}{\rho_0})^N$, $\forall$ N > 0.

3) We now want to show that even though the functions $\underline{A}_\infty$, $\underline{z}_\infty$ are defined on a set "filled with holes", so that one cannot define their derivatives, they nevertheless have a Taylor series at every point of W whose coefficients are just the values of the functions defined in (4.14).

This means that:

$$\left| \frac{\partial^{|\underline{a}_0|+|\underline{b}_0|} \underline{A}_\infty}{\partial \underline{A}^{\underline{a}_0} \partial \underline{z}^{\underline{b}_0}} (\underline{A}_1,\underline{z}_1) - \sum_{|\underline{a}|+|\underline{b}|\le p} \frac{\partial^{|\underline{a}_0|+|\underline{b}_0|+|\underline{a}|+|\underline{b}|} \underline{A}_\infty}{\partial \underline{A}^{\underline{a}} \partial \underline{z}^{\underline{b}}} (\underline{A}_2,\underline{z}_2)(\underline{A}_1-\underline{A}_2)^{\underline{a}}(\underline{z}_1-\underline{z}_2)^{\underline{b}} \right| \tag{4.15}$$

$$\le C(|\underline{a}_0|,|\underline{b}_0|,p)\,(|\underline{A}_2-\underline{A}_1|\rho_0^{-1}+|\underline{z}_2-\underline{z}_1|)^{p+1}$$

and a similar relation can be written for the $\underline{z}_\infty$. Here the constant $C(x,y,p)$ is $(\underline{A}_1,\underline{z}_1)$, $(\underline{A}_2,\underline{z}_2)$-independent.

The reason why a relation like (4.15) holds should be considered to be due to the enormous speed of convergence of $(\underline{A}_n,\underline{z}_n)$ (as well as of their derivatives) to $\underline{A}_\infty,\underline{z}_\infty$ and to the fact that the sets W_n are not too thin, if $\Gamma_\infty \neq \emptyset$.

More precisely we first show that if $(\underline{A},\underline{z}) \in \Gamma_\infty$ then if μ is a suitable constant W_n contains a sphere of radius $\xi_0\rho_n/4\mu\rho_0$ around $(\underline{A},\underline{z})$ provided the sphere is defined with respect to the metric in C^2 given by

$$|(\underline{A},\underline{z})-(\underline{A}',\underline{z}')| \equiv |\underline{A}-\underline{A}'|\rho_0^{-1}+|\underline{z}-\underline{z}'| \ . \tag{4.16}$$

In fact given two phase space points $(\underline{A},\underline{z})$, $(\underline{A}',\underline{z}')$ in $W(\frac{\tilde\rho_p}{4},\xi_p-3\delta_p;V^{(p)})$ with $(\underline{A},\underline{z}) \in V^{(p)}xT^\ell$ and $|(\underline{A},\underline{z})-(\underline{A}',\underline{z}')|<\tilde\rho_p/4\mu\rho_0$ then:

$$|\tilde{C}^{(p)}(\underline{A},\underline{z})-\tilde{C}^{(p)}(\underline{A}',\underline{z}')| < (1+\theta_p)\,|(\underline{A},z)-(\underline{A}',\underline{z}')|$$

with $\theta_n = B_{12}\frac{\varepsilon_n\rho_n C_n}{\rho_0}\delta_n^{-2\ell}$, $B_{12} = \ell(B_3'+3_3'')$. This is proved by dimensional estimates, see appendix G.

Let

$$\mu = \prod_{p=0}^{\infty} (1+\theta_p) > 0 .$$

Suppose now that $(\underline{A},\underline{z}) \in \Gamma_\infty$ and calling $(\underline{A}_n,\underline{z}_n) = \tilde{C}^{(n-1)}\ldots\tilde{C}^{(0)}(\underline{A},\underline{z}) \in V^{(n-1)}xT^\ell$ and $(\underline{A}'_n,\underline{z}'_n) = \tilde{C}^{(n-1)}\ldots\tilde{C}^{(0)}(\underline{A}',\underline{z}')$, one sees by induction that $|(\underline{A}',\underline{z}')-(\underline{A},\underline{z})| < \xi_0\rho_n/4\mu$ implies:

$$|\tilde{C}^{(n-1)}\ldots\tilde{C}^{(0)}(\underline{A},\underline{z}) - \tilde{C}^{(n-1)}\ldots\tilde{C}^{(0)}(\underline{A}',\underline{z}')|$$

$$\equiv |(\underline{A}_n,\underline{z}_n)-(\underline{A}'_n,\underline{z}'_n)| \leq \frac{\xi_0\rho_n}{4\mu\rho_0} \prod_{p=0}^{n-1} (1+\theta_p) \leq \frac{\xi_0\rho_n}{4\rho_0}$$

Hence, since $|(\underline{z}_n)_j| \equiv 1, \forall j = 1, \ldots, \ell, \forall n \geq 1$:

$$|\underline{A}_n-\underline{A}'_n| < \rho_n , \quad |(\underline{z}'_n)_j| \equiv |(\underline{z}'_n)_j-(\underline{z}_n)_j+(\underline{z}_n)_j| = \left|1+\frac{(\underline{z}'_n)_j-(\underline{z}_n)_j}{(\underline{z}_n)_j}\right|, \tag{4.17}$$

$$e^{-\xi_\infty} < 1 - \frac{\xi_0}{4} < 1 - \frac{\xi_0}{4}\frac{\rho_n}{\rho_0} \leq |(\underline{z}'_n)_j| \leq 1 + \frac{\xi_0\rho_n}{4\rho_0} \leq 1 + \frac{\xi_0}{4} < e^{\xi_\infty}$$

(recalling that the choice of δ_i in (3.6) implies $\xi_\infty > \xi_0/2$).

Therefore $(\underline{A}'_n,\underline{z}'_n) \in W(\rho_n,\xi_n;V^{(n-1)})$ and, consequently, $(\underline{A}',\underline{z}')$ is in W_n.

Let now $(\underline{A},\underline{z}), (\tilde{\underline{A}},\tilde{\underline{z}}) \in \Gamma_\infty$, $\tilde{\underline{z}} = e^{i\tilde{\underline{\phi}}}$, $\underline{z} = e^{i\underline{\phi}}$ and suppose that

$$\frac{\xi_0\rho_n}{4\mu\rho_0} < |(\underline{A},\underline{z}) - (\tilde{\underline{A}},\tilde{\underline{z}})| < \frac{\xi_0\rho_{n-1}}{4\mu\rho_0} \tag{4.18}$$

Then by the preceding argument the whole segment $\underline{A}(t) = \underline{A}t + \tilde{\underline{A}}(1-t)$, $\underline{\phi}(t) = \underline{\phi}t + \tilde{\underline{\phi}}(1-t)$, $t \in [0,1]$, is in W_n if we suppose, as we obviously can (without loss of generality), that the segment $t \longrightarrow \underline{\phi}(t)$ is the shortest segment joining $\underline{\phi}$ and $\underline{\phi}'$ in T^ℓ (notice that the angles $\underline{\phi}$ are defined only modulo 2π).

Therefore we can apply the Lagrange-Taylor formula to estimate $|\underline{A}_n(\underline{A},\underline{z}) - \underline{A}_n(\underline{A}',\underline{z}')|$ or, more generally to estimate the difference between two arbitrary derivatives of orders $\underline{a}_0$ with respect to the action variables and $\underline{b}_0$ with respect to the angle variables.

Given $M > 0$ it follows by (4.11) that

$$\left| \frac{\partial^{|\underline{a}_0|+|\underline{b}_0|} \underline{A}_n}{\partial \underline{A}^{\underline{a}_0} \partial \underline{z}^{\underline{b}_0}} (\underline{A},\underline{z}) - \sum_{\substack{\underline{a},\underline{b} \in Z_+^\ell \\ 0 \le |\underline{a}|+|\underline{b}| \le M}} \frac{\partial^{|\underline{a}|+|\underline{a}_0|+|\underline{b}|+|\underline{b}_0|} \underline{A}_n}{\partial \underline{A}^{\underline{a}_0+\underline{a}} \partial \underline{z}^{\underline{b}_0+\underline{b}}} (\tilde{\underline{A}},\tilde{\underline{z}}) \frac{(\underline{A}-\tilde{\underline{A}})^{\underline{a}}}{\underline{a}!} \frac{(\underline{z}-\tilde{\underline{z}})^{\underline{b}}}{\underline{b}!} \right| \tag{4.19}$$

$$\le \left[D(M) \max_{|\underline{a}|+|\underline{b}|=M+1} B_1(|\underline{a}_0|+|\underline{a}|, |\underline{b}_0|+|\underline{b}|) \; |(\underline{A},\underline{z})-(\tilde{\underline{A}},\tilde{\underline{z}})|^{M+1} \right]$$

where $D(M) = (M+1)^{2\ell} (M!)^{-2\ell}$.

So the r.h.s. of (4.19) does not explicitly depend on n and we shall take advantage of this.

Observe that because of (4.11) the ℓ.h.s. of (4.19) reaches its limit $n \to \infty$ uniformly in $(\underline{A},\underline{z})$ $(\tilde{\underline{A}},\tilde{\underline{z}})$ even though of course we cannot take the limit as $n \to \infty$ in (4.19), because of the constraint (4.18). We can still evaluate the difference between the ℓ.h.s. of (4.19) and the expression obtained from it by setting $n = +\infty$. (4.11) imply that this difference can be easily bounded. One should notice that

$$\underline{A}_\infty(\underline{A},\underline{z}) - \underline{A}_n(\underline{A},\underline{z}) = \sum_{k=n}^{\infty} \underline{\Xi}'_{(k)} (\underline{A}_k,\underline{z}_k)$$
$$\underline{z}_\infty(\underline{A},\underline{z}) - \underline{z}_n(\underline{A},\underline{z}) = \sum_{k=n}^{\infty} \underline{\Delta}'_{(k)} (\underline{A}_k,\underline{z}_k) \tag{4.20}$$

with $(\underline{A}_k(\underline{A},\underline{z}), \underline{z}_k(\underline{A},\underline{z}) = \tilde{\mathcal{C}}^{(k-1)} \dots \tilde{\mathcal{C}}^{(0)}(\underline{A},\underline{z})$. Then (4.8),(4.11) and the differentiation rules immediately yield bounds on (4.20), although of course, one gets involved in some conbinatorics. However one can also get bounds without any calculations by a dimensional estimate.

Recall that if $(\underline{A}',\underline{z}') \in C^{2\ell}$ and $|(\underline{A}',\underline{z}')-(\underline{A},\underline{z})| < \frac{\xi_0 \rho_{k+1}}{4\mu\rho_0}$ then by (4.17) $(\underline{A}_k(\underline{A}',z'),\underline{z}_k(\underline{A}',\underline{z}'))$ is in the domains of $\underline{\Xi}'_{(k)}$ and $\underline{\Delta}'_k$. This means that the functions $\underline{\Xi}'_{(k)}(\underline{A}_k(\underline{A}',\underline{z}'), \underline{z}_k(\underline{A}',\underline{z}'))$ are $\forall (\underline{A},\underline{z}) \in \Gamma_\infty$, holomorphic and bounded (by (4.8) with $\underline{h} = \underline{k} = 0$) in the poldisk:

$$|A_i'-A_i| < \frac{\xi_0 \rho_{k+1}}{4\mu} \frac{1}{2\ell} \qquad |z_i'-z_i| < \frac{\xi_0 \rho_{k+1}}{4\mu\rho_0} \frac{1}{2\ell} \; . \tag{4.21}$$

Therefore by dimensional bounds:

$$\left|\frac{\partial^{a+b}\Xi'_k(\underline{A}_k,\underline{z}_k)}{\partial \underline{A}^{\underline{a}}\partial \underline{z}^{\underline{b}}}\right| \le B''_3 C_k \varepsilon_k \rho_k \delta_k^{-2\ell}\left(\frac{8\ell\mu}{\xi_0}\frac{\rho_0}{\rho_{k+1}}\right)^{|\underline{a}|+|\underline{b}|}\rho_0^{-|\underline{a}|}\underline{a}!\underline{b}!$$

$$\left|\frac{\partial^{a+b}\Delta'_{(k)}(\underline{A}_k,\underline{z}_k)}{\partial \underline{A}^{\underline{a}}\partial \underline{z}^{\underline{b}}}\right| \le B'_3\varepsilon_k C_k E_k C_k \delta_k^{-3\ell}\left(\frac{8\ell\mu}{\xi_0}\frac{\rho_0}{\rho_{k+1}}\right)^{|\underline{a}|+|\underline{b}|}\rho_0^{-|\underline{a}|}\underline{a}!\underline{b}!. \tag{4.22}$$

Hence, calling

$$\frac{1}{2}\zeta_n(p,q) = \sum_{k=n}^{\infty} B''_3 C_k \varepsilon_k \rho_k \delta_k^{-2\ell}\left(\frac{8\ell\mu}{\xi_0}\frac{\rho_0}{\rho_{k+1}}\right)^{p+q} p!q! \tag{4.23}$$

we see that the difference between the (ℓ.h.s.) of (4.19) and its value for $n = +\infty$ is bounded by

$$\left(\sum_{0\le|\underline{a}|+|\underline{b}|\le M} \frac{\zeta_n(|\underline{a}_0|+|\underline{a}|,|\underline{b}_0|+|\underline{b}|)}{\underline{a}!\underline{b}!}\rho_0^{-|\underline{a}_0|}\left(\frac{\xi_0}{4\mu}\frac{\rho_{n-1}}{\rho_0}\right)^{|\underline{a}|+|\underline{b}|}\right) \equiv \tag{4.24}$$

$$\equiv \bar{\zeta}(|(\underline{A},\underline{z})-(\tilde{\underline{A}},\tilde{\underline{z}})|,|\underline{a}_0|,|\underline{b}_0|,M)$$

where the function $\bar{\zeta}(x;p,q,M)$ is defined as the ℓ.h.s. of (4.24) with $|\underline{a}_0| = p$, $|\underline{b}_0| = q$ when $x \in [\frac{\rho_{n-1}}{4\mu},\frac{\rho_n\xi_0}{4\mu})$. Clearly because of the bounds (3.57) $\bar{\zeta}(x;p,q,M) \xrightarrow[x\to 0]{} 0$ faster than any power of x.

7) Hence we deduce from (4.19):

$$\left|\frac{\partial^{|\underline{a}_0|+|\underline{b}_0|}\underline{A}_\infty}{\partial \underline{A}^{\underline{a}_0}\partial \underline{z}^{\underline{b}_0}} - \sum_{|\underline{a}|+|\underline{b}|<M}\frac{\partial^{|\underline{a}_0|+|\underline{a}|+|\underline{b}_0|+|\underline{b}|}\underline{A}_\infty}{\partial \underline{A}^{\underline{a}_0+\underline{a}}\partial \underline{z}^{\underline{b}_0+\underline{b}}}\frac{(\underline{A}-\tilde{\underline{A}})^{\underline{a}}}{\underline{a}!}\frac{(\underline{z}-\tilde{\underline{z}})^{\underline{b}}}{\underline{b}!}\right| \le$$

$$\le \{D(M)\max_{|\underline{a}|+|\underline{b}|<M+1} B_1(|\underline{a}_0|+|\underline{a}|,|\underline{b}_0|+|\underline{b}|)\}|(\underline{A},\underline{z})-(\tilde{\underline{A}},\tilde{\underline{z}})|^{M+1} \tag{4.25}$$

$$+\bar{\zeta}(|(\underline{A},\underline{z})-(\tilde{\underline{A}},\tilde{\underline{z}})|,|\underline{a}_0|,|\underline{b}_0|,M)$$

where on the ℓ.hs. the first derivative is computed at $(\underline{A},\underline{z})$ and the others at $(\tilde{\underline{A}},\tilde{\underline{z}})$.

An identical argument with similar conclusions can be carried out for $\underline{z}_\infty$ using the second line of (4.8), (4.22) instead of the first. Hence the functions $(A_\infty, \underline{z})$ are 2ℓ functions on Γ_∞. Γ_∞ as well as the sets W_∞ and V^∞ are closed (because the boundary of $V^{(n)}$ has no intersection with that of $V^{(n-1)}$). It is a general result that such functions can always be regarded as restrictions of $C^\infty(V^\ell x T^\ell)$-functions to Γ_∞, [10].

4) By construction the functions $\underline{z}_\infty$ have the form $\underline{z}_\infty = e^{i\underline{\phi}_\infty}$, $\underline{\phi}_\infty \in T^\ell$ and

$$\underline{A}_\infty = \underline{A} + \underline{\alpha}(\underline{A},\underline{\phi})$$
$$\underline{\phi}_\infty = \underline{\phi} + \underline{\beta}(\underline{A},\underline{\phi}) \tag{4.26}$$

with $\underline{\alpha},\underline{\beta}$ in $C^\infty(\Gamma^{(\infty)})$ i.e. if one wishes, $\underline{\alpha},\underline{\beta}$ have extensions to $C^\infty(VxT^\ell)$.

The functions $(\underline{A}_\infty,\underline{\phi}_\infty)$ being limits of canonical variables $(\underline{A}_n,\underline{\phi}_n)$, by construction verify the canonical commutation relations

$$\{A_{\infty i},A_{\infty j}\} = 0 \ , \ \{\underline{\phi}_{\infty i},\underline{\phi}_{\infty j}\} = 0 \ , \ \{A_{\infty i},\phi_{\infty j}\} = \delta_{ij} \tag{4.27}$$

on Γ_∞, which make sense since the $\underline{A}_\infty,\underline{\phi}_\infty$ admit derivatives on Γ_∞ defined as the restrictions to Γ_∞ of the derivatives of their C^∞-extensions (and coinciding with the ones defined in (4.14) above).

The map $(\underline{A},\underline{\phi}) \longleftrightarrow (\underline{A}_\infty,\underline{\phi}_\infty)$ is by construction one-to-one on Γ_∞ and furthermore its Jacobian determinant has value 1 on Γ_∞ because it is the limit of the Jacobian determinants of the maps $(\underline{A},\underline{\phi}) \longleftrightarrow (\underline{A}_n,\underline{\phi}_n)$ which are canonical.

This last remark can be used in conjunction with the ordinary theorem on implicit functions, to invert (4.26) in the vicinity of a torus $\{\underline{A}\}xT^\ell$ in the form

$$\underline{A} = \underline{A}_\infty + \underline{\alpha}'(\underline{A}_\infty,\underline{\phi}_\infty)$$
$$\underline{\phi} = \underline{\phi}_\infty + \underline{\beta}'(\underline{A}_\infty,\underline{\phi}_\infty) \tag{4.28}$$

with $(\underline{A}_\infty,\underline{\phi}_\infty)$ also varying in the vicinity of a torus $\{\overline{\underline{A}}\}xT^\ell \in V^{(\infty)}xT^\ell$.

Of course one still should have the doubt that $\Gamma_\infty = \emptyset$:, a case which would make the whole construction trivial.

Therefore we proceed to estimate from below the measure of Γ_∞. For this we need to analyze another family of functions: the free pulsations $\underline{\omega}_n(\underline{A})$ of the n-th order free Hamiltonian $h_n(\underline{A})$:

$$\underline{\omega}_n(\underline{A}) = \frac{\partial h_n}{\partial \underline{A}}(\underline{A}) = \underline{\omega}_{n-1}(\underline{A}) + \frac{\partial f_{(n-1)\underline{0}}(\underline{A})}{\partial \underline{A}} \quad . \tag{4.29}$$

The analysis of such functions (defined only in $V^{(n)}$) is identical to that of the functions $\underline{A}_n, \underline{z}_n$ and we do not repeat it.

The conclusions are: denote $\underline{\omega}_n(\underline{A}) = \underline{\omega}_0(\underline{A}) + \underline{\tau}_n(\underline{A})$. Then:

$$\lim_{n\to\infty} \underline{\omega}_n(\underline{A}) \equiv \underline{\omega}_\infty(\underline{A}) = \underline{\omega}_0(\underline{A}) + \underline{\tau}_\infty(\underline{A}) \tag{4.30}$$

exist $\forall\, \underline{A} \in V^{(\infty)}$ and $\underline{\omega}_\infty$ is a C^∞-function in the same sense in which the $\underline{A}_\infty$, $\underline{z}_\infty$ were such.

Furthermore, $\forall\, n \geq 1$, $\forall\, \underline{A} \in V^{(n-1)}$

$$\left|\frac{\partial^{\underline{a}}(\underline{\omega}_n(\underline{A})-\underline{\omega}_0(\underline{A}))}{\partial \underline{A}^{\underline{a}}}\right| \leq K_{\underline{a}}\, \varepsilon_0(E_0C_0)^{b_1^{\underline{a}}}\, \xi_0^{-b_2^{\underline{a}}}\, \rho_0^{-|\underline{a}|}$$

$$|\tau_n(\underline{A})-\tau_n(\underline{A}')| \leq \tilde{K}\, \varepsilon_0(E_0C_0)^{\tilde{b}_1}\, \xi_0^{-\tilde{b}_2}\, |\underline{A}-\underline{A}'|\rho_0^{-1} \tag{4.31}$$

where $K_{\underline{a}}$, $b_1^{\underline{a}}$, $b_2^{\underline{a}}$, $\tilde{K}$, $\tilde{b}_1$ $\tilde{b}_2$ are constants depending only on ℓ. It turns out that $\tilde{b}_1 < 5$, $\tilde{b}_2 < 4\ell$, $\tilde{K} < \sqrt{B_{11}}$. As an example we prove in appendix H the second (4.31).

(4.31) will now be used to estimate the multiplicity of the map $\underline{A} \longrightarrow \underline{\omega}_n(\underline{A})$ on $V^{(n-1)}$.

Repeating the argument in (2.14) we choose $\hat{r} = \frac{1}{2}\rho_0(E_0\eta_0\rho_0^{-1})^{-1}$ and notice that, by (2.14), (4.31): $\forall\, \underline{A},\underline{A}' \in V^{(n-1)}, |\underline{A}-\underline{A}'|$ $\hat{r}$:

$$\begin{aligned}
|\underline{\omega}_n(\underline{A})-\underline{\omega}_n(\underline{A}')| &= |\underline{\omega}_0(\underline{A})-\underline{\omega}_0(\underline{A}')+\tau_n(\underline{A})-\tau_n(\underline{A}')| \\
&\geq |\underline{\omega}_0(\underline{A})-\underline{\omega}_0(\underline{A}')| - |\underline{\tau}_n(\underline{A})-\underline{\tau}_0(\underline{A})| \\
&\geq \eta_0^{-1}(1-\tfrac{1}{4})|\underline{A}-\underline{A}'| - \tilde{K}\, \varepsilon_0(E_0C_0)^{\tilde{b}_1}\xi_0^{-\tilde{b}_2}\rho_0^{-1}|\underline{A}-\underline{A}'| \\
&= \eta_0^{-1}(\tfrac{3}{4} - K\, \varepsilon_0\eta_0\rho_0^{-1}(E_0C_0)^{\tilde{b}_1}\, \xi_0^{-\tilde{b}_2})|\underline{A}-\underline{A}| \geq \frac{\eta_0^{-1}}{2}|\underline{A}-\underline{A}'|
\end{aligned} \tag{4.32}$$

because the remarks on the size of the constants $\tilde{K}$, $\tilde{b}_1$, $\tilde{b}_2$ and the hugeness of the constants in (3.59) automatically imply that the term in the last parenthesis is $< \frac{1}{4}$ (in fact enormously smaller). So the map $\underline{A} \longrightarrow \underline{\omega}_n(\underline{A})$ is one-to-one in $V^{(n)} \cap$ (sphere of radius $\hat{r}$). Hence since $V^{(n-1)} \subset V$ we can estimate the multiplicity T of this map on the whole $V^{(n-1)}$ by the same estimate (2.15).

It is now easy to estimate the volume of $\Gamma^{(n)}$. In fact the measure of $\Gamma^{(n)}$ is the same as that of $V^{(n)} x T^{\ell}$ because the map C_{n-1} is canonical. To estimate vol $V^{(n)}$ we just notice that $V(n) \supset \overline{V}^{(n)}$ and $\overline{V}^{(n)}$ is obtained from $V^{(n-1)}$, which is a union of spheres of radius $\frac{\tilde{\rho}_{n-1}}{2}$ (and hence of radius $\rho_n = \frac{\tilde{\rho}_{n-1}}{8} < \frac{\tilde{\rho}_{n-1}}{2}$), by first excising from each such sphere an outer shell of width ρ_n and, afterwards, deleting from each such sphere the resonant points for h_{n-1}. In the first step we obtain a set V' whose volume is obviously (see appendix L):

$$\text{vol } V' \geq (1-\frac{\tilde{\rho}_n}{\rho_n})^{\ell} \text{ vol } V^{(n-1)} . \tag{4.33}$$

Next we must estimate the measure of the resonant points in V', i.e. the measure of the points in V' for which

$$|\underline{\omega}_{n-1}(\underline{A}')\cdot\underline{\gamma}| < C_{n-1}^{-1}|\underline{\gamma}|^{-\ell} \tag{4.34}$$

for some $\underline{\gamma}$, $0 < |\underline{\gamma}| \leq N_{n-1}$.

Calling this set V" and using (2.15), (3.58):

$$\text{vol } V'' \equiv \int_{V''} d\,\underline{A} < T \int_{\underline{\omega}_{n-1}(V'')} d\underline{\omega} \sup \left|\det\left(\frac{\partial \underline{\omega}_{n-1}}{\partial \underline{A}}\right)\right|^{-1} < T\,\eta_{n-1}^{\ell} \int_{\underline{\omega}_{n-1}(V'')} d\underline{\omega}$$

$$\leq (4(\eta_0 E_0 \rho_0^{-1})\,\rho_0^{-1})\,(4\eta_0)^{\ell} \sum_{\underline{\gamma}\neq\underline{0}} \int_{\substack{|\underline{\omega}\cdot\underline{\gamma}|<C_{n-1}^{-1}|\underline{\gamma}|^{-\ell-1}\\ \underline{\omega}\in\underline{\omega}_{n-1}(V'')}} d\underline{\omega} \leq 2^{4\ell} E_0^{-\ell} r^{\ell} (\eta_0 E_0 \rho_0^{-1})^{2\ell} . \tag{4.35}$$

$$\sum_{\underline{\gamma}\neq\underline{0}} \int_{\substack{|\underline{\omega}\cdot\underline{\gamma}|<C_{n-1}^{-1}|\underline{\gamma}|^{-\ell-1}\\ |\underline{\omega}|<E_{n-1}<4E_0}} d\underline{\omega} \leq \left(\frac{2^{\ell} r^{\ell}}{\ell!}\right) \ell!\, 2^{3\ell} E_0^{-\ell} (\eta_0 E_0 \rho_0^{-1})^{2\ell} \frac{(4E_0)^{\ell-1} 2^{\ell-1}}{(\ell-1)!} \frac{2}{C_0 n^2} \cdot$$

$$\cdot \left(\sum_{\underline{\omega}\neq\underline{0}} |\underline{\gamma}|^{-\ell-1} \right) \leq \ell\, 2^{6\ell} (\eta_0 E_0 \rho_0^{-1})^{2\ell} (E_0 C_0)^{-1} n^{-2} \left(\sum_{\underline{\gamma}\neq\underline{0}} |\underline{\gamma}|^{-1-\ell} \right) \text{ vol } V.$$

So

$$\text{vol } V^{(n)} \geq \left(1-\frac{\tilde{\rho}_n}{\rho_n}\right)^{\ell} \text{vol } V^{(n-1)} - \bar{B}_{12}\frac{(E_0\rho_0^{-1}\eta_0)^{2\ell}}{E_0C_0}\frac{1}{n^2}\text{ vol } V$$

$$\geq \left(\prod_{n=0}^{\infty}\left(1-\frac{\tilde{\rho}_n}{\rho_n}\right)^{\ell} - \bar{B}_{12}\left(\sum_{n=1}^{\infty}\frac{1}{n^2}\right)\left(\frac{(E_0\rho_0^{-1}\eta_0)^{2\ell}}{E_0C_0}\right)\right)\text{ vol } V \qquad (4.36)$$

$$\geq \left(1-B_{12}\frac{(E_0\rho_0^{-1}\eta_0)^{2\ell}}{E_0C_0}\right)\text{ vol } V$$

with $B_{12} = \ell\, 2^{8\ell+5}(\ell+1)!$, as elaborated in appendix M.

At this point we set

$$\lambda = B_{12}\frac{(E_0\eta_0\rho_0^{-1})^2}{E_0C_0} \qquad (4.37)$$

and eliminate C_0 by using condition (3.58) which becomes

$$B_{12}^{11}\, B_{11}\,\frac{\varepsilon_0}{E_0}\,\lambda^{-11}(E_0\eta_0\rho_0^{-1})^{5+22\ell}\;\xi_0^{-10(3\ell+1)+\ell} > 1 \qquad (4.38)$$

and $B_{12}^{11}\, B_{11} \leq (\ell+31)!^{31}$, for example.

Also notice that by construction

$$|\underline{\omega}_\infty(\underline{A})\cdot\underline{\gamma}|^{-1} \leq 4\, C_n|\underline{\gamma}|^{\ell} \qquad \text{if} \quad N_{n-1} \leq |\underline{\gamma}| < N_n. \qquad (4.39)$$

In fact

$$|\underline{\omega}_\infty(\underline{A})\cdot\underline{\gamma}|^{-1} = |\underline{\omega}_n(\underline{A})\cdot\underline{\gamma}|^{-1}\,|\,1-(\underline{\omega}_n(\underline{A})\cdot\underline{\gamma})^{-1}\,(\underline{\omega}_\infty(\underline{A})-\underline{\omega}_n(\underline{A}))\cdot\underline{\gamma}|^{-1}$$

$$\leq 2C_n|\underline{\gamma}|^{\ell}(1-2C_nN_n^{\ell+1}\sum_{k=n}^{\infty}\varepsilon_k)^{-1} < 4C_0|\underline{\gamma}|^{\ell}(1+n)^2 \qquad (4.40)$$

by (3.58) (which implies that the parenthesis is ≤ 2)

So since by (3.57) $N_n < (\frac{3}{2})^n N_0$, $N_{n-1} \leq |\underline{\gamma}| < N_n$ implies $(1+n)^2 \leq 2\log|\underline{\gamma}|$ and we rewrite (4.39)

$$|\underline{\omega}_\infty(\underline{A})\cdot\underline{\gamma}|^{-1} \leq 8\, C_0|\underline{\gamma}|^{\ell}(\log|\underline{\gamma}|)^2. \qquad (4.41)$$

5) It remains to check that the $\underline{A}'_\infty$ are prime integrals on Γ_∞. Clearly if we fix $(\underline{A},\underline{z}) \in \Gamma_\infty$ and $t > 0$ and if we call S the Hamiltonian flow for the Hamiltonian H_0 on Γ_n and $S^{(n)}$ the flow for H_n on $V^{(n)} x T^\ell$ then, if the ℓ.h.s. and r.h.s. make sense:

$$\tilde{C}^{(n-1)} \ldots \tilde{C}^{(0)}(S_t(\underline{A},\underline{z})) = S_t^{(n)}\tilde{C}^{(n-1)} \ldots \tilde{C}^{(0)}(\underline{A},\underline{z}) \equiv S_t^{(n)}(\underline{A}_n,\underline{z}_n) \quad (4.42)$$

From the form of the Hamilton equations it follows (see appendix N):

$$|S_t^{(n)}(\underline{A}_n,\underline{z}_n)-(\underline{A}_n,\underline{z}_n \ell^{i\underline{\omega}_n(\underline{A}_n)t})| < \ell[(e^{(1+2E_0t)\varepsilon_n t}-1) + \varepsilon_n t\rho_n\rho_0^{-1}] \quad (4.43)$$

at least as long as the motion $S_\tau^{(n)}(\underline{A}_n,\underline{z}_n)$ stays in $W(\rho_n,\xi_n;V^{(n-1)})$, $\forall\, \tau \in [0,t]$, i.e. as long as the r.h.s. of (4.35) is $< \rho_n/\rho_0$. This is certainly true for n large since $\varepsilon_n \longrightarrow 0$ much faster than ρ_n. Therefore for large enough n (4.42) makes sense and holds. Thus $S_t(\underline{A},\underline{z}) \in \Gamma_\infty$ (because the argument above yields that $S_t(\underline{A},\underline{z}) \in W_\infty$ and Γ_∞ is the "real part" of W_∞) and (4.42), (4.43) and the definition and properties of $\underline{A}_\infty$, $\underline{z}_\infty$ allow us to take the limit $n \longrightarrow \infty$ in (4.42) $\forall(\underline{A},\underline{z}) \in \Gamma_\infty$. One finds

$$(\underline{A}_\infty(S_t(\underline{A},\underline{z})),\underline{z}_\infty(S_t(\underline{A},\underline{z})) = (\underline{A}_\infty(\underline{A},\underline{z}),\underline{z}_\infty\, e^{i\underline{\omega}_\infty(\underline{A}_\infty)t}) \quad (4.44)$$

which clearly means that $\underline{A}_\infty$ is a prime integral on Γ_∞. Furthermore, the set described by (4.28) setting $\underline{A}_\infty \in V^{(\infty)}$ constant and letting $\underline{\phi}_\infty$ vary in T^ℓ is an invariant set (torus) and the motion on it is quasiperiodic with the pulsation $\underline{\omega}_\infty(\underline{A}_\infty)$.

In other words things go as if we had $V^{(\infty)} x T^\ell$ as phase space and h_∞ as Hamiltonian. The change of coordinates between the original variables and the new vairables is, however, only defined on Γ_∞, or it is interesting only there.

Finally notice that if $\tilde{A}(\underline{A},\underline{\phi})$ is another prime integral of class C^∞ on Γ_∞ then it can be expressed in the coordinate system $(\underline{A}_\infty,\underline{\phi}_\infty)$ via (4.28) as a function of $\underline{A}_\infty$, $\underline{\phi}_\infty$ near the torus $\{\underline{A}_\infty\} x T^\ell$:

$$\tilde{\underline{A}}(\underline{A},\underline{\phi}) = \underline{a}(\underline{A}_\infty(\underline{A},\underline{\phi}),\underline{\phi}_\infty(\underline{A},\underline{\phi})). \quad (4.45)$$

Then

$$\underline{\tilde{A}}(S_t(\underline{A},\underline{\phi})) = \underline{a}(\underline{A}_\infty(\underline{A},\underline{\phi}),\underline{\phi}_\infty(\underline{A},\underline{\phi})+\underline{\omega}_\infty(\underline{A}_\infty)t) \tag{4.46}$$

but by (4.41) the points $\underline{\phi}_\infty(\underline{A},\underline{\phi}) + \underline{\omega}_\infty(\underline{A}_\infty)t$ cover the whole torus T^ℓ densely as $t \longrightarrow \infty$ so that if $\underline{\tilde{A}}$ is t-independent it can only be a function of $\underline{A}_\infty$ alone: i.e.

$$\underline{a}(\underline{A}_\infty, \underline{\phi}_\infty) = \underline{\tilde{a}}(\underline{A}_\infty) \tag{4.47}$$

for some $\tilde{a}$.

§5. Some comments.

The reason why we get the horrible bounds (1.22) is to be found in the fact that we have proceeded rather carelessly in bounding the constants and in keeping track of the factors δ_0 popping out everywhere. Also and most importantly we have been **very** careless in treating (3.42) ÷ (3.46) and (3.36) to obtain the final result (3.57), (3.58).

An examination of the proof shows that the powers of E_0C_0, ε_0C_0, $E_0\eta_0\rho_0^{-1}$ appearing in the various inequalities up to (3.46) are almost certainly optimal and with some care one could also find the optimal powers to which δ_0 will be raised. After (3.46) we performed very bad estimates to finish quickly.

It is however clarifying to remark that there is an easy way to find the optimal results, i.e. to find the best possible values of α, β, γ in (1.14), starting from (3.41), (3.44) and (3.34).

The point is to set E_kC_k equal to E_0C_0, $E_k\eta_k\rho_k^{-1}$ equal to $E_0\eta_0\rho_k^{-1}$, δ_k equal to ξ_0 and delete everywhere the logarithms (appearing in $N_0^{\ell+1}$). In this way we get a recursion relation which works under milder conditions and since we have deleted logarithms the true recursion relation should work under a slightly stronger condition. If the modified recursion relation will work under a condition of the type (1.14) with $\bar{\alpha}$, $\bar{\beta}$, $\bar{\gamma}$ replacing α, β, γ then the true recursion relation should work (i.e. should be indefinitely iterable) under a condition of the type (1.14) with α, β, γ replaced by $\bar{\alpha}+$, $\bar{\beta}+$, $\bar{\gamma}+$ where $x+$ denotes a number larger (by any amount) than x.

The recursion relations (3.41) ÷ (3.43) and the appended constraints (3.34), (3.44) become

$$\begin{aligned} \varepsilon_{k+1} &= \text{const } C_0\varepsilon_k^2 (E_0C_0)^2 \xi_0^{-5\ell-1} \\ \rho_{k+1} &= \rho_k/\text{const } E_0C_0\xi_0^{-\ell-1} \end{aligned} \tag{5.1}$$

with the constraints:

$$\text{const } \varepsilon_k C_0E_0C_0\xi_0^{-\ell-1} < 1 \text{ , const } \varepsilon_k\eta_0\rho_k^{-1} < 1. \tag{5.2}$$

The discussion of the recursion relation (5.1) and of the constraint relation easily leads to the condition

$$\text{const}\ \varepsilon_0 C_0 (E_0 C_0)^{2+1/8} (E_0 \eta_0 \rho_0^{-1}) \xi_0^{-x} < 1 \tag{5.3}$$

$x = (5+2^{-6})\ell + (1+\frac{1}{4})$.

So recalling the relation between C_0 and λ ($\lambda = \text{const}(E_0\eta_0\rho_0^{-1})^{2\ell}/E_0C_0$) we expect that the best constants in (1.18) should be

$$\alpha = 3 + \frac{1}{8} + \ ,\ \beta = 1 + 6\ell + \frac{\ell}{4}\ell + \ ,\ \gamma = x + \tag{5.4}$$

which should be the best compatible with the bounds obtained up to (3.34).

As a final comment we can discuss another consequence of the preceding proof. Again we discuss this very informally.

It is often puzzling, particularly in the concrete applications, that the region Γ_∞ is hard to describe. Clearly the fact that it is the image of $V^{(\infty)} \times T^\ell$ is not a very satisfactory answer.

On the other hand the hypothesis that Γ_∞ be determined by the location of the resonances of h_0 cannot be true since clearly f_0 has to play a role.

A rather good description of Γ_∞ is the following:

Consider the functions $\underline{\omega}_0, \underline{\omega}_1, \underline{\omega}_2, \ldots$ defined on $V, V^{(0)}, V^{(1)}$, respectively. Consider the surfaces:

$$\Sigma_{\underline{\gamma}} = \{\underline{A} | \underline{\omega}_j(\underline{A}) \cdot \underline{\gamma} = 0, \underline{A} \in V^{(j-1)}\} \ \text{if}\ N_{j-1} \leq \gamma \leq N_j; \tag{5.5}$$

where we have set $V^{(-1)} \equiv V$, $N_{-1} = 0$.

These are smooth surfaces since the Jacobian matrix of ω_j has an inverse that by construction can be bound by $\eta_j < 6\eta_0$ in $V^{(j-1)}$.

Then we eliminate from V layers* of width

$$K\left(\frac{\varepsilon_0}{E_0}\right)^\sigma E_0 |\underline{\gamma}|^{-\ell-\frac{1}{2}} \tag{5.6}$$

around $\Sigma_{\underline{\gamma}}$.

It is not difficult to prove that if σ is small enough then the set $\hat{V}$ of the points found in ths way is such that $\hat{V} \times T^\ell \subset \Gamma_\infty$ (however $\hat{V} \times T^\ell$ will in general not be invariant. If one wants an invariant set one should just let $\hat{V} \times T^\ell$ "flow", i.e. replace it by $\cup_t S_t(\hat{V} \times T^\ell)$).

---

* a "layer" being $\{\underline{A} \,|\, |\underline{\omega}(\underline{A}) \cdot \frac{\underline{\gamma}}{|\underline{\gamma}|}| < K(\frac{\varepsilon_0}{E_0})^\sigma E_0 |\underline{\gamma}|^{-\ell-\frac{1}{2}}\}$

The proof involves the ideas of the proof of §3, §4, mainly those necessary to prove proposition 4 which however has not been discussed here. Also for ε_0 small the set $\hat{V}\times T^\ell$ has large measure. The value of σ can essentially be found by replacing C_0 in (1.14) with $(\frac{\varepsilon_0}{E_0})^\sigma E_0$. In this way, using the estimates (1.18), one gets that $\sigma < \frac{1}{12}$. However if one uses (5.4) one gets $\sigma < \frac{1}{3}$.

In any case it is clear that to be sure that a point $(\underline{A},\underline{z})$ is in Γ_∞ one has to perform the full perturbative algorithm to construct the functions ω_j.

However if one excises from $V\times T^\ell$ only the "0-th order" resonances, i.e. those determined by $\underline{\omega}_0$ (i.e. the layers around $\sum_{\underline{\gamma}}$ with $|\underline{\gamma}| \in N_0$) one finds a set of points which can be described in the new canonical coordinates associated with the first map (since these are the points on which $\tilde{C}_0$ is defined). Such points will move quasiperiodically (with perturbed frequencies given by the $\underline{\omega}_1$'s) to a much better approximation than the points of $V\times T^\ell$ as a whole: the points of $V\times T^\ell$ move quasiperiodically up to a time $O(1/\varepsilon_0)$ (i.e. neglecting f_0), while those of $V\times T^\ell$ do move quasiperiodically up to a time of $O(1/\varepsilon_0{}^2 N_0{}^{\ell+1})$ (i.e. neglecting, in the new coordinates, f_1).

So with finitely many operations one can reach a high degree of accuracy.

Appendix A.

Let $(\underline{A},\underline{z}) \in W(\tilde{\rho}_0,\xi_0;V^{(0)})$. Then there is $\underline{A}_0 \in \overline{V}^{(0)}$ such that $|A_i-A_{oi}| < \frac{3}{2}\tilde{\rho}_0$ and, since $\tilde{\rho}_0 < \rho_0/2$:

$$|\underline{\omega}_0(\underline{A})\cdot\underline{\gamma}|^{-1} = |\underline{\omega}_0(\underline{A}_0)\cdot\underline{\gamma}|^{-1}\ |1+(\underline{\omega}_0(\underline{A}_0)\cdot\underline{\gamma}|^{-1}(\underline{\omega}_0(\underline{A})-\underline{\omega}_0(\underline{A}_0))\cdot\underline{\gamma}|^{-1}$$

$$\leq C_0|\underline{\gamma}|^{\ell}(1-C_0N_0^{\ell+1}|\underline{\omega}_0(\underline{A})-\underline{\omega}_0(\underline{A}_0)|)^{-1}.$$

There is a path Λ leading from $\underline{A}_0$ to $\underline{A}$, entirely contained in $W(\tilde{\rho}_0,\xi_0;V^{(0)})$ and consisting of 2ℓ coordinate* segments of length $\frac{\tilde{\rho}_0}{2}$ or at most $\tilde{\rho}_0$ and such that every coordinate A_i of a point $\underline{A} \in \Lambda$ has a distance $> \rho_0 - \tilde{\rho}_0$ from the boundary of $W(\rho_0,\xi_0;V) \cap \{i\text{-th coordinate hyperplane}\}$. This follows the very definition of $W(\tilde{\rho}_0,\xi_0;V^{(0)})$. We do not explicitly use the fact that V is a sphere as this property will not be preserved in the later applications of (3.12). We only use the fact that $V^{(0)}$ is a union of spheres of radius $\tilde{\rho}_0/2$. So

$$|\underline{\omega}_0(\underline{A})-\underline{\omega}(\underline{A}_0)| = \left|\int \frac{\partial\underline{\omega}_0}{\partial\underline{A}}(\tilde{\underline{A}})\cdot d\underline{A}\right| \leq \frac{3}{2}\tilde{\rho}_0\ 2\ell\frac{E_0}{\rho_0-\tilde{\rho}_0} \leq 6\ell E_0\frac{\tilde{\rho}_0}{\rho_0}$$

Thus $B_2 = 2^3\ell$, as we shall bound all the constants by powers of 2 (for simplicity).

---

*i.e. segments along which only one coordinate varies.

Appendix B. (3.15), (3.16).

Using (3.5), (3.11) we find, $\forall\ (\underline{A}',\underline{z}) \in W(\tilde{\rho}_0,\xi_0-\delta_0;V^{(0)})$:

$$\left|\frac{\partial\Phi_0}{\partial\underline{\phi}}(\underline{A}',\underline{z})\right| \le \sum_{|\underline{\gamma}|\le N_0} \varepsilon_0\rho_0\ 2C_0|\underline{\gamma}|^{\ell} e^{-\delta_0|\underline{\gamma}|}$$

$$\equiv\ 2\varepsilon_0 C_0\rho_0\ \ell!(\frac{2}{\delta_0})^{\ell} \sum_{|\underline{\gamma}|\le N_0} (\frac{\delta_0|\underline{\gamma}|}{2})^{\ell}\frac{1}{\ell!}\ e^{-\delta_0|\underline{\gamma}|}$$

$$\le 2^{\ell+1}\varepsilon_0 C_0\rho_0\delta_0^{-\ell}\ \ell! \sum_{|\underline{\gamma}|\le N_0} e^{-\frac{\delta_0}{2}|\underline{\gamma}|} \le 2^{\ell+1}_{\ell}\varepsilon_0 C_0\rho_0\delta_0^{-\ell}\ \ell!(\frac{4}{8})^{\ell}\ .$$

Also recalling that $\tilde{\rho}_0 < \rho_0/2$ we can dimensionally estimate $\left|\frac{\partial\underline{\omega}_0}{\partial\underline{A}'}(\underline{A}')\right|$ for $\underline{A}' \in \pi_1(W(\rho_0,\xi_0-\delta_0;V^{(0)}))$ as $E_0/(\rho_0-\tilde{\rho}_0)$ and:

$$\left|\frac{\partial\Phi_0}{\partial\underline{A}'}(\underline{A}',\underline{z})\right| \le \sum_{0<|\underline{\gamma}|\le N_0} \varepsilon_0 2C_0|\underline{\gamma}|^{\ell} e^{-\delta_0|\underline{\gamma}|}$$

$$+ \left|\sum_{0<|\underline{\gamma}|\le N_0} \frac{f_{0\underline{\gamma}}(\underline{A}')}{i(\underline{\omega}_0(\underline{A}')\cdot\underline{\gamma})^2}\ \frac{\partial\underline{\omega}_0(\underline{A}')}{\partial\underline{A}'}\cdot\underline{\gamma}\ \underline{z}^{\underline{\gamma}}\right|$$

$$\le \varepsilon_0 C_0 2^{3\ell+1}\ell!\ \delta_0^{-2\ell} + \sum_{\underline{\gamma}} 4C_0^{\ 2}|\underline{\gamma}|^{2\ell}\ \varepsilon_0\rho_0\frac{E_0}{\rho_0-\tilde{\rho}_0} e^{-\delta_0|\underline{\gamma}|}$$

$$\le 2^{3\ell+1}\ell!\varepsilon_0 C_0\delta_0^{-2\ell} + 8C_0^{\ 2}\varepsilon_0 E_0\ 2^{2\ell}(2\ell)!\ \delta_0^{-2\ell}\ (\frac{4}{\delta_0})^{\ell}$$

$$\le 2^{4(\ell+1)}(2\ell)!\ \varepsilon_0 C_0 E_0 C_0\ \delta_0^{-3\ell}\ .$$

Appendix C. Dimensional implicit function theorem.

Consider the equations

$$\underline{z}' = \underline{z} \exp \underline{F}(\underline{A}',\underline{z}) \quad \text{and} \quad \underline{A} = \underline{A}' + \underline{G}(\underline{A}',\underline{z}) \tag{C.1}$$

where $\underline{F}$ and $\underline{G}$ are holomorphic in a region of the form $W(\rho,\xi;S)$ where S is some union of spheres of radius $\rho/2$, i.e. S is not too thin compared to its complex extension $W(\rho,\xi;S)$. Let $\xi \leq 1$.

We look for conditions of invertibility of the first equation with respect to $\underline{z}$ and of the second with respect to $\underline{A}'$.

Consider the first and search for a condition guaranteeing that $\underline{z}_1 \neq \underline{z}_2 \Longrightarrow \underline{z}_1' \neq \underline{z}_2'$. Given $\underline{z}_1$, $\underline{z}_2$ such that $e^{-(\xi-\delta/2)} < |z_{\alpha,j}| < e^{\xi-\delta/2}$, $\alpha = 1, 2$; $j = 1, \ldots, \ell$, and denoting $\|F\| = \sup|F(\underline{A}',\underline{z})|$ in $W(\rho,\xi;S)$ we consider a coordinate path Λ joining $\underline{z}_1$ to $\underline{z}_2$ staying in the annulus $\{e^{-(\xi_0-\delta/2)} < |z_i| < e^{\xi_0\delta/2}, i = 1, \ldots, \ell\}$, and of length bounded by $\pi|\underline{z}_1 - \underline{z}_2|$. Then

$$|z_{1j}'-z_{2j}'| = |(z_{2j}-z_{2j})\, e^{F_j(\underline{A}',\underline{z}_1)} + z_{2j}(e^{F_j(\underline{A}',\underline{z}_1)} - e^{F_j(\underline{A}',z_2)})| \geq$$

$$\geq |z_{1j}-z_{2j}|e^{-\|F\|} - e^{\xi}\, e^{\|F\|}\, |F_j(\underline{A}',\underline{z}_1) - F_j(\underline{A}',\underline{z}_2)| \geq$$

$$\geq |z_{1j}-z_{2j}|e^{-\|F\|} - e^{\xi}\, e^{\|F\|}\, \left|\int_\Lambda \frac{\partial F_j}{\partial \underline{z}}(\underline{A}',\underline{z})\cdot d\,\underline{z}\right| \geq$$

$$\geq |z_{1j}-z_{2j}|e^{-\|F\|} - e^{\xi+\|F\|}\pi|\underline{z}_1-\underline{z}_2|\, \sup\left|\frac{\partial F_j}{\partial z_k}(\underline{A}',\underline{z})\right|$$

the supremum being taken over $W(\rho,\xi-\frac{\delta}{2};S)$. It is therefore dimensionally bounded by $\|F\|\, e2\delta^{-1}$.

Hence, summing over j:

$$|\underline{z}_1-\underline{z}_2| > |\underline{z}_1-\underline{z}_2|\ e^{-\|F\|}\ (1-e^{2+2\|F\|}\pi\ell 2^{\ell-1} e\|F\|)\ .$$

Hence if $e\pi\ell 2\delta^{-1}\ e^{2+\|F\|}F < \frac{1}{2}$, and a fortiori , if

$$2^{11}\ \ell\|F\|\delta^{-1} < 1$$

(since $4\pi\ e^4 < 2^{11}$) we have $\underline{z}_1' \neq \underline{z}_2'$ when $\underline{z}_1 \neq \underline{z}_2$. (C.2) is sufficient to guarantee also that the Jacobian matrix $T_{ij} = \frac{\partial z_i}{\partial z_j}$ has nonvanishing determinant. In fact writing $J = 1 + \sigma$ one finds, by (C.2):

$$|\sigma_{ij}| = |\delta_{ij}(e^F-1)+z_i e^F \frac{\partial F}{\partial z_j}| \leq \delta_{ij}\ e^{\|F\|}\|F\| + e^{\xi+\|F\|}\|F\|\ 2\ e\delta^{-1}$$

$$\leq 4e^{2+\|F\|}\|F\|\ \delta^{-1} < 2^6\|F\|\delta^{-1} < \frac{1}{2\ell}$$

so since $\sum_j|\sigma_{ij}| < \frac{1}{2}$ it follows from well known algebraic results that $\det(1+\sigma) \neq 0$.

Everything being holomorphic it is clear that the inverse of the first map of (C.1), which is defined in the image of $W(\rho,\xi-\frac{\delta}{2};S)$ by the map $(\underline{A}',\underline{z}) \longrightarrow \underline{z}\exp \underline{F}(\underline{A}',\underline{z})$, will be holomorphic in its domain of definition.

Denote the inverse map $\underline{z} = \underline{z}'\ \exp\ \underline{\Delta}\ (\underline{A}',\underline{x}')$. Then if $\underline{z}' = \underline{z}\exp \underline{F}(\underline{A}',\underline{z})$ we have necessarily:

$$\underline{\Delta}(\underline{A}'.\underline{z}') \equiv -\underline{F}(\underline{A}',\underline{z}) \tag{C.3}$$

Therefore

$$\|\underline{\Delta}\| \leq \|\underline{F}\| < \frac{\delta}{2^{11}\ell} < \frac{\delta}{2} \tag{C.4}$$

so that is clear that as $(\underline{A}',\underline{z})$ varies in $W(\rho,\xi-\frac{\delta}{2};S)$ the point $(\underline{A}',\underline{z}')$ will vary covering at least $W(\rho,\xi-\delta;S)$.

Hence $\underline{\Delta}$ is defined and holomorphic in $W(\rho,\xi-\delta;S)$ and varifies (C.3), (C.4).

The argument for the theory of the second equation in (C.1) is, of course, essentially the same.

Let $\|\underline{G}\| = \sup |\underline{G}|$ in $W(\rho,\xi;S)$. Let $\underline{A}_1', \underline{A}_2' \in W(\frac{2}{3}\rho,\xi;S)$ and look for a condition guaranteeing that $\underline{A}_1 \neq \underline{A}_2$.

Consider first the case when $|A_{1j}'-A_{2j}'| \geq \rho/4$ for some $j = 1, \ldots, \ell$. Then

$$|A_{1j}-A_{2j}| \geq \rho/4 - 2\|\underline{G}\|$$

so $\underline{A}_1 \neq \underline{A}_2$ if

$$\|\underline{G}\| < \rho/8. \tag{C.5}$$

Let then $|A_{1j}-A_{2j}| < \rho/4$, $\forall\, j = 1, \ldots, \ell$, and notice that if $\underline{A}_1 \in C(\frac{2}{3}\rho,\xi;\{\underline{A}_0\})$ for $\underline{A}_0 \in S$ then $\underline{A}_2 \in C(\frac{3}{4}\rho,\xi;\{\underline{A}_0\})$ and we can join $\underline{A}_1$ to $\underline{A}_2$ by a path Λ which lies entirely in $W(\frac{3}{4}\rho,\xi;S)$ with length $\leq |\underline{A}_1-\underline{A}_2|$. Therefore

$$\begin{aligned}
|A_{1j}-A_{2j}| &\geq |A_{1j}'-A_{2j}'| - \left|\int_\Lambda \frac{\partial G_j}{\partial \underline{A}'}(\underline{A}',\underline{z})\, d\,\underline{A}'\right| \\
&\geq |A_{1j}'-A_{2j}'| - \sup_{i,k}\left|\frac{\partial G_i}{\partial A_k'}\right| |\underline{A}_1'-\underline{A}_2'| \\
&\geq |A_{1j}'-A_{2j}'| - \frac{\|\underline{G}\|}{\rho-\frac{3}{4}\rho} |\underline{A}_1'-\underline{A}_2'|
\end{aligned}$$

where the supremum is taken over i,k and $(\underline{A}',\underline{z}) \in W(\frac{3}{4}\rho,\xi;S)$ and is dimensionally bounded in the last step. Therefore

$$|\underline{A}_1-\underline{A}_2| \geq |\underline{A}_1'-\underline{A}_2'|\,(1-4\ell\rho^{-1}\|\underline{G}\|).$$

Hence if

$$\|\underline{G}\| \leq \tfrac{1}{2}\frac{\rho}{4\ell} \tag{C.6}$$

$\underline{A}_1 \neq \underline{A}_2$ and we can invert the second equation in the form $\underline{A}' = \underline{A} + \underline{\Xi}'(\underline{A},\underline{z})$ with

$$\underline{\Xi}'(\underline{A},\underline{z}) = -\underline{G}(\underline{A}',\underline{z}) \tag{C.7}$$

for $\underline{A} = \underline{A}' + \underline{G}(\underline{A}',\underline{z})$ (which is uniquely defined for $\forall(\underline{A}',\underline{z}) \in W(\frac{2}{3}\rho,\xi;S)$). Since, as before, the condition (C.6) also implies that the Jacobian determinant $J_{ij} = \delta_{ij} + \frac{\partial G_i}{\partial A_j'}(\underline{A}',\underline{z})$ does not vanish in $W(\frac{2}{3}\rho,\xi;S)$ we see that $\underline{\Xi}'$ will be holomorphic in its domain of definition which is the image of $W(\frac{2}{3}\rho,\xi;S)$ under $(\underline{A}',\underline{z}) \longrightarrow \underline{A}' + \underline{G}(\underline{A}',\underline{z})$. Also $\underline{\Xi}'$ will be bounded in its domain by

$$\|\underline{\Xi}'\| < \|\underline{G}\| < \frac{\rho}{8\ell} < \frac{\rho}{8} \; . \tag{C.8}$$

(C.8) implies that the image of $W(\frac{2}{3}\rho,\xi;S)$ under the map $(\underline{A}',\underline{z}) \longrightarrow \underline{A}' + \underline{G}(\underline{A}',\underline{z})$ covers $W(\frac{2}{3}\rho-\frac{\rho}{8},\xi;S) \supset W(\frac{\rho}{2},\xi;S)$.

(C.3), (C.7) show that $\underline{\Xi}',i\underline{\Delta}$ are real on $V \times T^{\ell}$ if $i\underline{F}$, $\underline{G}$ are real. In our application $\underline{F}(\underline{A}',\underline{z}) = i\frac{\partial\Phi_0}{\partial \underline{A}'}(\underline{A}',\underline{z})$ and $\underline{G}(\underline{A}',\underline{z}) = \frac{\partial\Phi_0}{\partial \underline{\phi}}(\underline{A}',\underline{z})$, $\rho = \tilde{\rho}_0$, $\xi = \xi_0 - \delta_0$, $\delta = \delta_0$ and $S = V^{(0)}$.

So by (C.6) and (3.16) the conditions for invertibility of the first equation of (3.17) become:

$$8\ell\, B_3'' \, \varepsilon_0 \rho_0 C_0 \delta_0^{-2\ell} \tilde{\rho}_0^{-1} > 1$$

and by (C.8), (3.15) the conditions for invertibility of the second equation of (3.17), become:

$$2^{11}\ell \, B_3' \, \varepsilon_0 C_0 E_0 C_0 \delta_0^{-3\ell} \, \delta_0^{-1} < 1$$

having required that the functions $\underline{\Xi}'$ and $\underline{\Delta}$ have, respectively, domains at least as large as $W(\frac{\rho_0}{2},\xi_0-\delta_0;V^{(0)})$ and $W(\rho_0,\xi_0-2\delta_0;V^{(0)})$. The above two conditions can be implied by (recalling the definitions of $\tilde{\rho}_0$ and N_0):

$$(8\ell\, B_3'' B_2 + 2^{11}\ell\, B_3')\, \varepsilon_0 C_0 E_0 C_0 N_0^{\ell+1} \delta_0^{-2\ell} < 1$$

so by (3.15), (3.16) one can take:

$$B_4 = 2^{12+4\ell}(2(\ell+1))!$$

(3.36) follow from (C.4), (C.8) and (3.30) from (C.7), (C.3).

Appendix D.

(3.42) is trivial from (3.38) and from the definitions of E_0, ε_0. To prove (3.43) we notice that for any two $\ell \times \ell$ matrices R,S we have $|RS| < |R|\,|S|$ so that, defining

$$M_1(\underline{A}) = \frac{\partial^2 h_0}{\partial \underline{A} \partial \underline{A}}(\underline{A}) + \frac{\partial^2 f_{\underline{00}}}{\partial \underline{A} \partial \underline{A}}(\underline{A}) \equiv M_0(\underline{A}) + \sigma(\underline{A})$$

we find that $\forall\,(\underline{A},\underline{z}) \in W(\frac{\rho_0}{2},\xi_0;V)$:

$$|\sigma(\underline{A})| = \sum_{i,j} |\sigma_{ij}(\underline{A})| = \sum_{i,j} \left|\frac{\partial^2 f_{\underline{00}}}{\partial \underline{A} \partial A_j}(\underline{A})\right| \le \ell^2 \frac{\varepsilon_0}{\rho_0 - \frac{\rho_0}{2}} = 2\ell^2\,\varepsilon_0 \rho_0^{-1}$$

since $f_{\underline{00}}(\underline{A}) = (2\pi)^{-\ell} \int_{T^\ell} f_0(\underline{A},\underline{\phi})\, d\underline{\phi}$ implies

$$\left|\frac{\partial f_{\underline{00}}}{\partial \underline{A}}(\underline{A})\right| \le \sup_{\underline{\phi}} \left|\frac{\partial f_0}{\partial \underline{A}}(\underline{A},\underline{\phi})\right| \le \varepsilon_0 .$$

Hence if $\eta_0 |\sigma(\underline{A})| < \frac{1}{2}$, which can be implied by

$$4\ell^2\, \eta_0 \varepsilon_0 \rho_0^{-1} < 1 \quad , \tag{D.1}$$

we find:

$$|M_1(\underline{A})^{-1}| \equiv |M_0(\underline{A}) + \sigma(\underline{A})|^{-1} \equiv |(M_0(\underline{A})(1 + M_0(\underline{A})^{-1}\sigma(\underline{A}))^{-1}|$$

$$\equiv |(1+M_0(\underline{A})^{-1}\sigma(\underline{A}))^{-1} M_0(\underline{A})^{-1}| \equiv |[(1+M_0(\underline{A})^{-1}\sigma(\underline{A}))^{-1} - 1] M_0(\underline{A})^{-1} +$$

$$+ M_0(\underline{A})^{-1}| \le \eta_0 + \eta_0\, |(1+M_0(\underline{A})^{-1}\sigma(\underline{A}))^{-1} - 1|$$

$$\le \eta_0 + \eta_0 \frac{\eta_0 |\sigma(\underline{A})|}{1 - \eta_0 |\sigma(\underline{A})|} \le \eta_0 (1 + 4\ell^2 \varepsilon_0 \rho_0^{-1} \eta_0)$$

if (D.1) holds. So $\overline{B}_5 = 4\ell^2$.

Appendix E. (3.44)

Notice that f_1 is defined on $W(\tilde{\rho}_0,\xi_0-2\delta_0;V^{(0)})$, i.e. on the domains of $\underline{\Xi}$, $\underline{\Delta}$. Also (see (3.37), (3.38)):

$$f_1(\underline{A}',\underline{z}') = f_0(\underline{A}'+\underline{\Xi}(\underline{A}',\underline{z}'),\underline{z}'e^{i\underline{\Delta}(\underline{A}',\underline{z}')}) - f_{\underline{00}}(\underline{A}') + h_0(\underline{A}'+\underline{\Xi}(\underline{A}',\underline{z}')) - h_0(\underline{A}') \,. \tag{E.1}$$

Recalling that (by definition) Φ_0 is such that:

$$\underline{\omega}_0(\underline{A}')\cdot\frac{\partial\Phi_0}{\partial\underline{\phi}}(\underline{A}',\underline{z}) + f_0^{[\leq N_0]}(\underline{A}',\underline{z}) - f_{\underline{00}}(\underline{A}') = 0 \tag{E.2}$$

i.e. by (3.30):

$$-\underline{\omega}_0(\underline{A}')\cdot\underline{\Xi}(\underline{A}',\underline{z}') + f_0^{[\leq N_0]}(\underline{A}',\underline{z}'\,e^{i\underline{\Delta}(\underline{A}',\underline{z}')}) - f_{\underline{00}}(\underline{A}') = 0 \tag{E.3}$$

we can rewrite f_1 as $f_1^{I} + f_1^{II} + f_1^{III}$ with:

$$f_1^{I}(\underline{A}',\underline{z}') = h_0(\underline{A}'+\underline{\Xi}) - h_0(\underline{A}')-\underline{\omega}_0(\underline{A}')\cdot\underline{\Xi}$$

$$f_1^{II}(\underline{A}',\underline{z}') = f_0^{[\leq N_0]}(\underline{A}'+\underline{\Xi},\underline{z}'e^{i\underline{\Delta}}) - f_0^{[\leq N_0]}(\underline{A}',\underline{z}'e^{i\underline{\Delta}}) \tag{E.4}$$

$$f_1^{III}(\underline{A}',\underline{z}') = f_0^{[\geq N_0]}(\underline{A}'+\underline{\Xi},\underline{z}'e^{i\underline{\Delta}})$$

which we write as

$$f_1^{I}(\underline{A}',\underline{z}') = \int_0^1 (1-t)dt \sum_{i,j=1}^{\ell} \frac{\partial^2 h_0}{\partial A_i\,\partial A_j}(\underline{A}'+t\,\underline{\Xi})\;\Xi_i\;\Xi_j$$

$$f_1^{II}(\underline{A}',\underline{z}') = \int dt \sum_{j=1}^{\ell} \frac{\partial f_0^{[\leq N_0]}}{\partial A_j}(\underline{A}'+t\,\underline{\Xi},\underline{z}'e^{i\underline{\Delta}})\underline{\Xi}_j \tag{E.5}$$

$$f_1^{III} = f_0^{[\leq N_0]}(\underline{A}'+\underline{\Xi},\underline{z}'e^{i\underline{\Delta}})$$

Let $(\underline{A}',\underline{z}') \in W(\frac{\tilde{\rho}_0}{4},\xi_0-3\delta_0;V^{(0)})$. From (3.31) we see that $(\underline{A}'+t\underline{\Xi},\underline{z}'e^{i\underline{\Delta}}) \in W(\frac{\tilde{\rho}_0}{2},\xi_0-2\delta_0;V^{(0)})$ for $t \in [0,1]$. So we can easily bound f_1^{I} by dimensional estimates, and f_1^{II} f_1^{III} by pure brute force (i.e. by (3.5)) in $W(\frac{\tilde{\rho}_0}{4},\xi_0-3\delta_0;V^{(0)})$. In fact

$$\left|\frac{\partial^2 h_0}{\partial A_i \partial A_k}(\underline{A}')\right| \le \frac{E_0}{\rho_0 - \frac{\tilde{\rho}_0}{2}} < 2E_0\rho_0^{-1}$$

$$\left|\frac{\partial f_0^{[\le N_0]}}{\partial \underline{A}'}(\underline{A}',\underline{z})\right| \le \sum_{|\underline{\gamma}|\le N_0} \varepsilon_0 e^{-\delta_0|\underline{\gamma}|} \le \delta_0^{-\ell}\varepsilon_0 2^{\ell}$$

$$\left|f^{[>N_0]}(\underline{A}',\underline{z})\right| \le \sum_{|\underline{\gamma}|>N_0} |f_{0\underline{\gamma}}(\underline{A}')|\, e^{(\xi_0-2\delta_0)|\underline{\gamma}|}$$

$$\le \sum_{|\gamma|>N_0} |\underline{\gamma}|\, |f_{0\underline{\gamma}}(\underline{A}')|\, e^{(\xi_0-2\delta_0)|\underline{\gamma}|} \le \varepsilon_0\rho_0 \sum_{|\underline{\gamma}|>N_0} e^{-2\delta_0|\underline{\gamma}|}$$

$$\le 2^{\ell}\, \varepsilon_0\rho_0\, e^{-\delta_0 N_0}\, \delta_0^{-\ell} \le 2^{\ell}\varepsilon_0^{2}\rho_0 C_0$$

so that (E.4) and (3.36) imply

$$|f_1^{I}| \le \tfrac{1}{2}(2E_0\rho_0^{-1})(B_3''\varepsilon_0\rho_0 C_0\delta_0^{-2\ell})^2 \le (B_3'')^2\, E_0 C_0\varepsilon_0^{2} C_0\rho_0\, \delta_0^{-4\ell}$$

$$|f_1^{II}| \le 2^{\ell}\varepsilon_0\delta_0^{-\ell}(B_3''\varepsilon_0\rho_0 C_0\delta_0^{-2\ell}) \tag{E.6}$$

$$|f^{III}| \le 2^{\ell}\varepsilon_0^{2}\rho_0 C_0 \quad .$$

So that, since $(B_3'') + 2^{\ell}B_3'' + 2^{\ell} < 2(B_3'')^2$ (see line after (3.16)):

$$|f_1| \le 2(B_3'')^2\, E_0 C_0\varepsilon_0^{2} C_0\rho_0^{-4\ell}\rho_0 \quad \text{in} \quad W(\frac{\tilde{\rho}_0}{4},\xi_0-3\delta_0;V^{(0)}). \tag{E.7}$$

(E.7) allows the following dimensional estimate in $W(\frac{\tilde{\rho}_0}{8},\xi_0-4\delta_0;V^{(0)}) \equiv W(\rho_1,\xi_1;V^{(0)})$:

$$\left|\frac{\partial f_1}{\partial \underline{A}'}\right| + \frac{1}{\rho_1}\left|\frac{\partial f_1}{\partial \underline{\phi}}\right| < 2(B_3'')^2\, E_0 C_0\varepsilon_0^{2} C_0\delta_0^{-4\ell}\rho_0\left(\left(\frac{\tilde{\rho}_0}{4}-\rho_1\right)^{-1}+\rho_1^{-1}\frac{e}{\delta_0}\right)$$

and recalling the values of B''_3, B_2 and the expressions for $\tilde{\rho}_0$ and $\rho_1 = \tilde{\rho}_0/8$ we find that:

$$\left|\frac{\partial f_1}{\partial \underline{A}'}\right| + \frac{1}{\rho_1}\left|\frac{\partial f_1}{\partial \underline{\phi}'}\right| \le 2\; 2^{8(\ell+1)}(2\ell)!^2\; E_0 C_0 \varepsilon_0{}^2 C_0 \delta_0{}^{-4\ell} \rho_0 \tilde{\rho}_0{}^{-1}\; 8\cdot 4\cdot \delta_0{}^{-1} \tag{E.8}$$

$$\le 2^6 2^{10(\ell+1)} B_2\; \ell!^3\; \varepsilon_0{}^2 C_0 N_0{}^{\ell+1} (E_0 C_0)^2\; \delta_0{}^{-4\ell-1} = \bar{\varepsilon}_1$$

so

$$\bar{B}_6 = 2^{-2} 2^{10(\ell+1)} \ell\; \ell!^2 \qquad \bar{x} = 4\ell + 1. \tag{E.9}$$

Also using $N_0 = 2\;\delta_0{}^{-1}\;\log(\varepsilon_0 C_0 \delta_0{}^{\ell})^{-1}$ and

$$\sqrt{\mu}\,(\lg \mu^{-1})^a \le 2^{2a}(2a)! \qquad \forall\, \mu \in (0,1), \forall\, a > 0 \tag{E.10}$$

we see that

$$\bar{\varepsilon}_1 \le 2^{\ell+1} \bar{B}_6\; 2^{\ell+1}(\ell+1)!\; (\varepsilon_0 C_0 \delta_0{}^{\ell})^{\frac{3}{2}}\; \delta_0{}^{-7\ell-2}\; E_0{}^2 C_0 \equiv \varepsilon_1 \equiv \tag{E.11}$$

$$= B_6 (\varepsilon_0 C_0 \delta_0{}^{\ell})^{\frac{3}{2}}\; \delta_0{}^{-7\ell-2}\; E_0{}^2 C_0$$

with

$$B_6 = 2^{14(\ell+1)}\; 2^{-2} (\ell+1)!^4\;. \tag{E.12}$$

ppendix F.

Suppose $E_k \le 4E_0$. Then (3.56) imply (since $_6(\frac{k+1}{k+2})^{2\ell+2} \ge B_6 2^{-2\ell-2} \ge 1$, $\frac{C_{k+1}}{C_k} = \frac{k+2}{k+1} \le 2$):

$$\mu_{k+1} \ge \mu_k^{\frac{3}{2}}$$

$$\mu_{k+1} \le 4^2 B_6 (E_0 C_0)^2 \ \delta_k^{-2(3\ell+1)} \cdot 4(\mu_k)^{\frac{3}{2}} \tag{F.1}$$

$$\left(\frac{\rho_{k+1}}{\rho_k}\right)^{-1} \le 4\, B_8\ E_0 C_0 \delta_k^{-\ell-1} (\log \mu_k^{-1})^{\ell+1}\ (k+1)^2$$

so that

$$\mu_k \ge (\mu_0)^{(\frac{3}{2})^k} \tag{F.2}$$

and bounding $\prod_{j=0}^{\infty} (1+j)^{(\frac{3}{2})^{-j}} \le 2^4$:

$$\mu_k \le (2^6 B_6 16^{2(3\ell+1)} (E_0 C_0)^2\ \xi_0^{-2(3\ell+1)})^{2[(\frac{3}{2})^k - 1]}$$

$$\mu_0^{(\frac{3}{2})^k} \left(\prod_{j=1}^{k} (1+j)^{(\frac{3}{2})^{k-j}} \right)^{2(3\ell+1)+2} \tag{F.3}$$

$$\le (B_9 (E_0 C_0)^4\ \xi_0^{-4(3\ell+1)} \mu_0)^{(\frac{3}{2})^k}$$

$$\left(\frac{\rho_{k+1}}{\rho_k}\right)^{-1} \le (B_{10}\ \xi_0^{-\ell-1} E_0 C_0 (\log \mu_0^{-1})^{\ell+1})^k\ k!^{2(\ell+2)} (\tfrac{3}{2})^{(\ell+1)k\frac{k+1}{2}}$$

with

$$B_9 = 2^{12} B_6^2\ 16^{4(3\ell+1)} (2^4)^{2(3\ell+1)+2} = 2^{100\ell+68} (\ell+1)!^8 \tag{F.4}$$

$$B_{10} = 4\ B_8\ 16^{\ell+1} = 2^{5\ell+14}\ \ell \quad .$$

Consider now the condition (3.57) assuming $E_k \le 4E_0$, $\eta_k \le 4\eta_0$. We find that the (ℓ.h.s.) of (3.52) is bounded by;

$$2^{\ell+1} B_7 4^3\, 16^{4\ell+1}\, E_0 C_0 (\eta_0 E_0 \rho_0^{-1})^2\, \xi_0^{-4\ell-1}\, \mu_k \left(\frac{\rho_k}{\rho_0}\right)^{-2} (\log \mu_k^{-1})^{\ell+1}$$

$$\le 2^{\ell+1} 2^6\, 2^{16\ell+4} B_7\, 2^{\ell+1} (\ell+1)!\, E_0 C_0 (\eta_0 E_0 \rho_0^{-1})^2\, \xi_0^{-4\ell-1} \sqrt{\mu_k} \left(\frac{\rho_k}{\rho_0}\right)^{-2}$$

$$\le 2^{18\ell+12} (\ell+1)!\ B_7\, E_0 C_0 (\eta_0 E_0 \rho_0^{-1})^2\, \xi_0^{-4\ell-1}$$

$$\cdot\, (B_9 (E_0 C_0)^4\, \xi_0^{-4(3\ell+1)} \mu_0)^{\frac{1}{2}(\frac{3}{2})^k}\ (B_{10}\, E_0 C_C\, \xi_0^{-\ell-1} (\log \mu_0^{-1})^{\ell+1})^k$$

$$\cdot\, k!^{2(\ell+2)} \left(\frac{3}{2}\right)^{(\ell+1)\frac{k(k+1)}{2}}$$

$$\le 2^{18\ell+12} (\ell+1)!\ B_7\, E_0 C_0 (\eta_0 E_0 \rho_0^{-1})^2 \xi_0^{-4\ell-1}$$
$$(B_9 B_{10}^{\ 2} (E_0 C_0)^2 \xi_0^{-2\ell-2} (\log \mu_0^{-1})^{2(\ell+1)} (E_0 C_0)^4 \xi_0^{-4(3\ell+1)} \xi_0)^{\frac{1}{2}(\frac{3}{2})^k}$$
$$\cdot\, 2^{2(\ell+2)(\frac{3}{2})^k}\ 2^{8(\frac{3}{2})^k (\ell+1)}$$

where the last step holds $\forall\, k \ge 0$ if it holds for $k = 0$.

$$(2^{18\ell+12} (\ell+1)! B_7)^2\ (E_0 C_0)^2 (\eta_0 E_0 \rho_0^{-1})^4\ \xi_0^{-8\ell-2}\ B_9 B_{10}^{\ 2}\ 2^{4\ell+8} 2^{16\ell+16}$$

$$(E_0 C_0)^6\ \xi_0^{-2\ell-2-12\ell-2}\ \mu_0 (\log \mu_0^{-1})^{2(\ell+1)} < 1.$$

Hence recalling $B_7 = 2^{6\ell+5} (\ell+1)!^2$

$$2^{142\ell+190} (\ell+1)!^{14} (E_0 C_0)^8 (\eta_0 E_0 \rho_0^{-1})^4\ \xi_0^{-22\ell-8}\ \mu_0 (\log \mu_0^{-1})^{2(\ell+1)} < 1$$

which is implied by

$$2^{142\ell+190} (\ell+1)!^{14}\ (E_0 C_0)^8 (\eta_0 E_0 \rho_0^{-1})^4\ \xi_0^{-22\ell-8}\ \mu_0^{\frac{8}{10}}\ 5^{2(\ell+1)} (2(\ell+1))!$$

$$\le 2^{150\ell+198} (\ell+1)!^{16} (E_0 C_0)^8 (\eta_0 E_0 \rho_0^{-1})^4\ \xi_0^{-22\ell-8}\ \mu_0^{\frac{8}{10}} < 1$$

i.e. by

$$2^{175\ell+250} (\ell+1)!\ (E_0 C_0)^{10}\ (\eta_0 E_0 \rho_0^{-1})^5\ \xi_0^{-28\ell-10}\ \mu_0 < 1$$

which follows from

$$(\ell+20)!^{20}(E_0C_0)^{10}\,(\eta_0E_0\rho_0^{-1})^5\,\xi_0^{-10(3\ell+1)} < 1\,. \qquad (F.5)$$

If one wants a better estimate one should avoid estimating $\mu(\log\mu)^{\ell+1}$ by $\sqrt{\mu}$.

If we impose (F.5) we find, in particular, that, $((1+k) < 2^{(\frac{3}{2})^k})$:

$$\mu_k\,\delta_k^{-\ell} < \xi_0^{-\ell}2^{4\ell}\,(1+k)^{2\ell}\,(B_9(E_0C_0)^4\,\xi_0^{-4(3\ell+1)}\,\mu_0)^{(\frac{3}{2})^k}$$
$$<(2^{4\ell}\xi_0^{-\ell}B_9\,2^{2\ell}\,(E_0C_0)^4\,\xi_0^{-(3\ell+1)4}\,\mu_0)$$

but in the course of the above inequalities we required that the term in parenthesis to be such that $B_9\,B_{10}{}^2(E_0C_0)^4\,\xi_0^{-(3\ell+1)4}\,\mu_0 2^{8(\ell+1)2}\xi_0^{-4(3\ell+1)} <$

Hence

$$\mu_k\delta_k^{-\ell} \le (\frac{1}{4^4})^{(\frac{3}{2})^k}$$

also

$$\varepsilon_k \le \frac{1}{C_0}\,\mu_k\,\delta_k^{-\ell} \le E_0(\frac{1}{4^4})^{(\frac{3}{2})^k}$$

so

$$E_k \le E_0 + \sum_{k=0}^{\infty}\varepsilon_k \le E_0 + E_0\sum_{k=0}^{\infty}(\frac{1}{4^4})^k \le 2\,E_0 \le 4E_0$$

$$\eta_k \le \eta_0\prod_{k=0}^{\infty}(1+\sqrt{\mu_k\delta_k^{-\ell}}) \le \eta_0\,\exp\sum_{k=0}^{\infty}(\frac{1}{4^2})^k \le \eta_0\,e^{\frac{16}{15}} < 4\eta_0\,.$$

Hence (F.6) is sufficient to guarantee the possibility of an infinite induction.

Appendix G.

Let Λ be a path in $W(\tilde{\rho}_p, \xi_p - 3\delta_p; V^{(p)})$ joining $(\underline{A}_1, \underline{z}_1) \in V^{(p)} \times T^{\ell}$ to $(\underline{A}_2, \underline{z}_2)$ and entirely contained in $C(\frac{\tilde{\rho}_p}{4}, \xi_p - 3\delta_p; \{\underline{A}_1\})$.

We can take Λ to be a path consisting of a piece Λ_1 joining $(\underline{A}_1, \underline{z}_1)$ to $(\underline{A}_2, \underline{z}_1)$ at constant $\underline{z}$ and continue by a path Λ_2 joining $(\underline{A}_2, \underline{z}_1)$ to $(\underline{A}_2, \underline{z}_2)$ at constant $\underline{A}$. Furthermore the length of the first path can be taken $< \ell \tilde{\rho}_p/4$ and that of the second $< 2\pi \tilde{\rho}_p/4\rho_0$ (this can easily be seen remarking that $\frac{\tilde{\rho}_k}{4\rho_0} < \delta_k^{\ell+1}$ as is implied by the presence of $N_k^{\ell+1}$ in the denominator of the relation defining $\tilde{\rho}_k$: $\tilde{\rho}_k = \rho_k / 8B_2 E_k C_k N_k^{\ell+1}$ (and by (3.58)).

Expressing the variations $\tilde{C}^{(p)}(\underline{A}_1, \underline{z}_1) - \tilde{C}^{(p)}(\underline{A}_2, \underline{z}_2)$ as an integral over $\Lambda_1 \cup \Lambda_2$ one finds

$$\frac{1}{\rho_0} \left| \pi_1 (\tilde{C}^{(p)}(\underline{A}_1, \underline{z}_1) - (\underline{A}_1, \underline{z}_1)) - \pi_1 (\tilde{C}^{(p)}(\underline{A}_2, \underline{z}_1) - (\underline{A}_2, \underline{z}_2) \right|$$

$$\leq \ell \frac{\tilde{\rho}_p}{4\rho_0} \quad \text{(r.h.s. of the first line of (4.8) with } |\underline{k}| = 1, |\underline{h}| = 0)$$

$$+ \frac{2\pi}{4} \frac{\tilde{\rho}_p}{4\rho_0} \cdot \text{(r.h.s. of the first line of (4.8) with } |\underline{k}| = 0, \; |\underline{h}| = 1)$$

$$\leq \frac{\theta_p}{2}$$

with θ_1 given e.g. by (4.15). (When $\tilde{\rho}_p$ does not drop out of the estimates we have bounded it by $\tilde{\rho}_p / 8E_p C_p (2\delta_p^{-1})^{\ell+1}$ discarding the logarithmic term (≥ 1)).

Similarly, one treats π_2 and finally one gets (4.15).

Appendix H. (4.31)

Let $\underline{A}, \underline{A}' \in V^{(n)}$ and $\rho_{k-1} < |\underline{A}-\underline{A}'| < \rho_k$. Then

$$|\underline{\tau}_n(\underline{A})-\underline{\tau}_n(\underline{A}')| \equiv |\sum_{p=0}^{n-1} (f_{p\underline{0}}(A)-f_{p\underline{0}}(\underline{A}'))| .$$

Suppose $k \leq n-1$. Then

$$|\underline{\tau}_n(\underline{A})-\underline{\tau}_n(\underline{A}')| \leq \sum_{p=0}^{k} |f_{p\underline{0}}(\underline{A})-f_{p\underline{0}}(\underline{A}')| + 2\sum_{p=k+1}^{n} \varepsilon_p$$

since $|f_{p\underline{0}}^{(\underline{A})}| = |(2\pi)^{-\ell}\int_{T^\ell} f_p(\underline{A},\underline{\phi})d\underline{\phi}| \leq \varepsilon_p$. Since $|\underline{A}-\underline{A}'| < \rho_k$ the whole segment joining $\underline{A}$ to $\underline{A}'$ will be in $V^{(p)}$, $(p \leq k)$, and therefore

$$|\underline{\tau}_n(\underline{A})-\tau_n(\underline{A}')| \leq |\underline{A}-\underline{A}'| \sum_{p=0}^{k} \frac{\varepsilon_p}{\rho_p} + 2|\underline{A}-\underline{A}'| \sum_{p=k+1}^{n} \frac{\varepsilon_p}{|\underline{A}-\underline{A}'|}$$

$$\leq |\underline{A}-\underline{A}'|\ 3\sum_{p=0}^{\infty} \frac{\varepsilon_p}{\rho_p} = 3\frac{|\underline{A}-\underline{A}'|}{\rho_0\delta_0^{\ell}} C_0^{-1} \sum_{p=0}^{\infty} (1+p)^{2\ell}\mu_p\rho_p^{-1}$$

$$\leq |\underline{A}-\underline{A}'|\ K\ \varepsilon_0\rho_0^{-1}(E_0C_0)^a\xi_0^{-b}$$

where K, a, b are constants which can easily be estimated from (3.57), (3.58).

If $k > n-1$ nothing essentially changes, of course.

Appendix L.

Notice that one in fact builds a subset of V' by excising a boundary layer of width $\tilde{\rho}_n$ from each sphere of radius ρ_n in $V^{(n-1)}$.

The volume of the set obtained in this way can be bounded below observing that if we bring the centers of the new spheres closer to each other by a factor $(1 - \frac{\tilde{\rho}_n}{\rho_n})$ (without any change of the radii) we obtain a set with smaller volume. Its volume, however, is exactly $(1 - \frac{\tilde{\rho}_n}{\rho_n})^{\ell} \text{ vol } V^{(n-1)}$ as the two above operations are equivalent to a homothety by a factor $(1 - \tilde{\rho}_n/\rho_n)$.

Appendix M.

We bound $\sum_{\underline{\gamma}\neq\underline{0}} |\underline{\gamma}|^{-1-\ell} \leq 2^{-2(\ell+1)}(\ell+1)!$ and $\sum_{n=1}^{\infty} n^{-2} \leq 4$.

Thus $\overline{B}_{12} = \ell 2^{8\ell+4}(\ell+1)!$. Also

$$\prod_{n=0}^{\infty} (1-\frac{\tilde{\rho}_n}{\rho_n})^{\ell} \geq \exp -2\ell \sum_{n-0}^{\infty} \frac{\tilde{\rho}_n}{\rho_n} \geq 1 - 4\ell \sum_{n=0}^{\infty} \frac{\tilde{\rho}_n}{\rho_n}$$

if $4\ell \sum_n \frac{\tilde{\rho}_n}{\rho_n} < 1$. But by (3.54), (3.57), (3.58):

$$4\ell \sum_{n=0}^{\infty} \frac{\tilde{\rho}_n}{\rho_n} \leq \frac{4\ell}{B_8} \frac{1}{E_0C_0} \sum_{n=0}^{\infty} \frac{(\frac{3}{2})^{-n},}{(\log \mu_0^{-1})} \leq \frac{1}{E_0C_0}$$

since $(4\ell/B_8)^3 < 1$ and $(\log \mu_0^{-1})^{-1} < 1$ (by (3.59), and the large value of B_{11}). Hence if

$$B_{12} = 2\overline{B}_{12} = \ell 2^{8\ell+5}(\ell+1)!$$

we have, (recalling that $E_0\eta_0\rho_0^{-1} \geq 1$):

$$\text{vol } V^{(n)} > (1-B_{12}\frac{(E_0\eta_0\rho_0^{-1})^{2\ell}}{E_0C_0}) \text{ vol } V \text{ .}$$

Appendix N.

We argue as follows: the equations of motion are

$$\underline{A}' = -\frac{\partial f_n}{\partial \underline{\phi}'}\,,\quad \dot{\underline{\phi}}' = \underline{\omega}_n(\underline{A}') + \frac{\partial f_n}{\partial \underline{A}'} \equiv \underline{\omega}_n(\underline{A}_n) + (\underline{\omega}_n(\underline{A}') - \omega_n(\underline{A}_n)) + \frac{\partial f_n}{\partial \underline{A}'}$$

so that by the definition of ε_n

$$|\underline{A}' - \underline{A}_n| < \varepsilon_n \rho_n t$$

$$|\underline{\phi}' - \underline{\phi}_n - \underline{\omega}_n(\underline{A}_n)t| \le \frac{E_n}{\rho_n} \int_0^t |\underline{A}'(\tau) - A_n|\,d\tau + \varepsilon_n t$$

$$\le E_n \rho_n^{-1} \rho_n \varepsilon_n t\,t + \varepsilon_n t < (1+2E_0 t)\varepsilon_n t .$$

REFERENCES

[1] N. Kolmogorov: Dokl. Akad. Nauk. 98, 527, 1954.

[2] J. Moser: Nach. Akad. Wiss. Gottinger, II a, 1, 1962.

[3] V. Arnold: Russ. Math. Surv. 18, n° 5, 9, 1963 and 18, n° 6, 85, 1963.

[4] J. Pöschel: "Ueber differenzierbare Faserungen invarianter Tori", 1981, preprint, ETH-Zürich.

[5] L. Chierchia, G. Gallavotti: "Smooth prime integrals for quasi-integrable Hamiltonian systems", Il Nuovo Cimento, B67, 277, 1982.

[6] L. Kadanoff: Proc. Int. School of Physics, E. Fermi, Corso LI, ed. M. Green, Acad. Press, N.Y., 1971.
L. Kadanoff: Physics 2, 263, 1966.
L. Kadanoff et Rev. Mod. Phys. 39, 395, 1967, and
C. Di-Castro, G. Jona-Lasinio: Phys. Lett. 29A, 322, 1969, and
C. Di-Castro, G. Jona-Lasinio, L. Peliti: Ann. Phys. 87, 327, 1974.
K. Wilson: Phys. Rev. 179, 1499, 1969 and Phys. Rev. B4, 3174, 1971 and Phys. Rev. B4, 3184, 1971.
K. Wilson, J. Kogut: Physics Reports 12, 75, 1974.

[7] G. Gallavotti, A. Martin-Lof: Nuovo Cimento, 25B, 425, 1975.
G. Gallavotti, M. Cassandro: Nuovo Cimento, 25B, 691, 1975.
G. Benfatto, M. Cassandro, G. Gallavotti, F. Nicolò; E. Olivieri
E. Presutti, E. Scacciatelli; Comm. Math. Phys. 59, 143, 1978.
R. Griffiths, R. Pearce; J. Stat. Phys. 20, 499, 1979.
M. J. Westwater: Comm. Math. Phys. 72, 131, 1980.
K. Gawedski, A. Kupiainen: Comm. Math. Phys. 77, 31, 1980.
J. Fröhlich, T. Spencer: Comm. Math. Phys. 81, 527, 1981.
M. Cassandro, E. Olivieri: Comm. Math. Phys. 80, 255, 1981.

[8] H. Poincaré: "Les Methodes Nouvelles de la Mécanique Celeste", Gauthier-Villars, Paris, 1892, Vol. I, Ch. V, p. 233.
J. Moser: "Stable and random motions in dynamical systems", Ann. Math. Studies, Princeton Univ. Press, 1973, Princeton.

[9] For an interesting (non-rigorous) analysis see D. Escande, F. Doveil: J. Stat. Phys. 26, 257, 1981.
D. Escande: "Renormalization approach to non integrable Hamiltonians", Austin Workshop, March 1981, preprint Ecole Politecnique, Lab. Phys. Milieux Ionisés, Palaiseau, 1982
I learned some not too bad rigorous estimates in the forced pendulum case (considered in the above papers) from L. Chierchia (private communication).

REFERENCES (continued)

[10] H. Whitney: Trans. Am. Math. Soc. 36, 63, 1934.

[11] H. Rüssmann: Celestial Mech. 14, 33, 1976 and Comm. Pure Appl. Math. 29, 755, 1976, and Lecture Notes in Physics, Vol. 38, 1975, ed. J. Moser.

Progress in Mathematics

Edited by J. Coates and S. Helgason

Progress in Physics

Edited by A. Jaffe and D. Ruelle

- A collection of research-oriented monographs, reports, notes arising from lectures or seminars
- Quickly published concurrent with research
- Easily accessible through international distribution facilities
- Reasonably priced
- Reporting research developments combining original results with an expository treatment of the particular subject area
- A contribution to the international scientific community: for colleagues and for graduate students who are seeking current information and directions in their graduate and post-graduate work.

Manuscripts

Manuscripts should be no less than 100 and preferably no more than 500 pages in length.

They are reproduced by a photographic process and therefore must be typed with extreme care. Symbols not on the typewriter should be inserted by hand in indelible black ink. Corrections to the typescript should be made by pasting in the new text or painting out errors with white correction fluid.

The typescript is reduced slightly (75%) in size during reproduction; best results will not be obtained unless the text on any one page is kept within the overall limit of 6x9½ in (16x24 cm). On request, the publisher will supply special paper with the typing area outlined.

Manuscripts should be sent to the editors or directly to:
Birkhäuser Boston, Inc., P.O. Box 2007, Cambridge,
Massachusetts 02139

PROGRESS IN MATHEMATICS
Already published

PM 1 Quadratic Forms in Infinite-Dimensional Vector Spaces
Herbert Gross
ISBN 3-7643-1111-8, 432 pages, paperback

PM 2 Singularités des systèmes différentiels de Gauss-Manin
Frédéric Pham
ISBN 3-7643-3002-3, 346 pages, paperback

PM 3 Vector Bundles on Complex Projective Spaces
C. Okonek, M. Schneider, H. Spindler
ISBN 3-7643-3000-7, 396 pages, paperback

PM 4 Complex Approximation, Proceedings, Quebec, Canada, July 3-8, 1978
Edited by Bernard Aupetit
ISBN 3-7643-3004-X, 128 pages, paperback

PM 5 The Radon Transform
Sigurdur Helgason
ISBN 3-7643-3006-6, 207 pages, hardcover

PM 6 The Weil Representation, Maslov Index and Theta Series
Gérard Lion, Michèle Vergne
ISBN 3-7643-3007-4, 348 pages, paperback

PM 7 Vector Bundles and Differential Equations
Proceedings, Nice, France, June 12-17, 1979
Edited by André Hirschowitz
ISBN 3-7643-3022-8, 256 pages, paperback

PM 8 Dynamical Systems, C.I.M.E. Lectures, Bressanone, Italy, June 1978
John Guckenheimer, Jürgen Moser, Sheldon E. Newhouse
ISBN 3-7643-3024-4, 305 pages, hardcover

PM 9 Linear Algebraic Groups
T. A. Springer
ISBN 3-7643-3029-5, 314 pages, hardcover

PM 10 Ergodic Theory and Dynamical Systems I
A. Katok
ISBN 3-7643-3036-8, 346 pages, hardcover

PM 11 18th Scandinavian Congress of Mathematicians, Aarhus, Denmark, 1980
Edited by Erik Balslev
ISBN 3-7643-3040-6, 526 pages, hardcover

PM 12 Séminaire de Théorie des Nombres, Paris 1979-80
Edited by Marie-José Bertin
ISBN 3-7643-3035-X, 404 pages, hardcover

PM 13 Topics in Harmonic Analysis on Homogeneous Spaces
Sigurdur Helgason
ISBN 3-7643-3051-1, 152 pages, hardcover

PM 14 Manifolds and Lie Groups, Papers in Honor of Yozô Matsushima
Edited by J. Hano, A. Marimoto, S. Murakami, K. Okamoto, and H. Ozeki
ISBN 3-7643-3053-8, 476 pages, hardcover

PM 15 Representations of Real Reductive Lie Groups
David A. Vogan, Jr.
ISBN 3-7643-3037-6, 776 pages, hardcover

PM 16 Rational Homotopy Theory and Differential Forms
Phillip A. Griffiths, John W. Morgan
ISBN 3-7643-3041-4, 258 pages, hardcover

PM 17 Triangular Products of Group Representations and their Applications
S. M. Vovsi
ISBN 3-7643-3062-7, 142 pages, hardcover

PM 18 Géométrie Analytique Rigide et Applications
Jean Fresnel, Marius van der Put
ISBN 3-7643-3069-4, 232 pages, hardcover

PM 19 Periods of Hilbert Modular Surfaces
Takayuki Oda
ISBN 3-7643-3084-8, 144 pages, hardcover

PM 20 Arithmetic on Modular Curves
Glenn Stevens
ISBN 3-7643-3088-0, 236 pagers, hardcover

PM 21 Ergodic Theory and Dynamical Systems II
A. Katok, editor
ISBN 3-7643-3096-1, 226 pages, hardcover

PM 22 Séminaire de Théorie des Nombres, Paris 1980-81
Marie-José Bertin, editor
ISBN 3-7643-3066-X, 374 pages, hardcover

PM 23 Adeles and Algebraic Groups
A. Weil
ISBN 3-7643-3092-9, 138 pages, hardcover

PM 24 Enumerative Geometry and Classical Algebraic Geometry
Patrick Le Barz, Yves Hervier, editors
ISBN 3-7643-3106-2, 260 pages, hardcover

PM 25 Exterior Differential Systems and the Calculus of Variations
Phillip A. Griffiths
ISBN 3-7643-3103-8, 349 pages, hardcover

PM 26 Number Theory Related to Fermat's Last Theorem
Neal Koblitz, editor
ISBN 3-7643-3104-6, 376 pages, hardcover

PM 27 Differential Geometric Control Theory
Roger W. Brockett, Richard S. Millman, Hector J. Sussmann, editors
ISBN 3-7643-3091-0, 349 pages, hardcover

PM 28 Tata Lectures on Theta I
David Mumford
ISBN 3-7643-3109-7, 254 pages, hardcover

PM 29 Birational Geometry of Degenerations
Robert Friedman and David R. Morrison, editors
ISBN 3-7643-3111-9, 410 pages, hardcover

PM 30 CR Submanifolds of Kaehlerian and Sasakian Manifolds
Kentaro Yano, Masahiro Kon
ISBN 3-7643-3119-4, 223 pages, hardcover

PM 31 Approximations Diophantiennes et Nombres Transcendants
D. Bertrand and M. Waldschmidt, editors
ISBN 3-7643-3120-8, 349 pages, hardcover

PM 32 Differential Geometry
Robert Brooks, Alfred Gray, Bruce L. Reinhart, editors
ISBN 3-7643-3134-8, 267 pages, hardcover

PM 33 Uniqueness and Non-Uniqueness in the Cauchy Problem
Claude Zuily
ISBN 3-7643-3121-6, 185 pages, hardcover

PM 34 Systems of Microdifferential Equations
Masaki Kashiwara
ISBN 0-8176-3138-0
ISBN 3-7643-3138-0, 182 pages, hardcover

PM 35 Arithmetic and Geometry Papers Dedicated to I. R. Shafarevich on the Occasion of His Sixtieth Birthday Volume I Arithmetic
Michael Artin, John Tate, editors
ISBN 3-7643-3132-1, 373 pages, hardcover

PM 36 Arithmetic and Geometry Papers Dedicated to I. R. Shafarevich on the Occasion of His Sixtieth Birthday Volume II Geometry
Michael Artin, John Tate, editors
ISBN 3-7643-3133-X, 495 pages, hardcover

PM 37 Mathématique et Physique
Louis Boutet de Monvel, Adrien Douady, Jean-Louis Verdier, editors
ISBN 0-8176-3154-2
ISBN 3-7643-3154-2, 454 pages, hardcover

PM 38 Séminaire de Théorie des Nombres, Paris 1981-82
Marie-José Bertin, editor
ISBN 0-8176-3155-0
ISBN 3-7643-3155-0, 359 pages, hardcover

PM 39 Classical Algebraic and Analytic Manifolds
Kenji Ueno, editor
ISBN 0-8176-3137-2
ISBN 3-7643-3137-2, 644 pages, hardcover

PM 40 Representation Theory of Reductive Groups
P. C. Trombi, editor
ISBN 0-8176-3135-6
ISBN 3-7643-3135-6, 308 pages, hardcover

PM 41 Combinatorics and Commutative Algebra
Richard P. Stanley
ISBN 0-8176-3112-7
ISBN 3-7643-3112-7, 102 pages, hardcover

PM 42 Théorèmes de Bertini et Applications
Jean-Pierre Jouanolou
ISBN 0-8176-3164-X
ISBN 3-7643-3164-X, 140 pages, hardcover

PROGRESS IN PHYSICS

Already published

PPh 1 Iterated Maps on the Interval as Dynamical Systems
Pierre Collet and Jean-Pierre Eckmann
ISBN 3-7643-3026-0, 256 pages, hardcover

PPh 2 Vortices and Monopoles, Structure of Static Gauge Theories
Arthur Jaffe and Clifford Taubes
ISBN 3-7643-3025-2, 294 pages, hardcover

PPh 3 Mathematics and Physics
Yu. I. Manin
ISBN 3-7643-3027-9, 112 pages, hardcover

PPh 4 Lectures on Lepton Nucleon Scattering and Quantum Chromodynamics
W.B. Atwood, J.D. Bjorken, S.J. Brodsky, and R. Stroynowski
ISBN 3-7643-3079-1, 574 pages, hardcover

PPh 5 Gauge Theories: Fundamental Interactions and Rigorous Results
P. Dita, V. Georgescu, R. Purice, editors
ISBN 3-7643-3095-3, 406 pages, hardcover

PPh 6 Third Workshop on Grand Unification, 1982
P.H. Frampton, S.L. Glashow, H. van Dam, editors
ISBN 3-7643-3105-4, 388 pages, hardcover